Marine Ecology and
Fisheries

Marine Ecology and Fisheries

D.H.CUSHING
Fisheries Laboratory, Lowestoft

CAMBRIDGE UNIVERSITY PRESS
CAMBRIDGE
LONDON · NEW YORK · MELBOURNE

Published by the Syndics of the Cambridge University Press
The Pitt Building, Trumpington Street, Cambridge CB2 IRP
Bentley House, 200 Euston Road, London NW1 2DB
32 East 57th Street, New York, NY 10022, USA
296 Beaconsfield Parade, Middle Park, Melbourne 3206, Australia

Library of Congress Catalogue Card Number: 74-82218

hard covers ISBN: 0 521 20501 8
paperback ISBN: 0 521 09911 0

First published 1975

Printed in Great Britain
at the Alden Press, Oxford

Contents

Preface

The study of fish populations that sustains the fisheries is at present a narrow but intense one. For the purposes of conservation single stocks are examined in considerable detail, and information on stock, growth, mortality and year class strength has been collected for decades. But the interaction between exploited stocks in the form of competition has not been studied by fisheries scientists in anything but the most general terms. As the collections of information lengthen into decades it has become clear that quite independently of exploitation numbers slowly rise and fall in the course of time. The mechanisms by which the annual year classes are generated are more or less obscure largely because they are rooted in the lives of larvae and juveniles and fishery biologists have tended to examine the lives of adolescent and adult fishes.

The study of marine ecology is a broader discipline in which processes are studied as parts of an ecosystem, and attempts have been made to elicit principles of some general application. Indeed, much of our detailed knowledge of ecosystems and how they work is based on work at sea, or perhaps more generally, in water. Production processes, as for example in the classical description of the spring outburst of algae in the sea, are well understood even if argument will continue on the details of mechanism. There is a considerable contrast between the analysis of the single stock of fish like the plaice in the southern North Sea or the yellowfin tuna of the eastern tropical Pacific and the broad predator–prey relationship that comprises most of the planktonic algae and most of the small animals that eat them.

During the twenties and thirties of the present century, the two disciplines were not noticeably separate, and indeed most people working in fisheries laboratories called themselves marine biologists. Since the Second World War the fisheries biologists have been very busy with the detailed work needed to support the international commissions on conservation and to make them successful. At the same time there has been a resurgence of interest in oceanography like that in the seventies of last century when the *Challenger* put to sea. Although the main source is military, as the missile submarine has become the supreme weapon, the search for minerals and for protein has provided a hopeful source of funds. Fisheries biologists work in government laboratories whereas marine ecologists work in the stations of Marine Biological Associations, universities and oceanographic institutes. Any divergence between the two sorts of marine biologists should not be insisted upon too much because they talk to each other readily and often.

One of the reasons for this book is to suggest that the time of divergence is drawing to an end and that the practitioners in each field can learn a great deal from each other. Another reason, perhaps obvious to some, is that my own interests have spread in both fields during my working life. There are good scientific reasons for drawing the two disciplines closer together. In fisheries biology the mechanisms of population control remain unrevealed, that is the generation of year class strength, the mechanism of stabilization and the maintenance of competition. To discover them a considerable population study is needed of life in the larval stages and on the nursery grounds. Traditionally, the plankton and inshore environment has been studied more by marine ecologists (but not exclusively) rather than by the fisheries biologists who have been more concerned with those age groups represented in the fisheries. It is no accident that models of the production cycle are being developed within the fisheries laboratories, at least in Europe.

The study of marine ecology, particularly that of marine ecosystems, does not proceed very quickly largely because the information needed for advance is very great. As the food webs are investigated further they become more complicated and each animal tends to grow out of the role allotted to it by the examining biologist. Except in fishes, the death rates and growth rates of such animals are unknown and will not be known until the temperature coefficients of their development have become established much more fully than they are at present. The proportions of the components of an ecosystem can change profoundly as will be shown and the numbers of some populations may rise or fall by several orders of magnitude. The problems of stability in an ecosystem are those of the stabilization mechanisms. It is in this field that the two disciplines overlap.

The book is arranged in three parts. The first four chapters describe the mechanisms of production in the sea, the parameters of importance and the models that have been developed. The second group of four chapters deals with the biology of the abundant and commercial fish stocks, their population dynamics including the possible dependence of year class strength upon the variability of the production cycles. The third and last group of chapters describes more general themes, the use of food-chain studies to fisheries biologists, the nature of climatic change and the match in time of larval production and that of their food. It is in the latter systems that the methods hitherto used in marine ecology need to be applied to fisheries research. Because the processes involved are those of stabilization some of the modes of thought used by fisheries biologists might be useful to marine ecologists in their study of the changes in ecosystems.

Acknowledgments

In the summer of 1969 I gave a series of lectures to postgraduate students in the Department of Oceanography, Oregon State University in Corvallis, Oregon, USA. I am grateful to Professor John Byrne who invited me to Corvallis. I am also grateful to Dr H. A. Cole, CMG for allowing me to travel; he has read and criticized the text and I am grateful to him for the help he has given me. As specialists, I must thank Mr T. Wyatt for reading Chapters 1–4, Dr F. R. Harden Jones for reading Chapter 5, Mr D. J. Garrod and Mr A. C. Burd for reading Chapter 7 and Dr R. R. Dickson for reading Chapters 10 and 11. Their criticisms were most valuable, but the errors that remain are mine.

The sources of all previously published figures are indicated in the legends in the normal way. My thanks are due to all those who gave permission to reproduce their original figures.

Symbols and abbreviations

A' constant in migration equation
A constant; coefficient of eddy diffusivity, g/cm per s
A_m maintenance
A_p relative speed of predator to prey
A_s assimilations, A_{s_n} in units of nitrogen
A area, A_r area; cross sectional
B biomass, B_0
C catch; C_E, C_S
C_p cruising speed
C_w specific growth rate
D true density
D_c compensation depth; ${}_ND_c$, noon compensation depth, m
D_{cr} critical depth, m
D_m depth of mixing
E exploitation rate, E_t
E_1, E_2, E_3, efficiencies
E_L echo level
E_x excretion in units of nitrogen
F instantaneous rate of fishing mortality
G grazing rate, G_0 initial grazing rate
G' growth rate
H degree of gut fullness; H_k, maximum H; H_0, initial H; H_{t0} quantity of food needed to elicit attack; H_l, quantity of food needed to satiate; H_{tl}, attack threshold
H' one-way transmission loss, dB
I irradiance, ly/min; I_0, subsurface irradiance; I_m, irradiance for maximum photosynthesis; T_o compensation irradiance
I' energy consumed by population
K growth coefficient
K_1, K_2 Ivlev's growth coefficients
K' a dimensionless coefficient
K_s half saturation constant
K_t transport coefficient ($= N_c'$)
L depth of photic layer, m
L' loss of nutrients
L_t length at time t
L_∞ asymptote length
M instantaneous mortality rate; M_0, initial mortality rate
N number, N_0
N' nutrient; N_0', N_1', μg-at./l, μM/l; mg/m^3; N_0', nutrient in lower layer
N'' limiting nutrient factor ($= (0.55-N')/0.55$)
N_c' nutrient/cell
P stock (algae or fish, etc.); P_0,P_1; numbers, g C/m^3, etc.
P_G quantity grazed
P' predatory mortality, constant
P'' constant in migration equation
P_N stock of algae in units of nutrient, and stock in numbers of fish
P_T stock at time T, time of maximum stock
P_R production; P_n
P_W stock in weight; P_m max. stock in weight
P^α constant
P_h photosynthesis
P_r ($= P_h/P$) R
Q secondary production
Q_a energy in activity
Q_{cc} energy in daily ration
Q_d specific dynamic activity
Q_g energy in increment of weight
Q_r energy of metabolism
Q_s standard metabolism in units of energy consumption
Q_w energy excreted
Q'' constant
Q_c total incoming radiation; mWh/cm^2

Q'_s total incoming radiation; ly/min
Q'_{ss} photosynthetic energy; ly/min
R algal reproductive rate; R_{max}
R' recruitment; R'_r at replacement
R_a ration, $R_{a_{max}}$
R_s resistance
R_d radius
R_e respiration
R_p attack speed of predator
R_N regenerated nutrient
S source level, dB
S' mean angular deviation, radians
S_s strike success
T absolute temperature
T_s time at maximum of spring outburst
T' ml O_2 consumed per g body wt
T_a target strength, d*b*
T_l density of larvae, *n*/l, or *n*/m^3
U prey speed
U_N uptake of nutrient
V vertical mixing
V' velocity, directional component of velocity
V''_p predator speed
V'_L loss rate of algae/unit stock
V_a attack speed
V_c cruising speed
V_p attack speed of predator
V_r predator speed, less prey speed
V_t virtual populations, V_λ, $V_{\lambda-1}$
V'' predator speed
W weight, W_t
W_∞ asymptotic weight
W_h rate of upwelling, m/d
Z total instantaneous mortality rate, Z
Y yield; Y_N, in numbers; Y_W in weight

a' constant
a ($= P_r/P_m$)
c constant (= 0.774)
c' interprey distance
c'' volume of algal cell in μ^3
c_g rate at which grazing mortality increases with time
d density as n/m^3 or n/m^2
e proportion regenerated (=0.85)
f fishing intensity, or effort per unit area

g grazing coefficient, l/d per g C per m^3; gravity
g_1, g_2 coefficients of efficiency
g' fishing effort
h herbivores, g C/m^3
k extinction coefficient, constant
k' saturation level
k_f feeding coefficient
l length, l_t length at time t
m mixing coefficient
m_0 mortality rate
m_1 rate of increase of mortality rate
n_e number eaten
n_f number of fish in a shoal
n_i number in *i*th age
p quantity produced in photosynthesis; p_m, maximum quantity, g C/g C per h or mg chlorophyll/m^2; p_i, total production
p food in Ivlev equation; p_n
p' phosphorus; p'_T total phosphorus; p'_R residual phosphorus; p'_u, phosphorus absorbed; $\alpha p'_u$, phosphorus in animal flesh; $\beta p'_u$, phosphorus regenerated; p'_p, phosphorus in algae; p'_0, p'_1 μg-at./l
p'' photosynthetic coefficient
q catchability coefficient
r sighting range (or perceptive range)
r' respiration
r_a radius of shoal
r_i rate of increase
r_n shortest distance travelled
t time
t_0 age at which growth starts
t_s time to search
t'_p age of entry into the fishery
t age of extinction of the cohort
t_p time spent in pursuit
v sinking rate, m/d
w P/P_r
w_g volume of grazer
w_i average weight in *i*th age group
z depth, *m*
z_p z_1 depth of photic layer

α constant; mean direction at which tags are recovered
α' coefficient of density-independent mortality

β search field constant
β' coefficient of density-dependent mortality
γ constant
δ sun's altitude; δ_N, at noon
ζ efficiency of nitrogen conversion
θ direction measured in angle from axis of the coastline
ρ density
Ω summation constant.

1

Production cycles in space and time

Introduction

The study of production cycles throughout the seasons began in Europe and North America in the early years of this century. Since then, they have been studied in other parts of the world, but until recently the observations have been limited to accessible stations. There is now a fair amount of information available on production cycles in different areas, even most inhospitable ones, on their variability and on their development in theoretical terms. Methods are not yet standardized, so as much information as possible is summarized from results with disparate means of collection. A few examples are examined in detail in which methods are more or less common and from these some conclusions are drawn.

The durations and amplitudes of production cycles

In general, samples in the sea have been taken once (or twice) a month, so if the outburst in spring lasts two, three or four months, it can only be barely described; if the plants or animals increase in numbers by an order of magnitude or more, the cycle appears to be well described with only a few samples and this appearance of accuracy is deceptive. Lund's (1949) sampling of *Asterionella formosa* Hass in Windermere produced the most complete description of a cycle yet made. He took samples once a week, with many more at the peak. The cycle is described as a roughly bell-shaped curve, symmetrical about its peak so that the magnitude at this point can be described as the amplitude of the cycle. The spread from trough across the peak in time to trough again can be described as the duration. Nearly all the production cycles are in fact cycles of standing stock and the true production is of much greater magnitude. Such a cycle in stock expresses the means by which production in the water takes place, so to describe a stock cycle as a production cycle is perhaps not perverse. In subsequent pages, the distinction between stock and production will be maintained.

The simplest way of counting the algae in the sea is to let them settle from a water sample (5–100 ml) on to the base of a special capsule and to count them from underneath with an inverted microscope (Utermöhl,

1933,1958). This method has two disadvantages. First, some of the smallest flagellates do not sink very quickly and may disintegrate on preservation. Secondly, there are many species in the phytoplankton and it takes a long time to count them if an adequate number of samples has been taken. An extension of this method is to measure the sizes of cells and to express the total quantity in mm^3/l.

A much quicker method is to estimate the quantity of chlorophyll in a water sample; the water is filtered and the residue is dissolved in 90% acetone in water and then the chlorophyll is estimated by absorption spectrophotometry or by excitation fluorometry. There is an extensive literature on methods (Strickland & Parsons, 1968; Loftus & Carpenter, 1971).

Most of the methods detect in addition to chlorophyll, its breakdown products free in the water as phaeophytins or chlorophyllides, which are produced predominantly by the destruction of algal cells by grazing (Currie, 1962; Lorenzen, 1967). But some breakdown can occur in the dark independently of grazing (Yentsch, 1965). If the production cycle is discontinuous (i.e. stops in winter), the initial stages are obviously well estimated, because grazing is low. The peak may be overestimated, because it includes grazed cells and duration might be overestimated also. If the production cycle is continuous all the year round, the same biases apply but they may be distributed equally throughout the cycle. If that is the case, the differences in amplitude and duration may be well estimated by chlorophyll in tropical and subtropical waters. In the future, the cycles may be properly estimated by fluorescence methods to separate the chlorophyll from its degradation products after acidification (Lorenzen, 1966); such a method would be competent to detect those small flagellates inaccessible to earlier methods.

A third method is the radiocarbon method by which an increment of algal carbon is measured during a period of hours (Nielsen, 1952); since Wolfe & Schelske (1967) have adapted the method to scintillation counting, it is an absolute one. Of all methods of estimating quantities of algae, it is the most reliable because it must estimate those flagellates inaccessible to counting and is not vulnerable to those biases in the chlorophyll method due to degradation products. Because the method measures increments in carbon, a production cycle in one area or water mass can be properly described with frequent samples in time, whether the cycle is continuous or discontinuous. But differences between samples scattered across the ocean may indicate differences in stock, because differences in productivity or rate of production may be less than the considerable differences in stock that can occur.

The same principles apply to the production cycles observed in freshwater, with the following provisos. Blue-green algae sometimes dominate the lake plankton in high summer. Because they contain gas bubbles, they are often concentrated at the very surface or close to it. The estimation of chlorophyll is more complicated under such conditions because of the masking pigments. The radiocarbon method is used successfully in fresh water, but the quantity of carbon dioxide must be estimated at the same time. In the open sea, a standard quantity of 90 mg/l is used.

The study of production cycles includes animal production because the greater proportion of algal material is transferred to animal flesh. To catch the animals, the meshes of the nets should be small enough (about 105 μm) to retain the smallest juvenile stages (for example, nauplii) of the herbivores. There are smaller grazers, such as ciliates and tintinnids, which are sometimes important, but they can be conveniently counted directly in water samples. Larger herbivores like euphausids are not caught properly in the daytime by slowly hauled nets, because they dodge the mouths of such nets (Bridger, 1958; Fleming & Clutter, 1965). High-speed nets (hauled at about five knots) sample euphausids and fish larvae adequately at any time of day and night. Such high-speed nets have not yet been used to sample the production of animals in oceanic waters and so the cycles have not so far been fully sampled there. But one cycle, at least, was fully sampled in the North Sea (Cushing *et al.*, 1963).* Elsewhere, two forms of net have been used, a Hensen net (mesh size 420 μm) and the American metre net (mesh size 210 μm; its Russian analogue has a slightly smaller mesh size). The Hensen net samples older herbivores only and the metre net catches both juvenile and adult animals. Neither net catches euphausids properly in the daytime. With such nets, the duration of the animal production cycle is probably well estimated, but the amplitude may be underestimated if the juvenile herbivores and euphausids are absent from the samples.

Table 1 shows the amplitudes and durations of production cycles in marine and freshwater algae and in zooplankton. For marine algae, in numbers and volume, the amplitudes and perhaps the durations of the cycles are greater in higher latitudes. In chlorophyll, there is no trend in latitude across a limited sample, but the amplitudes are very much less than those in numbers; in chlorophyll sampled from 84 μm mesh nets, the amplitudes are higher, perhaps because these very limited samples are mainly composed of diatoms. The radiocarbon measurements in time series represent true estimates of production whereas those in numbers and in chlorophyll are estimates of stock. Although there are series of observations in tropical and temperate waters, the latter are spread too far apart in time to describe the cycle properly.

In the freshwater phytoplankton, the amplitude in volume (or weight) is

* Cushing *et al.* (1963) throughout the text refers to a group; Cushing & Tungate (1963); Cushing (1963); Cushing & Većetiċ (1963); Cushing & Nicholson (1963).

Table 1 *The amplitude and duration of production cycles*

Region	Units	Amplitude	Duration (months)	Author
A. *Marine phytoplankton*				
Canadian arctic	*n*/l	?	3	Grainger (1959)
N. Atlantic	*n*/l	×70–100	4	Friedrich (1954)
Norwegian Sea	*n*/l	×1000	5–6	Halldall (1953)
Baltic Sea (Kiel Bay)	*n*/l	×15	4–6	Lohmann (1908)
W. Scotland (Loch Striven)	*n*/l	×100–1000	3–4	Marshall & Orr (1929)
NE. USA (Block Is.)	*n*/l	×20	2–3	Riley (1949)
NE. USA (Narragansett Bay)	*n*/l	×7; ×12	5	Martin (1965)
Malabar Coast, India	*n*/l	×29	3	Subramanyan (1959)
Sargasso Sea	*n*/l	×7	3	Riley (1957)
NE. USA (Block Is.)	Chlorophyll	×2	2–3	Riley (1949)
NE. USA (Long Is.)	Chlorophyll	×5	1	Conover (1956)
NE. USA (Georges Bank)	Chlorophyll	×20	3	Riley (1946)
English Channel (Southampton Water)	Chlorophyll	×20	3	Raymont (1963)
E. Greenland	Chlorophyll		3	Digby (1953)
Sargasso	Chlorophyll	×10–20	3–4	Ryther & Menzel (1960)
Antarctic	Net chlorophyll	×200	8	Hart (1934)
English Channel (E1)	Net chlorophyll	×40	2	Harvey *et al.* (1935)
E. Greenland	Net chlorophyll	×9000	3	Digby (1953)
Mediterranean (Saronicos Gulf)	Radiocarbon	×2	4	Becacos-Kontos (1968)
W. Greenland	Radiocarbon	(0–1.2 g C/m^2 per day)	8	Nielsen (1958)
Baltic	Radiocarbon	×10	8–9	Nielsen (1965)
Denmark	Oxygen	×10		Nielsen (1958)

Table 1 *contd.*

Region	Units	Amplitude	Duration (months)	Author
Sargasso	Radiocarbon	×3.5–10	3–5	Menzel & Ryther (1960)
Sargasso	Radiocarbon	×20	3	Menzel & Ryther (1961*a*)
B. *Freshwater phytoplankton (as selected by Hutchinson, 1967; references quoted by him)*				
Schleinsee (Germany)	*n*/ml	×2	3–6	Vetter
Worthersee (Austria)	*n*/ml	×4, ×7, ×15, ×100	2, 5, 1, 3	Findenegg
Lunzer Untersee (Austria)	*n*/ml	×60, ×500, ×8000	4, 4, 2	Ruttner
Windermere (England)	*n*/ml	×100, ×4000	4, 3–4	Lund
Lake Michigan USA)	*n*/ml	×4	8	Dariby
Millstättersee (Austria)	mm^3/m^3	×3	7	Findenegg
Lake Constance (Germany)	mm^3/m^3	×10	7	Grim
Lake Erie (USA)	mm^3/ml	×20	4	Chandler
Lake Mendota (USA)	mg/l	×3		Birge & Juday
Linsley Pond (USA)	mg/l	×3	2	Hutchinson
Lake Waubesa (USA)	mg/l	×5	5	Birge & Juday
Lake Beasley (USA)	mg/l	×1.5	12	Pennak
Boulder Lake (USA)	mg/l	×5	2	Pennak
Lake Gaynor (USA)	mg/l	×6	8	Pennak
Lake Maggiore (Italy)	*n*/ml	×30	4	Berardi & Tonolli
Lenore Lake (USA)	mm^3/l	×25	2	Anderson *et al.*
Lake Maggiore (Italy)	mg/m^3 chlorophyll	×4	5	Berardi & Tonolli
Soap Lake (USA)	mg/m^3 chlorophyll	×30	3	Anderson, Comita & Engstrom-Heg
C. *Marine zooplankton*				
Arctic		×6–10	3–5	Grainger (1959)

Table 1 *contd.*

Region	Units	Amplitude	Duration (months)	Author
N. Atlantic		×7	6–7	Friedrich (1954)
Norwegian Sea (Ona)	Displ. vol.	×20	7	
(Skrova)	Displ. vol.	×4	12	Wiborg (1954)
(Eggum)	Displ. vol.	×40	8	
(OWS M)	Displ. vol.	×1.5	12	Ostvedt (1955)
North Sea (Flamborough)	Displ. vol.	×3.5	7	Wimpenny (1966); Cattley (1950, 1954)
(Anholt Knob)	Displ. vol. (calc.)	×10.0	12	
(Smith's Knoll)	Displ. vol. (calc.)	×10.0	8	
(Borkumriff)	Displ. vol. (calc.)	×8.0		ICES (1912–14)
English Channel (Varne)	Displ. vol. (calc.)	×10.0	4	
(7 Stones)	Displ. vol. (calc.)	×40.0	8	
N. Pacific	mg/m^3	×200.0	3	McAllister (1961)
Norwegian Sea (OWS M)	n/m^3	×50–100	2.5	Ostvedt (1955)
NE. USA (Block Is.)	Displ. vol.	×10–20	3	Deevey (1952)
NE. USA (Long Is.)	Displ. vol.	×5	5	Deevey (1956)
NE. USA (Narragansett Bay)	n/m^3	×3, ×7	7	Martin (1965)
Irish Sea	n/m^3	×25	3–5	Herdman (1918)
English Channel (E1)	n/m^3	×30	6–7	Harvey *et al.* (1935)
N. Pacific (OWS X)	?	×3	?	Kitou (1958)
Great Barrier Reef	n/m^3	×3	6	Russell & Colman (1934)
Bermuda	n/m^3	×2	9 (?)	Moore (1949)
Mediterranean (Algiers)	n/m^3	×7	3	Bernard (1955)
Hawaii Windward	Displ. vol.	×10	2	King & Hida (1957)
Leeward			12	King & Hida (1957)

always very much less than that in numbers. The amplitude in Lake Windermere, in the English Lake District, when *Asterionella* is dominant, is three and a half orders of magnitude in numbers per litre (*n*/l) or in mg/l; in many lakes, however, there are sequences of production of different species, the amplitude of any one of which is high, but when they are summed, the total amplitude is lower. In the sea the division rates are much the same as those in lakes and there is a similar succession of species. Both in lakes and in the sea, the duration ranges from one to twelve months. Outside high latitudes in the sea, however, the average duration is about three months; in tropical waters, the cycle may be ill-defined when it is not very prominent. In lakes, the average duration tends to be longer, perhaps because the cycle, although often determined by physical conditions, does not suffer such rigorous physical limits as it does in the sea. If the marine and freshwater cycles are considered in the same way, the average outburst lasts about three months and its amplitude ranges from less than one to three orders of magnitude in chlorophyll or in algal volume.

Because the animals in the plankton have not always been completely sampled, the animal cycles shown in Table 1 may yield less information than the algal ones. The duration, which may be well estimated, ranges from two to twelve months, with an average of six or seven months. Most of the observations in quantity are in displacement volume and estimates of amplitude range from less than one to one or two orders of magnitude. The general conclusion is that the amplitude of animal production is less than that of the algae, but it lasts longer. No differences with latitude can be detected, but there may be differences between the continental shelf and the deep ocean. The differences between sampling methods have not been entered in the table, because they add no information, nor do they invalidate the conclusions drawn from the data.

Production cycles have not often been compared from year to year. Davidson (1934) sampled the cycles of algal stock with a net in Passamaquoddy Bay (NB, Canada) for five years and showed that the peaks in different years did not vary by more than a factor of three, although the increase in numbers in each year amounted to three orders of magnitude. Lund (1964) has published samples of the production cycle of *Asterionella* in Windermere for sixteen years (1945–60) and the results are remarkably consistent in the peak numbers, varying by less than a factor of three (except 1960). Table 2 shows the numbers of six copepod species sampled at the peak with a Hensen net by Herdman (1921) over a period of ten years in the Irish Sea. The table also shows the average biomass, as mg/m^3 dried weight for six areas off California sampled with a one-metre net, for a period of nine years (Thrailkill, 1959,1961,1963). In Herdman's results, the coefficients of variation for *Temora* and *Calanus* are 44% and 80%;

Table 2 *Time series in numbers and volumes of zooplankton*

Numbers in the Irish Sea

	1907	1908	1909	1910	1911	1912	1913	1914	1915	1916	Mean
Oithona	3503	3028	4152	8556	10795	10767	10501	12267	8567	8741	8171
Pseudo-calanus	1472	1909	2768	3788	3420	6566	5270	5025	7864	7076	4583
Acartia	678	1225	566	3340	3025	4600	4224	2443	3060	2320	2571
Paracalanus	335	980	483	2134	4281	3713	2067	5863	1180	1443	2276
Temora	1350	220	559	1442	994	2148	1552	1572	1170	1523	1234
Calanus	277	211	191	152	55	810	244	438	130	188	266

(Herdman, 1921)

Mg/m³ dried wt off California

	1949	1950	1951	1952	1953	1954	1955	1956	1957	Mean
45°N.–55°N.	35	36								
40°N.–45°N.	39	53	59	124				111		
35°N.–30°N.	22	94	41	65	60	50	91	92	64	64
30°N.–25°N.	11	90	18	28	61	23	89	58	27	45
25°N.–20°N.	7	12	10	20	23	17	38	79	21	25
20°N.–15°N.		5	16	12	13	9	11	17	18	13

(Thrailkill, 1959, 1961, 1963)

but for the other animals, there is an increasing trend with time, by a factor of three or four. If the true variation in numbers is represented by that of *Temora* and *Calanus*, the trend exceeds the normal variation by a factor of two to three. There are no trends in dried weight with time off California, but over fifteen degrees of latitude the mean weight increases by a factor of four from south to north. Colebrook & Robinson (1961,1965) and Robinson (1970) have examined the variability of production cycles in the waters around the British Isles, but their work will be examined in more detail in a later chapter.

The quantities in the production cycle sometimes lead one to suppose them to be so variable as to be unruly. The survey sketched in Tables 1 and 2 shows that some order can be extracted from the data: the annual variations in amplitude are restricted to two or three orders of magnitude in algal numbers, one or two orders of magnitude in algal volume, chlorophyll and animal biomass. The duration of the algal cycle is about three months, whereas that of the animals is on average about six or seven months. The variation between years in plants and animals is much less than the annual increase in amplitude; but secular and latitudinal trends may be much greater.

Some quantitative aspects of the production cycle

So far, the production cycle has only been described in the general terms of amplitude and duration. A typical production cycle in temperate waters, or the annual spring increase, will now be examined in more detail. The complete cycle in stock of both plants and animals in mm^3/l off the north-east coast of England as sampled in 1954 is illustrated in Figure 1 (Cushing & Vućetič, 1963). The study started in the third week of March and ended in the second week of June and the peak of algal stock occurred towards the end of April and the peak of animal stock took place between three weeks and a month later. The same patch of *Calanus finmarchicus* Gunn, was followed for the whole period by two ships working alternately. Algal cells were counted and measured in a settling cell with an inverted microscope; some flagellates were counted with this method, but some must have been inaccessible. The animals were sampled with a Fine International net (125 μm mesh size) for nauplii and juvenile herbivores, a Hensen net (420 μm mesh size) for adult herbivores and a Gulf III high-speed net for euphausids and fish larvae; ciliates and tintinnids were counted in the water samples.

The stock of algae increased from about 0.02 mm^3/l to just over 0.90 mm^3/l, nearly fifty times in about thirty days. The stock of animals increased from about 0.01 mg/l to about 0.32 mg/l (wet weight $\simeq mm^3/l$), an increase of thirty times in just over fifty days. If the curve of algal stock in time was symmetrical about the peak, the spread of the curve in time would be about twenty days. That of the stock of animals would be more than thirty days. The apparent rate of production of the algal stock was about 0.03 mm^3/l per day and that of animal stock about 0.006 mg/l per day. The production of algae was very much higher, because their reproductive rates at the height of the spring increase were between 0.5 and 1.0 divisions per day (Cushing, 1963). The increment in stock between time t_0 and t_1 is $(P_1 - P_0)$, where P_0 and P_1 are the stocks at times t_0 and t_1.

$$P_1 = P_0[\exp(R-G)(t_1-t_0)] \qquad (1)$$

where R is the instantaneous algal reproductive rate; G is the instantaneous rate of grazing mortality.

$$P_1 - P_0 = P_0[\exp(R-G)(t_1-t_0)-1]. \qquad (2)$$

The quantity produced, $P_R = [P_0R/(R-G)][\exp(R-G)(t_1-t_0)-1]$. (3)

The quantity grazed, $P_G = [P_0G/(R-G)][\exp(R-G)(t_1-t_0)-1]$. (4)

When $R>G$, production is not much greater than stock, but when $R \simeq G$, the increment of production can be very much greater than that of stock.

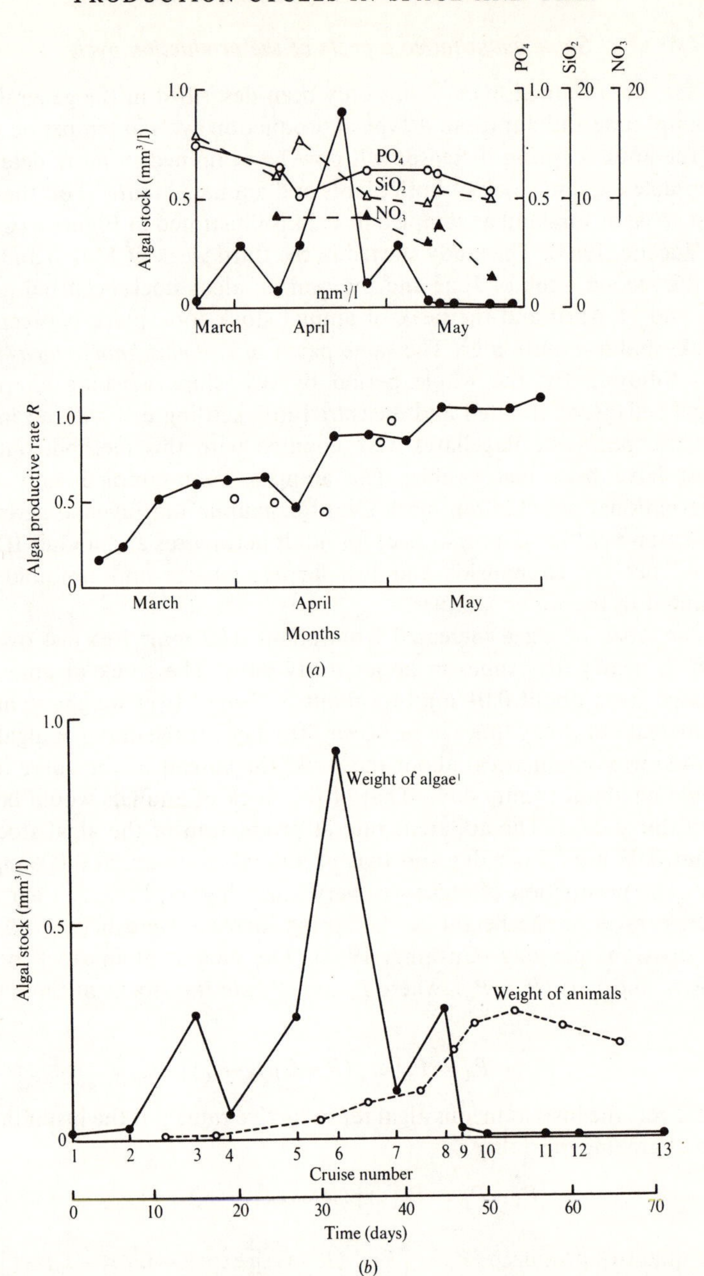

Figure 1 (*a*). The production cycle off the north-east of England in 1954. The upper figure shows the changes in algal stock in mm³/l, in phosphate, silicate and nitrate in μg-at/l. The lower figure shows the development of the algal reproductive rate estimated theoretically (●) and with the cell size decrease method (○) (Cushing *et al.*, 1963). (*b*) Stock of plants in mm³/l and weight of animals in mg/l.

When $R<G$, production is greater than stock and the same arguments apply to the quantity grazed. The algal stock is the quantity observed and sampled during the production cycle. The quantity produced is usually of much greater magnitude, although in time it traces a bell-shaped curve, which resembles that of the stock. The quantity grazed is the difference between the production and the stock; it peaks just after the peak in production because then stock is consumed and transferred to the quantity grazed. In this brief formulation, it was assumed that the losses by deaths other than grazing and by sinking are small compared with the changes due to the major exponents, R and G. At times other than during the spring outburst, other forms of mortality may become important. The method of estimating the algal reproductive rate (Cushing, 1963), that of the cell size decrease of diatoms as diatoms divide (the transapical width decreases and so the decrease in cell size indicates the algal reproductive rate) included losses due to sinking; an estimate of deaths, other than

Table 3 *Algal stock, algal reproductive rates, production and quantities grazed off the north-east coast of England in the spring of 1954* (*Cushing* et al. *1963c*)

Cruise	Stock (mm^3/l)	Cruise duration (days)	R	P_R (mm^3/l)	P_G
2	0.025				
		7	0.53	2.80	2.64
3	0.286				
		7	0.55	0.33	0.16
4	0.055				
		3	0.48	0.60	0.43
5	0.279				
		8	0.88	7.87	6.98
6	0.898				
		5	0.88	1.11	0.61
7	0.107				
		6	0.86	5.06	4.86
8	0.298				
		6	0.88	0.30	0.14
9	0.022				
		3	1.04	0.31	0.20
10	0.001				
		3	1.04	0.02	0.17
11	0.005				
		6	1.04	0.03	0.026
12	0.004				
		4	1.04	0.03	0.026
13	0.005				

grazing, was made by counting the numbers of dead cells and empty frustules, which were small.

Table 3 shows the stock, the quantity produced (P_R) and the quantity grazed (P_G), by cruises. The maximum production was 7.9 mm^3/l, about nine times the peak in algal stock, so the average rate of production was 0.26 mm^3/l per day or about forty times the rate of production of animal stock. Because the daily rate of animal production is rather low, there is less difference between herbivore stock and production, at least in the early part of the cycle before the older copepodite stages are eaten, so it always is an underestimate. The algal reproductive rate increased during the period from 0.25–1.0 divisions per day and the effective animal reproductive rate was 0.012–0.024 per day.

The amplitude of algal production was about 400 times. The peak was reached thirty-two days after the start of production. There was a delay of twenty-two days between the start and the beginning of effective grazing as indicated by the point of inflection of a smooth curve to the data. Presumably, effective grazing is generated by the appearance of the first medium copepodites about ten days after the spawning of the overwintering generation at the end of March or beginning of April. About ten days after the cycle had got under way, an interpolated value of stock is 0.075 mm^3/l and of production is 0.3 mm^3/l and these are presumably the food levels at which the overwintering generation spawned. The effective delay period of about three weeks determines the amplitude of production and the total quantity transferred to animal flesh. Presumably the differences between production cycles shown in Table 1 are generated by differences in delay periods.

The production cycles in different regions

In this section, the production cycles will be compared from different regions, those of high latitudes, temperate waters and those of subtropical and tropical seas. Figure 2 shows the algal production cycle from net hauls in the Southern Ocean south of the antarctic convergence; Currie (1964) converted Hart's data (1934) with a fixed chlorophyll–carbon ratio (0.009 PPU = 0.05 g C/m^2; the plant pigment unit, PPU was a rough unit of chlorophyll, established by Harvey, 1934). The amplitude of this cycle of algal stock is about forty times and it developed for about three months to reach the peak. The whole cycle lasted for eight months right into the antarctic winter.

The long duration of the cycle is attributed to two conditions peculiar to antarctic (and perhaps arctic) regions. The algae are exposed to nearly continuous daylight and so their daily reproductive rate may be higher, which delays the onset of effective grazing. The herbivorous euphausids

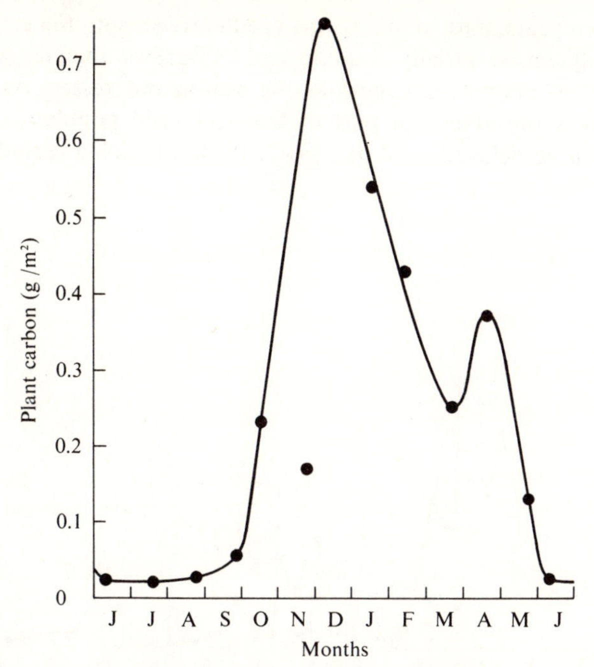

Figure 2. The production cycle in the antarctic (Currie, 1964); Hart's (1942) data were converted from chlorophyll to carbon.

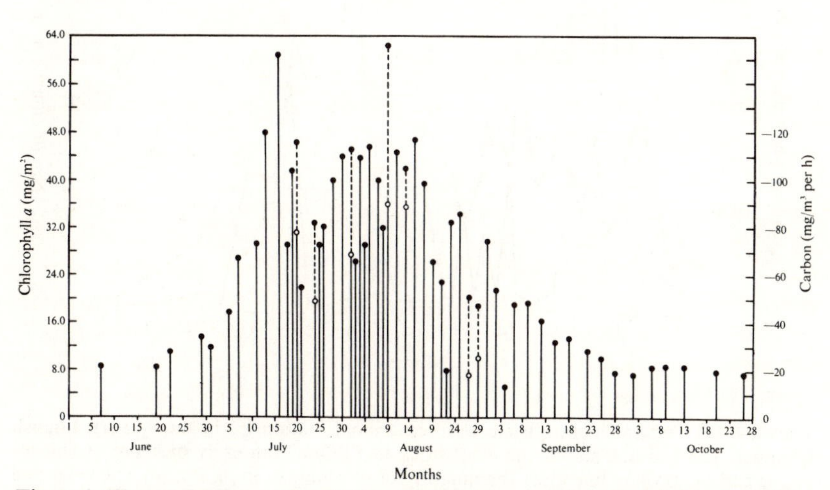

Figure 3. The production cycle of chlorophyll (●) in mg/m² carbon (○) in mg/m³ per hour at Drifting Station Alpha in the high arctic in 1958 (English, 1959).

live for two years, and so in any one productive season, the grazing population will consist initially of adults and half-grown animals, with a new generation of animals emerging as the season progresses. As this new generation is the abundant part of the euphausid population, effective grazing will be delayed until its arrival. If the full delay period is longer

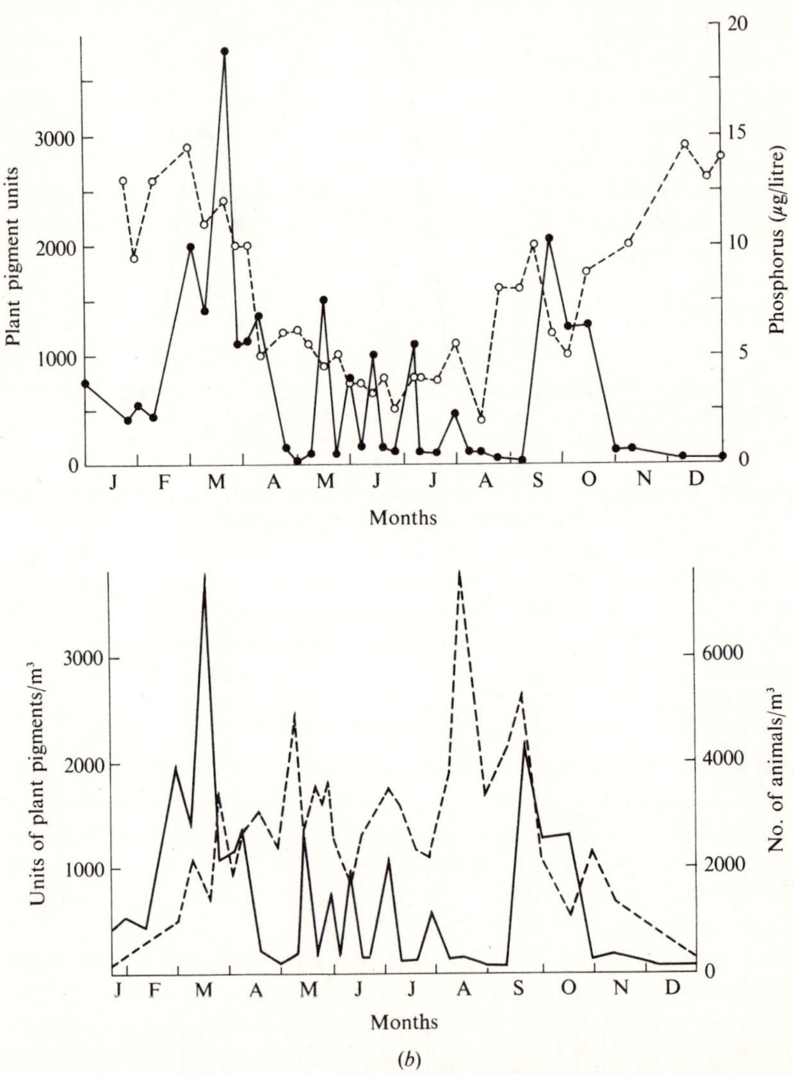

(*b*)

Figure 4. (*a*) The production cycle at International Station E1 in the western English Channel. The full line shows the algal stock in PPU/m³ (an early measure of chlorophyll) and the broken line gives the quantity of phosphorus in μg P/l. (*b*) The solid line shows the algal stock and the broken line shows the numbers m³ of animals (Harvey *et al.*, 1935).

than that of the temperate cycle, that is up to six weeks, as opposed to three weeks, then the whole cycle will last for longer and generate a larger stock.

In the high arctic, beneath the ice, the production cycle is slight, in contrast with that in the open antarctic. Figure 3 illustrates the cycle in chlorophyll and radiocarbon at 'Drifting Station' Alpha, at about 83°N. (English, 1959). The amplitude of the cycle is low in both factors, compared with that in the antarctic, with a factor of about six or seven and it takes about six weeks to develop; the whole cycle endures for less than three months. There is a small population of herbivores which probably controls the algae, unless there is another mechanism of reducing numbers after the peak. The samples were taken under the ice and show that a low-amplitude production is possible even in this inhospitable ocean. The low and persistent daylight probably determines the nature of the production cycle, as it does under the different condition in the Southern Ocean. A thin replica of the great antarctic cycle under the arctic ice shows the potentiality for the development of large standing stocks and heavy production in both the arctic and the antarctic.

In temperate waters, there is a different form of cycle and Figure 4 shows a complete annual cycle there in PPU/m^3 of algae from net hauls, in phosphorus and animals in n/m^3 at International Station E1, off Plymouth in the western English Channel. There is an autumn outburst of algae as well as one in the spring, separated by a low level of stock in mid-summer. The spring outburst was completed in 1933 at the end of March about two months after it had started before any real decline of phosphorus had taken place; the autumn cycle reached a peak in September and faded away in October. The cause of the autumn outburst was formerly attributed to the regeneration of nutrients as the equinoctial gales broke up the thermocline (Atkins, 1925*b*). If Figure 4 is compared with Figure 14, the autumn outburst starts in August and peaks in September before any surface phosphorus has been regenerated. The figure shows that the autumn outburst occurs as the zooplankton density declines. The amplitude of the autumn cycle, relative to that in the spring, varies considerably and presumably such differences are related to differences in the decline of zooplankton populations.

In the deep tropical ocean, there are not many records of seasonal cycles but Menzel & Ryther (1960,1961*a*) have described an annual cycle in radiocarbon production off Bermuda (Figure 5). There is a mid-winter or very early spring peak (with an amplitude of ten times at most and perhaps as low as three times). Because the cycle appears to develop rather slowly from October to February, the delay period may be quite short and effective grazing might start relatively early in the cycle. The cycle in the winter of 1958–9 is possibly the normal one; in the winter of 1957–8,

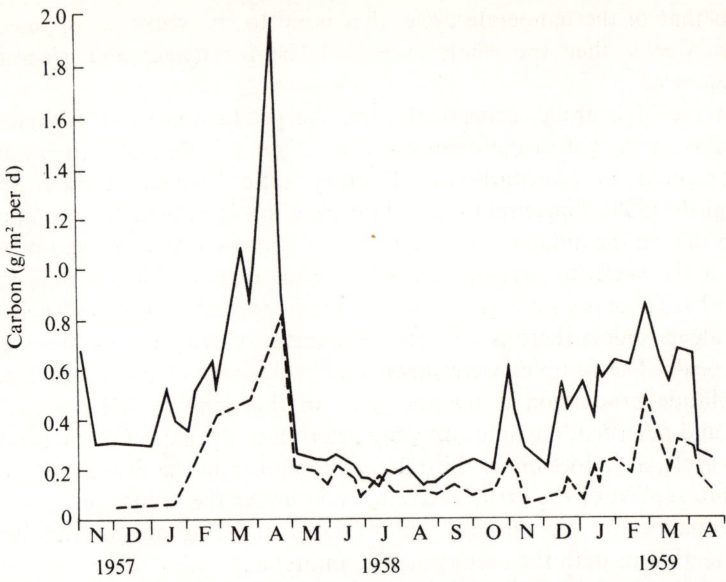

Figure 5. The production cycle in the Sargasso Sea in g C/m^2 per day (Ryther & Menzel, 1960); the net production (- - -) and the gross production (—) are shown.

there was a storm that may have increased the depth of mixing and forced the animals into deeper water, so perhaps introducing a longer delay. Absolutely, the delay would have still been short. If the cycle of three times in 1958–9 is typical of the tropical ocean outside the upwelling areas and the divergences, it is of low amplitude and it continues through all seasons.

The evidence selected here from the arctic, antarctic, English Channel and Sargasso Sea shows that the amplitude of the production cycle increases with latitude. If the production cycle is a function of a predator–prey relationship, then a high amplitude would indicate a long delay. It can also be seen that the delay period increases with latitude partly because there is a continuous production short of 40°N. (or S.) in latitude and a discontinuous one beyond it. Then the delay period increases with seasonal difference. It is unlikely that the slight differences in seasonal light intensity in tropical waters can generate the amplitude of the production cycle there. Such small differences may, however, affect the seasonal pattern of mortality. In higher latitudes, the production cycles are seen to be discontinuous, with no production at all in mid-winter. After production has started, the overwintering generation of herbivores spawns and then, later in the season, the grazing capacity of the growing generation of new herbivores appears, i.e. where the production cycles stops in mid-winter, there is a long delay period.

The range of seasonal cycle with latitude is summarized in Figure 6 in pictorial form. There is a single peak in the arctic mid-summer, a double peak in temperate waters and a continuous trend in the tropics. The seasonal differences increase with latitude and so does the delay period. In the arctic or antarctic, the herbivores have to live for the greater part of

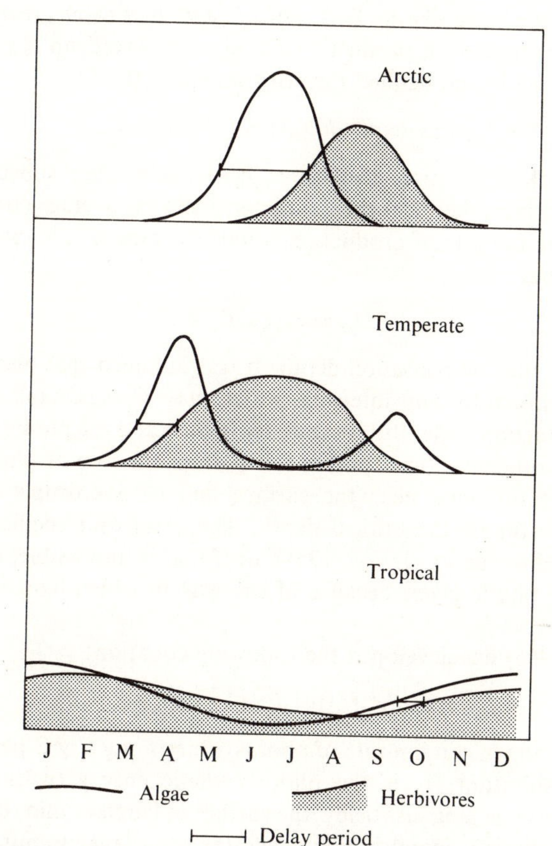

Figure 6. A diagrammatic representation of production cycles in different regions (Cushing, 1959).

the year with little food, usually in deep water (e.g. *Calanus hyperboreus* Krøyer) (Conover, 1962) and the animals tend to be bigger and to carry large fat stores. Because they are rather larger, their life history is longer in the cooler water and in itself this must increase the length of the delay period. Steuer (1910) compared the appearance of plankton in Greenland, in the North Sea and in the Mediterranean and he arrived at much the same conclusion as that summarized in Figure 6.

Variability in the production cycle

In temperate and high latitude waters, production stops in winter and so the start of the cycle in spring each year is a most important part of the process. Gran & Braarud (1935) showed that there was a depth to which photosynthesis, integrated in the water column, equalled respiration summed to the same depth. This *critical depth* had to be greater than the depth of mixing before production could start. Sverdrup (1953) put the problem formally and defined the critical depth, D_{cr}:

$$D_{cr}/[1-\exp(-kD_{cr})] = (1/k)(I_0/I_c), \qquad (5)$$

where k is the extinction coefficient; I_0 is the average subsurface light intensity, during a day; I_c is the light intensity at the average compensation depth (where the rate of production equals the rate of respiration). It can be shown that

$$D_{cr} = \exp(k)D_c/k,$$

where D_c is the compensation depth. It was assumed that photosynthesis was proportional to light intensity, which may be reasonable at the start of any production cycle. But if the water is clear and the photic layer (from surface to compensation depth) is fairly deep, the rate of photosynthesis can become inhibited near the surface and so Sverdrup's assumption might overestimate the critical depth. The extinction coefficient should be measured in energy (Jerlov, 1951), or if that is impossible, in the lower part of the photic layer, because of the way in which light is filtered in seawater.

Steele (1965) has developed the following equation:

$$D_{cr} = (p_m/r')\{\exp(1/k)\}[1-\exp(-I_0/I_m)], \qquad (6)$$

where p_m is the maximum rate of photosynthesis in g C/g C per h; r' is the rate of respiration; I_0 is the photosynthetic energy (within the band 380–720 nm) reaching just below the surface in langleys/min (ly/min); I_m is the photosynthetic energy which generates p_m in langleys/min. Nielsen & Hansen (1959) showed that the rate of respiration amounts to about 0.1 p_m so in a day, $p_m/r' = 5$. This method has two advantages (1) that the compensation depth need not be measured, and (2) that the dependence of photosynthesis on light intensity is well described.

As the production cycle develops in temperate waters, the magnitude of the algal reproductive rate depends upon the ratio of D_c to D_m, the depth of mixing. The compensation depth increases as the average daily solar radiation increases in the season and the depth of mixing decreases as the average daily wind strength decreases during the same period. Because radiation increases and wind strength decreases as the season proceeds, the production ratio (D_c/D_m) increases. So, the algal reproductive rate increases from a low value (about 0.3) to a high value (about 1.0), when $D_c/D_m \simeq /1$. The increase in the algal reproductive rate is a function of two environmental factors, the increase in radiation and the decrease in wind strength and, as will be shown later, it is through this agency that climatic factors affect production cycles and at a second stage, the fortunes of the fisheries.

An estimate of the compensation depth can be made from estimates of radiation and cloud cover distributions (Cushing, 1972*a*). From Laevastu (1960);

$$Q_c = 135 \sin \delta \, (a + b \sin \delta) \text{ mWh/cm}^2, \tag{7}$$

where Q_c is the total incoming radiation; δ is the sun's altitude; a and b are constants.

Lumb (1964) tabulates values of a and b for a variety of conditions.

$$Q'_s = (Q_c/69.6)\text{ly/min} \tag{8}$$

$$Q'_{ss} = 0.5\, Q'_s - (3\, Q'_s/\delta_N)\text{ly/min},$$

where Q'_{ss} is the photosynthetic energy reaching the surface at noon in the band 380–720 nm (Strickland, 1960, suggests that half the total energy is used in the photosynthetic band). Laevastu (1960) recommended the expression ($3/\delta$) to describe the losses due to reflection at the altitude, δ. Then,

$${}_N D_C = (1/k) \ln (3.5 \; 10^{-3}/Q'_{ss}), \tag{9}$$

where ${}_N D_c$ is the noon compensation depth. Strickland (1960) suggested that the energy reaching the compensation depth was 3.5 10^{-3} ly/min. From Jerlov (1951), energy extinction coefficients can be used. Thus, it is possible to estimate the compensation depth from meteorological observations. The depth of mixing can sometimes be measured with bathythermograph observations and sometimes from measurements of wind strength, but on occasion the physical structures are complex enough to preclude an estimation of the depth of mixing.

Colebrook (1965) has shown that production cycles can vary in timing, amplitude and spread. Cushing (1973*a*) suggested that all three sources of variation originate in differences in the rate at which the production ratio (D_c/D_m) develops. By the same token, there are variations in production in different areas. It was shown earlier that differences in the character of production cycles between latitudes have their sources in the delay periods. The latter may be modified by differences in temperature. The production ratio is controlled by the increase in radiation and the decrease in wind strength, which may be linked in any single environment in a number of subtle ways. As wind and sun vary seasonally in different ways over quite short distances, the production ratio is the most important single factor connecting climatic change with the recruitment of year classes to the fisheries. Colebrook & Robinson (1961,1965) have examined the colour of plankton recorder samples as degrees of 'greenness' by statistical squares for about fifteen years in British waters. Figure 7 shows four different forms of production cycle in Marsden squares in an area between Iceland, Norway and the northern coast of Spain. The four types

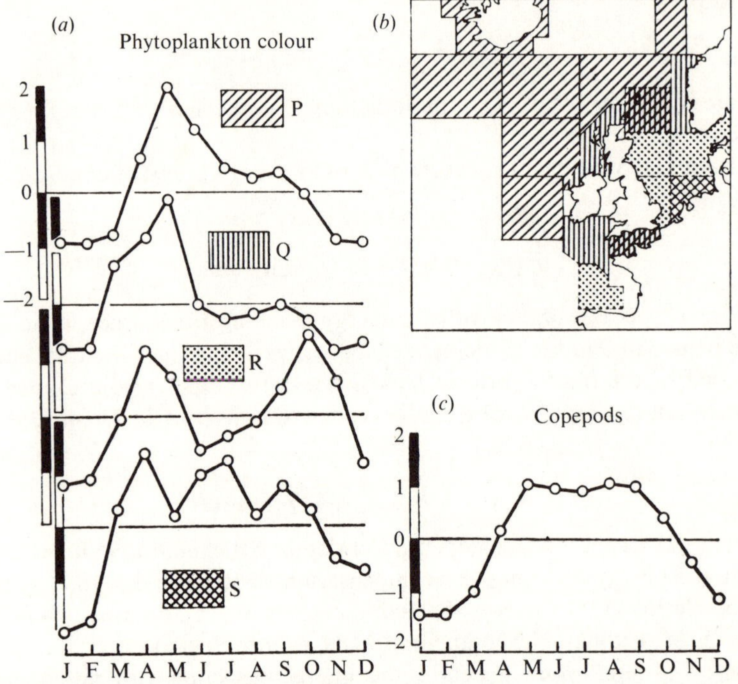

Figure 7. The production cycles, in greenness, estimated from the plankton recorder network in the waters around the British Isles (Colebrook & Robinson, 1965); see text for explanation.

shown in the figure belong to characteristic areas and are indicated by letters. The oceanic cycle *P*, peaks in May, persists for five months and there is a second low-amplitude cycle in August; the cycle is restricted to waters beyond the continental shelf. There is a shelf cycle, *Q*, which starts earlier than the oceanic one and there is also a low amplitude autumn peak. In the central North Sea, there is a 'Bank' cycle, *R*, which reaches a peak in April and there is a pronounced autumn peak of the same magnitude as that in the spring; this form of cycle is also found in the Bay of Biscay. In the Southern Bight and the German Bight of the North Sea, there is an early production cycle, *S*, persisting in a variable manner between March and October. In the English Channel and in the northern North Sea, the cycle appears to be intermediate in character between *P* and *Q*. The oceanic cycle occurs over the greatest depth of water and the inshore cycle is found in shallow water and the two middle cycles occur in waters of intermediate depth. Hence, the differences between the types of cycle are mainly differences in depth.

In shallower water, the depth of mixing is obviously limited by the depth of water. This apparently direct effect of water depth on a production cycle is not, however, as simple as it would at first appear. Shallow waters are often relatively sheltered, whereas the ocean is exposed to the wind from all directions and from great distances; so the depth of mixing in the ocean is greater than that in sheltered water and it decreases in depth more slowly as spring progresses. Near the coast, the wind strength is often reduced by its passage over the land, but a more important factor is that the fetch of the wind is much reduced. There is a marked association between the timing of the production cycle and the depth of water. In very shallow water, as in the southern North Sea, a further complexity is introduced. In this area, the tidal streams are swift and they scour the sea bed and make the water turbid. They are strong enough to mix the algal cells from top to bottom in a calm sea and so the depth of mixing is the depth of the sea. There is some evidence that, as the wind rises, the turbidity increases and the compensation depth decreases. So the development of the production ratio (in this case, D_c/Z, where Z is the depth of water) depends upon the effect of the wind strength directly upon the compensation depth (Cushing, 1972*a*). The effect of tides is greatest between mid-Channel and 53°N. (Dietrich, 1950,1954*a*) and outside this area, west and north, the water is clearer and thermoclines become established in summer. From these areas out to the ocean, the production cycle is governed by depth of water, as the mixing depth is determined by the strength of the wind. Further, the start time of the cycle in deeper water tends to be later because differences in the critical depth are controlled by the depth of mixing. The oceanic cycle, the last in this trend, starts last and persists in summer in a single peak.

An interesting point about the oceanic cycle is that in the autumn, the overwintering stages of copepods sink to below 600 m (Ostvedt, 1955; Conover, 1962), where they winter in cool water. In spring, they rise to the surface at about the right time to create the first generation of effective grazers. It was suggested above that the spring outburst in some temperate and high-latitude waters owed its nature to the delay to the spawning of the overwintering generation at a fixed food density. The onset of effective grazing by the new generation and the length of the delay period were inversely related to temperature, because of the rate of development. In the ocean, the overwintering generation climbs to the surface perhaps in response to the increased rate of change of light intensity during the day, perhaps independently of the production cycle itself. Yet the oceanic cycle is probably the most variable; perhaps its development in time is slow enough to allow independent emergence of grazers from the depths to be as effective as the emergence linked to food production in shallower water. Heinrich (1962) has pointed out that some copepods, like *Calanus*, depend upon the food available during the spring outburst for spawning and that others, like *Pseudocalanus*, do not. There may be two forms of control, one linked to the production cycle and the other independent of it. The study of the variability of the production cycle is complex because many processes are fitted together. Those that act directly, with obvious links, must be distinguished from those which only do so in an evolutionary sense, and for which the links must appear coincidental.

Conclusion

The production cycle in any region, excluding upwelling areas and areas of divergence discussed in the next chapter, depends upon three factors:

(1) the critical depth, which governs the start of production;

(2) the delay period, which depends upon the appearance of the first new zooplankton generation of the year – whether it is linked to the cycle or not;

(3) the rate at which the production ratio, D_c/D_m, approaches unity as the production cycle develops.

Each of the three factors is indirectly affected by the depth of the sea and the extinction coefficient of the water and the delay period is governed partly by temperature. The light intensity at the compensation depth is approximately 1% of the average daily surface irradiance and the depth depends on the magnitude of radiation and upon the clarity of the water. The depth of mixing is greater in deeper water, but the deeper water is often clearer. Differences in production cycle become less as the depth of water is increased.

If upwelling areas and the oceanic divergences are temporarily ignored, the noon compensation depth increases towards the deeper water and towards the equator; however, at or near the equator itself, it is not so deep because of the divergences in the equatorial currents. In the Sargasso Sea, in summer, which is the less productive season, the compensation depth can be as great as 120 m (Clarke & Denton, 1962), but at the height of the productive season in the winter it can be as little as 60 m, because some light is absorbed and scattered by the algal cells (Riley & Schurr, 1959). The production ratio is unity, or not far from it, for a large part of the year.

Hence, production is continuous and, because the delay period is short, the cycle is of low amplitude. In high latitudes the water may be clear, but the sun is low in the sky and much of its light is reflected, so the compensation depth is shallow. The depth of mixing tends to be deep except close to the ice edge where there is a stable layer of melt water (Marshall, 1958). The production ratio starts low, but increases in summer with the long day and failing winds. The production cycle is discontinuous; it starts late and the delay period is long, so the cycle is of high amplitude and it lasts for a long time.

Because the tropical cycle is of low amplitude and is continuous throughout the year in a deep photic layer, it is an efficient cycle. No algae are spilt on the ocean floor as in high latitudes or in upwelling areas, food is available all the year round and both growth and reproduction in the animals are probably continuous. The distance between algal cells in tropical waters outside the upwelling areas is much greater than that at the peak of the high latitude cycle, where the cells are so close to each other that a herbivore can eat a cell immediately it has finished the one before. The transfer of material from one food level to the next is higher than the average, and as a method of spreading energy continuously through all trophic levels, the tropical cycle is efficient. In the disbalanced, long delay period, high latitude system, the two important exponents, R and G, are equal only at the transient peak, but in the steady-state system they are nearly equal all the year round. So the contrast between the two systems is this: in the temperate and high latitude systems, a large quantity of material is transferred from one trophic level to the next at low efficiency, whereas in the quasi-steady state system less is transferred more efficiently. Cushing (1971*a*) compared the primary production and an underestimate of secondary production in the Indian Ocean and showed that the transfer coefficients in the clear oceanic water were three times greater than those in the upwelling areas (see Chapter 2).

Production cycles have been described in this chapter in two ways: (1) amplitudes and durations were estimated for a number of cycles; (2) a few cycles were selected to compare open sea conditions in different latitudes. One of the disadvantages of the first method is that seasonal studies have been executed at places which are convenient rather than representative. For example, more than half of the observations on marine phytoplankton in Table 1 were made in bays and fjords which are accessible even if distant from civilized centres. The first method leads to one very simple generalization that the amplitude of the algal cycle ranges from one to three orders of magnitude and that it endures for about three or four months and in one or two cases for longer. Another generalization is that the amplitude of the herbivore cycle ranges from one to two orders of magnitude and endures for seven or eight months. In Table 1 it is difficult to detect trends in amplitude or duration with depth or latitude. From the selected cycles illustrated in Figures 1 to 7, there is a trend in latitude from the continuous low amplitude production cycle in parts of the tropical and subtropical ocean to the discontinuous high amplitude cycle of higher latitudes. The boundary between the continuous and the discontinuous cycles is probably the poleward edge of the subtropical anticyclone. The difference between the two forms of cycle stems from the fact that production in the sea stops in winter in temperate and high latitudes.

Table 2 shows that the variation in biomass or numbers from year to year is low as compared with the variation in either in any one season. Differences between latitudes or trends in time appear to be more important than the average variation from year to year. In a different way Figure 7 illustrates considerable differences in algal cycles in British waters, a small area as compared with gross differences across the latitudes. The sources of variability were described in terms of the critical depth which governs the start of production in high and temperate latitudes and the production ratio (D_c/D_m) which controls the rate at which the production cycle develops. One of the most important factors in the cycle is the delay period, the time it takes for effective grazing to develop from the start of production. If the animals have not died off in mid-winter, as on the temperate or high latitude shelf, or have not migrated down into deep water in the high latitude ocean, the delay period is short as in low latitudes. In temperate and high latitudes, the overwintering generation spawns when food starts to become available or when the animals rise from deep water and it takes more time for the subsequent generation to develop, especially in cold water. The long delay is associated with high amplitude and long duration in high latitudes and the inefficient transfer of energy between trophic levels. The short delay period is associated with the continuous cycles of low amplitude with efficient transfer in low latitudes. Over an annual period, there may be less difference than expected between the two systems in the quantities produced.

2

Production cycles in upwelling areas

Introduction

In the previous chapter a comparison was made between the continuous low amplitude production cycles of the deep tropical ocean and the discontinuous high amplitude cycles of high latitudes. The differences in amplitude were attributed to differences in the length of the delay periods which were long in high latitude and short in low latitude. The upwelling areas in tropical and subtropical waters were explicitly excluded from this comparison because they were considered to be special cases of the high latitude form of cycle. Large quantities of living material are produced there as in high latitudes because upwelling imposes a long delay period on the cycle.

Physically, upwelling processes can and do occur at any latitude where the wind direction and the action of Corioli's force permit it. But the production cycles that occur in upwelling waters in high latitudes cannot be distinguished from the pattern of the high amplitude cycles found in the surrounding waters. In high latitudes the physical process of upwelling does not appear to be important to the development of the cycle. In tropical or subtropical waters the high amplitude cycles found in upwelling areas do differ from the steady-state cycles characteristic of low latitudes. Thus, though the upwelling process is widespread its biological significance is limited to the tropical and subtropical oceans.

The distribution of upwelling areas

In an upwelling area, the wind blows either parallel to, or at a slight angle to, the coast and by Corioli's force, the deflecting force of the earth's rotation, it drives the water offshore. The deflection is at right angles and to the right of the current in the northern hemisphere and the other way round in the southern hemisphere. Where the surface water is blown offshore, cool water upwells at the coast often not very far from the shoreline. This cool water is nutrient-rich and a production cycle of high amplitude occurs in it; as a consequence some of the world's great fisheries are found in upwelling areas. There are four main ones, in the California current, the Peru current, the Canary current and in the Benguela current and each is an eastern boundary current in a subtropical anti-cyclone in the Pacific or Atlantic oceans. The eastern boundary currents are broad, slow (about half a knot), a little freshened with rain, somewhat diffuse and they drift towards the equator and eventually join an equatorial current. The main high-pressure systems in the atmosphere over the ocean correspond to each subtropical anticyclone and they move slightly polewards as spring turns to summer. Consequently, the axes of the trade winds shift with season and, in each main area, the centre of upwelling shifts a little towards the pole as the season progresses.

The processes of upwelling, which may take place along a thousand miles or more of coastline in the main areas, are slow and the rate of upwelling is probably only of the order of 1–5 metres per day or less, on average; at the start of any upwelling cycle, or in some particular areas, the rate of upwelling may be higher, but it is the low average rate that generates the high production, as will be shown below. The water is rarely drawn up from depths greater than 200 m and in some lesser upwellings it originates from quite shallow water, 20–40 m. During the season of upwelling, a counter-current is generated below 200 m that travels towards the poles and sometimes surface counter-currents appear very close inshore also moving polewards. Off Peru, there is a major southbound counter-current at a considerable distance offshore.

About 100 km offshore in the four main areas, there is a dynamic boundary (Sverdrup, 1938) to the system, where the offshore movement of surface water is forced to sink by the inertia of the main mass of the oceanic surface waters. This convergence is associated with a divergence beyond it and Hart & Currie (1960) have described the whole system as a roller bearing, a cell of convergence and divergence. In principle, there is no difference between coastal upwellings and the divergences that may occur anywhere in mid-ocean (Yoshida, 1967). Beyond the dynamic boundary at 100 km, offshore divergences may persist for considerable distances, because of the general structure of the trade wind system. Divergences exist at various places around the rims of the oceanic anticyclones (Hidaka & Ogawa, 1958), but the processes are concentrated in the eastern boundary currents.

As noted above, upwelling is not very important from a biological point of view outside the tropical and subtropical regions. Poleward of the subtropical anticyclone lie the cyclones, on the boundaries of which water tends to sink rather than rise under the influence of the average winds. Hachey (1936–7) showed that upwelling occurs off Nova Scotia during a period of south-westerly winds; but the water sinks there when the wind blows from the north-east (Longard & Banks, 1952). The western boundary currents of the subtropical anticyclones, like the Gulf Stream or the Kuroshio, are too stable and too swift for normal upwelling; their speed and structure are effects of all winds blowing over all the ocean whereas upwelling is a local effect dependent upon intermittent winds. Again, the Somali current, the western boundary current in the Indian Ocean north of the equator, is twisted eastward by the Horn of Africa and the island of Socotra and as a result it tilts and cold water from below reaches the surface off the coast of Somalia.

In addition to the four major upwellings, there are a number of systems in the Indian Ocean and the Indonesian area dominated by the alternating monsoon system. In the eastern currents upwelling occurs on the Malabar coast of India, off the Andaman Islands, off Burma, Sumatra and off north-west Australia. That off the Malabar coast of India is the most important one of this group and its structure is a little mysterious. Darbyshire (1967) has suggested that it starts at the end of the south-west monsoon as the current tilts slightly. However, the eastern current upwellings occur during the period of the north-east monsoon, between October and March. The south-west monsoon blows from April to September and it is during this period that the geostrophic upwelling occurs off Somali when the Somali current flows. A most important upwelling occurs off southern Arabia during the south-west monsoon. The primary production in this area is among the highest in the world, although it is not yet exploited by fishermen to any substantial degree. A lesser upwelling occurs off the Orissa coast of India. In the Indonesian area, there are upwelling systems in the Flores and Banda Seas and off Vietnam there is an upwelling area north of the Mekong delta.

In the equatorial complex of currents in each of the major oceans, there are a number of systems of divergence. They lie along the edges of the north and south equatorial currents, above the undercurrent towards its eastern end and in the extensions to the eastern boundary currents, so each eastern tropical ocean is rich in production across a very widespread area. Where the equatorial counter-current turns into one of the equatorial currents, a dome is formed which is a large swirl within which the thermocline and cool water rises towards the surface, a form of upwelling. The most studied dome is that off Costa Rica, where the Pacific equatorial counter-current turns north and then east into the north equatorial current. In analogous positions in the equatorial system, lie the Guinea dome, the Angola dome and that off Sumatra. There is a host of minor upwellings which are gradually being discovered, for example, in the Gulfs of Panama, Nicoya and Tehuantepec in the eastern tropical Pacific, the area off Yucatan in the Gulf of Mexico and off Curaçao in the Gulf of Venezuela. In the Gulf of Tehuantepec, strong winds from the north are channelled into the bay: the same effect is found between Cape Town and the Cape of Good Hope (Andrews, Cram & Visser, 1970). This account of upwelling areas summarizes a more extensive one given by Cushing (1971*a*).

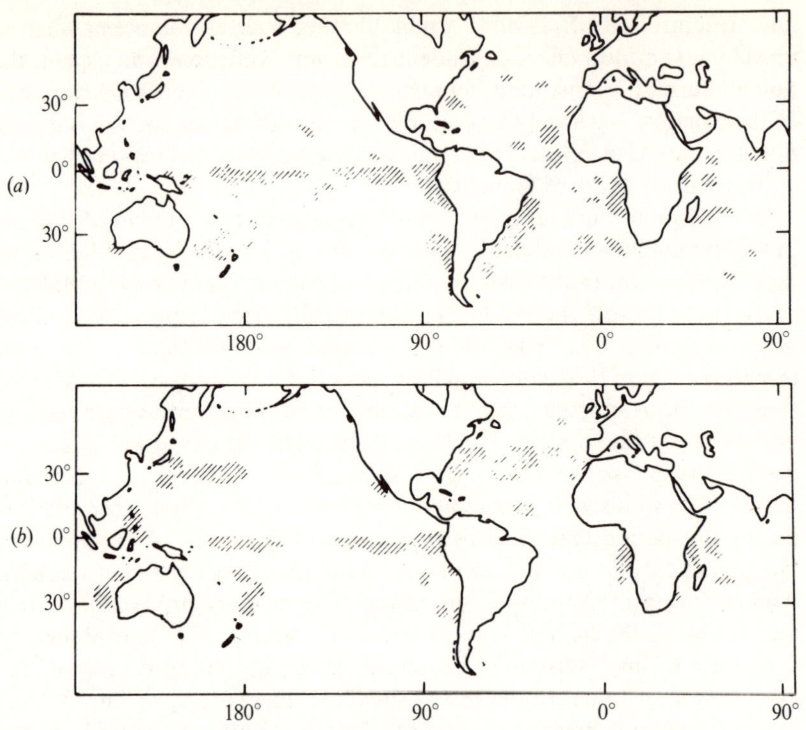

Figure 8. The distribution of sperm whale catches by Nantucket whalers between 1760 and 1926 (Townsend, 1935). (*a*) Northern summer; (*b*) northern winter.

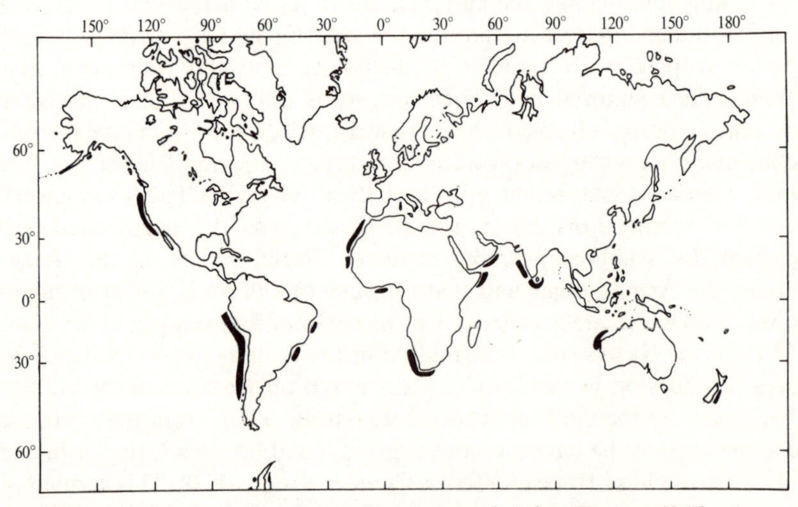

Figure 9. Phosphatic deposits on the continental shelves (Tooms, 1967).

The distribution of upwelling areas can be represented in a number of ways. Figure 8 shows the distribution of the catches of sperm whales made by Nantucket whalers between 1760 and 1926 (Townsend, 1935); Figure 9 shows the distribution of phosphatic deposits on the continental shelves; the distribution of guano islands is illustrated in Figure 10 (Hutchinson, 1950) and in Figure 11*a*,*b*,*c* and *d* are shown the phosphate, zooplankton (Reid, 1962) and radiocarbon distributions at or near the surface in the Pacific Ocean (Koblentz-Mishke, 1965) and that of radiocarbon in the world oceans (Koblentz-Mishke *et al.*, 1970). The four main upwelling areas, California, Peru, Canary and Benguela are shown in the whale

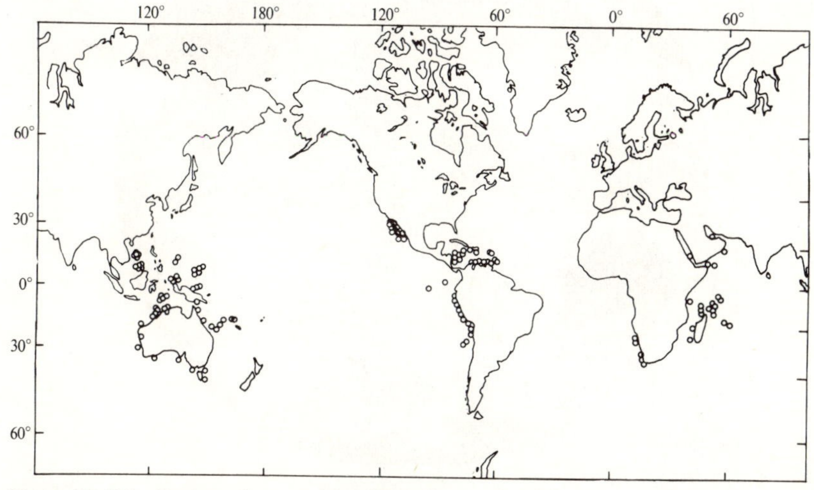

Figure 10. Distribution of guano islands on the continental shelves (after Hutchinson, 1950).

distributions and in the phosphatic deposits, but not in those of the guano islands. Some minor upwellings, like that south-east of Sri Lanka (Ceylon) and the wedge between the two equatorial currents north of Malagasy (Madagascar) are shown in the distribution of whale catches. An interesting point emerges from the distribution of phosphatic deposits and it is that the centre of upwelling off California is further north than it appears to have been in the last two decades (see Cushing, 1971*a*); again, the phosphatic deposits off South Africa extend east of Cape Agulhas, further east than the present regime of the Benguela current. As such deposits may represent the average position of upwelling for a long period of time, the present climate off California may be a deviation from the average, that is, the high pressure system over the North Pacific anticyclone may be, on average, a little further to the south.

The distributions of phosphate, zooplankton and radiocarbon show the equatorial complex in the Pacific Ocean very well. Not only are the two major systems, California and Peru, well shown, but also the divergences on the poleward edge of each equatorial current, the upwelling above the undercurrent in radiocarbon and the divergences of the eastern tropical ocean, together with the Costa Rica dome. It is even possible to trace the

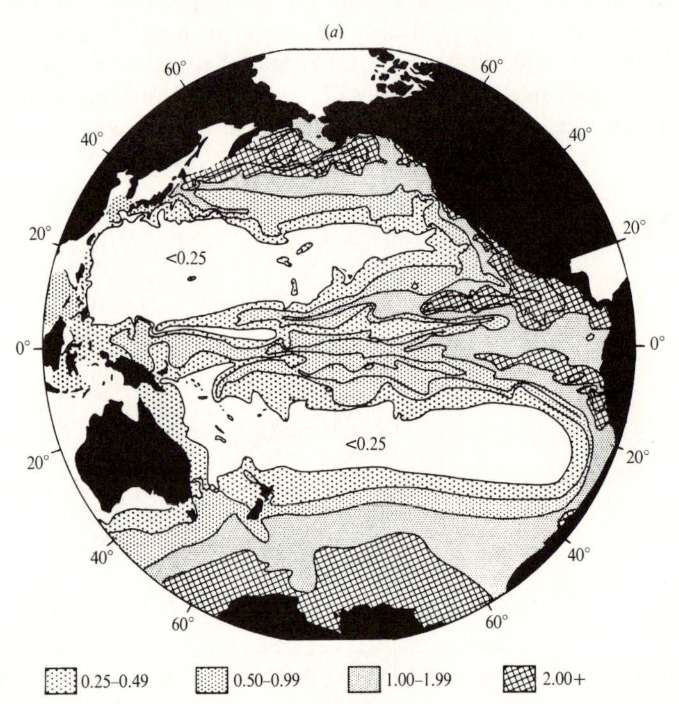

Figure 11. (*a*) Distribution of phosphates at 100 m in μg-at. PO_4-P/l in the Pacific (Reid, 1962). (*b*) Distribution of zooplankton between 200 m or 300 m and the surface in the Pacific (Reid, 1962); in ml displacement volume/1000 m^3. (*c*) Distribution of radiocarbon in the surface waters of the Pacific (Koblentz-Mishke, 1965); in mg C/m^3 per day; (1) < 100; (2) 100–150; (3) 150–250; (4) 250–650; (5) 650+. (*d*) Distribution of radiocarbon in the surface waters of the world oceans (Koblentz-Mishke *et al.*, 1970); in mg C/m^3 per day; (1) < 100; (2) 100–150; (3) 150–250; (4) 250–500; (5) 500+.

position of the equatorial counter-current in the distribution of the zooplankton from a position north of New Guinea almost to Costa Rica. A remarkable point is that the evidences of biological production extend so far away from the coastal upwellings and from equatorial divergences. It shows that the divergent processes are very widespread. The physical processes are limited to a band within 50–100 km of the coast in a coastal upwelling system, but the width of the biological system reaches hundreds of kilometres; but this is partly because the physical processes of divergence extend seaward and partly because the biological material drifts offshore.

The main conclusion from these diagrams is not that physical processes can be shown indirectly, but that the upwelling areas of various forms are very widespread and are of great biological importance.

It is often assumed that the production of living material is limited to the classical region of coastal upwelling within the dynamic boundary at 100 km from the shore. The distributions in Figure 11 show quite clearly that the production of the first two trophic levels spreads for considerable distances offshore. Fisheries in the main upwelling areas occur usually quite close to the shore, often within perhaps 50 km, where, of

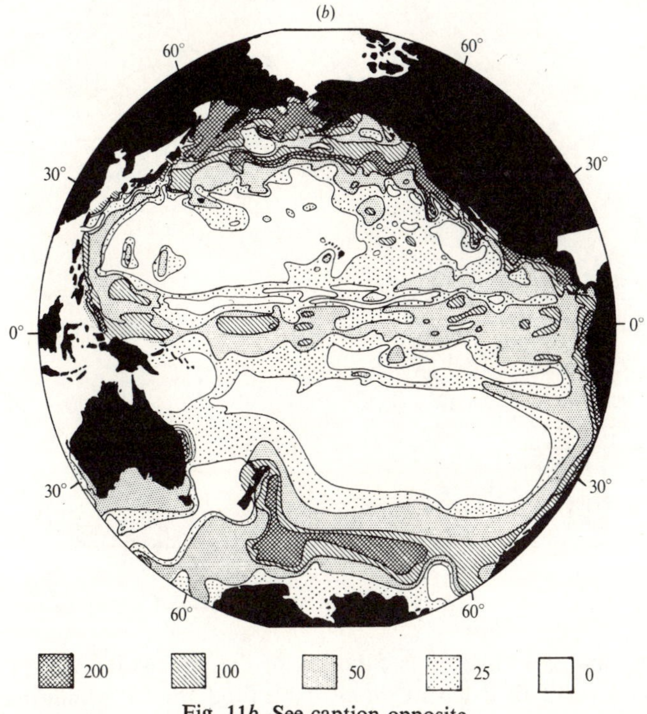

Fig. 11*b*. See caption opposite.

course, production is most intense and sardines, sardinellas and anchovies gather there. There are other stocks of mackerel, horse mackerel, hake and tuna-like fishes which depend either directly or indirectly upon the zones of production which extend up to 500 km offshore, or very much further in the eastern tropical ocean.

The production cycle in an upwelling area

In the first chapter the characteristics of the temperate production cycle were described. It starts in spring when the critical depth exceeds the depth

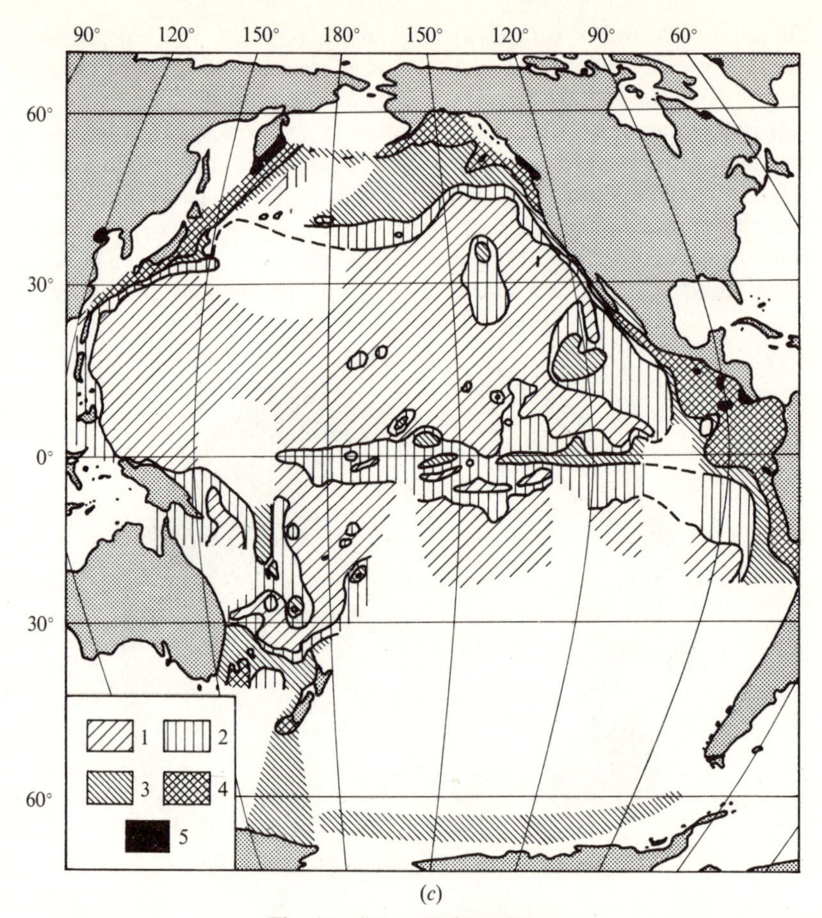

(c)

Fig. 11c. See p. 30 for caption.

of mixing and continues as a function of the production ratio (D_c/D_m). At a fixed level of algal density, or available food, the overwintering generation of herbivorous copepods spawns and, as the new generation grows, its grazing capacity increases. The delay period between the start of production and the onset of significant grazing mortality governs the form of the production cycle, its amplitude and spread (but not timing).

In an upwelling area, conditions that govern production resemble those in temperate waters. The water is drawn from 200 m in the main upwelling areas, which is below the depth to which most herbivorous zooplankton normally migrate in their diurnal vertical movements. The water rises slowly, on average at 1–5 m per day and contains a small resident population of algae and possibly a small residual population of herbivores. Let us suppose that the photic layer was 50 m deep, the production starts at the

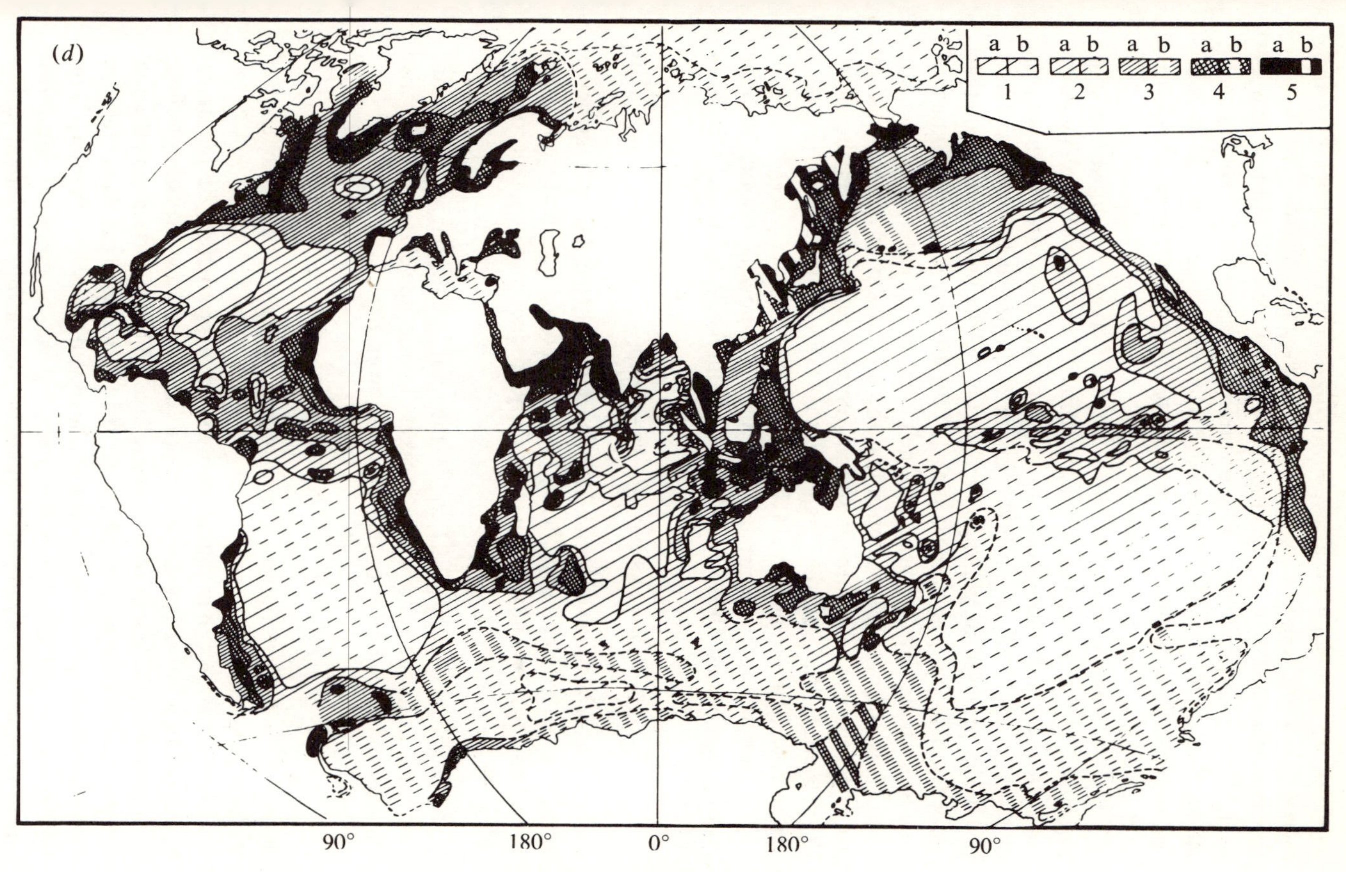

Fig. 11*d*. See p. 30 for caption.

depth of 50 m and increases exponentially in the rising water with the increase in radiation. When the water reaches the surface, somewhere between ten and fifty days later, the amplitude of the cycle must be near its peak. A residual population of herbivores in the rising water would generate grazing mortality and if it did not exist, active grazing would start as the water drifted offshore and mixed with water containing herbivores generated by earlier upwellings.

Thus, a simplified picture of the production cycle in an upwelling area presents a high amplitude cycle which attains its peak in the upwelling water as it reaches the surface. Again, in a simplification, the water drifts away at $\frac{1}{4}$–$\frac{1}{2}$ knot (6–12000 m per day) offshore as the production cycle declines under the pressure of grazing. Vinogradov (1970) has pointed out that vertical mixing plays an important part in the offshore drift and production may be sustained by divergences in the movement of water away from the shore. It is observed, for example, that the production in radiocarbon (in g C/m^2 per day) off southern Arabia remains high far from the shore (Ryther *et al.*, 1966), further than might be expected if the peak in a temperate production cycle was reached at the shoreline.

The mechanisms of the production cycle in the upwelling water, without grazing, can be described a little formally from Steele & Menzel (1962), as follows:

$$p = \alpha I_0[\exp-(-kz-2\exp(-kz))], \tag{10}$$

where I_0 is the average radiation at the surface in ly per day; z is the depth in m; k is the extinction coefficient (= 0.1, ignoring any increase with p); α is a constant (= 0.48); and p is the daily production in mg chlorophyll/m^2.

It can be shown (Cushing, 1971*a*) that the total production, p_t, in the water column during the time which it takes for the upwelling water to rise through the photic layer to the surface is given by:

$$p_t = [\alpha I_0/(2kW_h)]\{\exp[-2\exp(-kz_p)]-\exp(-2)\}, \tag{11}$$

where W_h is the ascending velocity in metres per day; z_p is the depth of the photic layer in metres.

Figure 12 shows the relation between total production during ascent and ascending velocity, with greater production at slow rates of upwelling, irrespective of the depth of the photic layer. The important point is that there is a critical upwelling velocity at about one metre per day and below it total production can be high, but above it there is less chance of high production. If the water rises too quickly it can be entrained into areas where production may become vulnerable to grazing, additional to that generated within the system. Off Peru, the wind blows more strongly than in other upwelling areas (Wooster & Reid, 1962). The average radiocarbon and zooplankton quantities are less off Peru than in the other three main areas, yet the anchoveta fishery there yielded the highest catches in the world, about 10 million tons per year, until recently.

The crux of the problem of production in an upwelling area is where and when grazing becomes effective. If the cycle is an exact analogue of the temperate one, grazing starts at about half the generation time of the zooplankton after the water enters the photic layer. Heinrich (1962) suggested that the generation time in the upwelling areas is about forty days, so

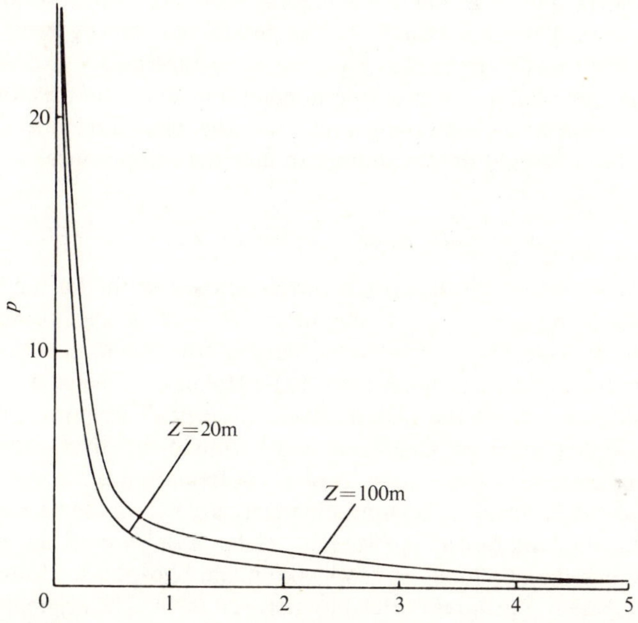

Figure 12. Relation between total production, p, during ascent and the rate of upwelling (Cushing, 1971), W_h, for two depths of the photic layer, 20 m and 100 m.

grazing might be expected to start after about twenty days. The whole system might be more complex for three reasons: (1) if the residual herbivores in the system are few or absent, grazing might not be effective in the upwelling water and at some stage, the cycle might become nutrient-limited; this is unlikely, however, because the traditional index of upwelling is nutrient-rich water at the surface near the shore which can only mean that the cycle was not nutrient-limited there; (2) when the water reaches the surface, the produced material might become vulnerable to older zooplankton by mixture; (3) euphausids might comprise a substantial proportion of the herbivore population, particularly a little way offshore. Jones

& Folkard (1968), in sections across the Canary current, show distributions of physical properties that might bring the vertically migrating zooplankton part of the way inshore.

When it is recalled that the productive mechanisms are reinforced hundreds of kilometres offshore by the divergences generated by wind stress, then it will be admitted that the full structure of the production cycle must be very complex. If the complications in the distributions of potential grazers are combined with those in the generation of primary production, the form of the cycle must differ in some degree from the simpler temperate one. Of course some such complexities do occur also in temperate waters, but in a different temporal order. Perhaps, the upwelling cycle is more flat-topped than the temperate one, as suggested above, because the transfer from the rising water to the surface drift offshore represents the change from a simple condition to a complex one. The analogy between the upwelling cycle and the temperate one is worth making, but it should not be allowed to hide the complexities.

The correlation between zooplankton and phosphorus

Figure 11 illustrates the surprising correlation across the Pacific between zooplankton and phosphorus in the surface layers. It was first observed by Hentschel & Wattenberg (1931) in the material from the *Meteor* expedition in the south Atlantic and more recently by Holmes, Schaefer & Shimada (1957) with data from the eastern tropical Pacific. The figure shows the main upwelling areas off California and Peru and the divergences in the equatorial system, in the eastern tropical ocean and on the borders of the subtropical anticyclones. The upwelling areas are rich in the production of algae and zooplankton and the water is rich in phosphorus. Seawards from the shoreline the quantities of zooplankton and phosphorus decrease and spatially there is a positive correlation between them. The original correlation, due to Hentschel and Wattenberg, led to the belief that the nutrients in the upwelled water caused the production there. Production cannot continue without nutrients, but as the phosphorus content is high all over the upwelling areas, it is unlikely that it is restrained by lack of nutrients. The algal reproductive rate starts to decline at rather low levels of nutrient (approx. 3.0 μM-NO_3; Eppley, Rogers & McCarthy, 1969).

An attempt will be made to account for the correlation observed between phosphorus and zooplankton. A correlation might be expected between the phosphorus used and the quantity of algae produced. As the algae are, in their turn, converted into animal flesh, the production of zooplankton might further be expected to depend upon the production of algae. Hence, there might be an inverse correlation between the quantity of zooplankton and the amount of phosphorus remaining in the water, the phosphorus observed. But there is no such correlation because the phosphorus observed is probably regenerated.

When phosphorus is used in the production cycle, it is converted into animal flesh, organic residue (dissolved and particulate) and a stock of algae, which is perhaps a transient component in the system (Harvey *et al.*, 1935). The phosphorus in the water, p'_R, is a residue from production processes and $p'_R < p'_T$, the initial quantity available for production, i.e. that available at 200 m in an upwelling area. Then the quantity used by the algae in production, $p'_u = p'_T - p'_R$. Let us suppose that p'_u is absorbed partly into a stock of algae (p'_p) and partly into animal flesh, ${}_{\alpha}p'_u$, and is partly regenerated, ${}_{\beta}p'_u$, and then the correlation is explained: for p'_R is perhaps composed of repeated increments of ${}_{\beta}p'_u$.

The high values of nutrients and the large quantities of zooplankton found in an upwelling area cannot be simply related because the zooplankton was not produced by the nutrients observed, except in a transient way. Hence, the correlation must be more complex, perhaps based on the regeneration of nutrients in the same processes which produced the zooplankton. The spatial extension of the correlation across the whole ocean from the anticyclones of low latitudes to the cyclones of high latitudes implies that regeneration must be rapid and continuous in all forms of the production cycle.

Wyrtki (1964) estimated production in the Costa Rica dome from the decrement of phosphorus during the period of upwelling and found that it corresponded fairly well with the radiocarbon measurements. In the next chapter it will be suggested that the decrement of nutrient,

$$\Delta N' = N'_c\,(0.85P_G - P_R), \qquad (12)$$

where N' is nutrient and N'_c is nutrient per cell.

It will be recalled that

$$P_G = \{P_0G/(R-G)\}\{[\exp(R-G)]-1\}$$

and

$$P_R = \{P_0R/(R-G)\}\{[\exp(R-G)]-1\}.$$

Then it can be shown that

$$\Delta N' = N'_c(P_1 - P_0)(0.85G - R)/(R-G). \qquad (13)$$

The radiocarbon measurement represents an increment to the algal population i.e. $(P_1 - P_0)$. So long as the R and G do not differ too much, as expected in a steady-state production cycle, $\Delta N'$ and $(P_1 - P_0)$ must be correlated.

The correlation between phosphorus and zooplankton extends across the whole Pacific from the Antarctic to the Aleutians and seaward from the upwelling areas to the centres of the subtropical anticyclones and it would be reasonable to extend it to the other oceans. Hence there is no real difference in character between the high amplitude cycles found in the upwelling areas and in the waters of high latitudes. Further, the phosphorus quantities observed in temperate seas during the late spring and summer may also be composed of regenerated phosphorus. The spatial correlation is one between the living material produced and the consequential nutrients regenerated. Similarly, the temporal correlation between the decline of nutrients and the rise of zooplankton observed, for example, in the English Channel may be of the same character, i.e. the observed decline of nutrients is the consequence of productive processes and not their cause.

The production cycle in an upwelling area starts from very low algal numbers, has a long delay period and hence a high amplitude and so is homologous with that in temperate waters. The water drifts away from the zone of coastal upwelling and mixes with the oceanic water offshore and the quantities of phosphorus and zooplankton decrease towards the centre of the deep ocean, until the whole cycle is in the quasi-steady state characteristic of that region. The transition from one cycle to another occupies the time the water takes to flow into the extension current towards the equatorial current. There is every stage between the two ideal cycles, the high amplitude one in the upwelling areas and the quasi-steady state one of the deep ocean.

The quantities produced in upwelling areas

In Cushing (1971*a*), a study was made of the production of living material in upwelling areas. From the distribution of surface temperatures, estimates were made of area and length of season; the area of biological production was considered to be two-and-a-half times as broad as the zone of coastal upwelling. The primary production was obtained from radiocarbon measurements in g C/m^2 per day and the total production was calculated by raising the average production by area and by length of season. The zooplankton has been sampled in many parts of the world as displacement volume from the top 100 or 200 m of the ocean, from standard nets; an average weight of secondary biomass was computed by area during the season of upwelling, in g C/m^2. A rough estimate was made of the generation times at different temperatures and so an underestimate of secondary production was obtained by raising the average biomass by the ratio of generation time to season. It is an underestimate because the average biomass during a generation represents the remainder after a fraction has been transferred by death to carnivores; if the generations overlap, the underestimate might be reduced. The tertiary production was derived by taking 1% of the primary production and 10% of the underestimated secondary production. The two estimates were correlated, which implies that a true ecological efficiency would be greater than 10% by a factor which is the ratio of true secondary production to the underestimated one.

With this rough method, common to all areas, it was found that the two major upwellings in the southern hemisphere, Peru and Benguela, were the most productive, each yielding 12–18 million tons wet weight of tertiary production each year. In the second rank were placed the two other main upwellings, Canary and California with annual tertiary productions of 8 and 3 millions tons, respectively, and that off southern Arabia with 2.5 million tons per year. There is a large number of less important systems, the Malabar coast of India (1.20 million tons per year), Orissa coast of India (1.0 million tons per year), Somali (1.25 million tons per year), Ghana (1.0 million tons per year), Java (2.3 million tons per year) and the Costa Rica dome (2.6 million tons per year). There are many small areas for which no calculations were made, like the gulfs in the eastern tropical Pacific. The extensive areas of production in equatorial complexes and the eastern tropical oceans have been excluded because they support only the tuna fisheries; in any case the production in such areas is not very intense, but is accumulated over broad areas. However valuable the oceanic tuna fisheries may be, their total yield in weight from the world ocean cannot be much more than 5 million tons wet weight per year (Gulland, 1970). The total yield from the world's upwelling areas is not yet fully exploited, but a total yield of about 100 million tons could be taken, of which a fair fraction (between one-third and one-half) is already being caught.

Another way of comparing the productivity of upwelling areas is to use the intensities (g C/m^2 per day) directly. Table 4 summarizes the results; in the original analysis each of the four main areas was divided into three or four sub-areas. As far as possible the boundaries are natural ones, often defined by the angle made by the coastline to the direction of the prevailing wind during the season of upwelling.

The richest areas in the intensity of production are off Peru, in the Canary current and in the Benguela current; the highest individual observations were recorded off southern Arabia (Ryther *et al.*, 1966), off Cape Blanco in the Canary current (Lloyd, 1970) and off Pisco in Peru (Strickland, Eppley & Rojas, 1969). The rank in Table 4 is very roughly the same as that given above, in that most of the main areas are also the most productive in intensity and that the lowest intensities are found in the open ocean. But there are certain anomalies; for example, the intensity of primary production in the Gulf of Thailand is as high as anywhere in the world. Further, those areas off Arabia, Somalia and Java are more intense in productivity than parts of the California current or parts of the Benguela. The differences in g C/m^2 per day across the whole world system are little more than

Table 4 *Some upwelling areas ranked in order of production intensity*

The major upwelling areas were divided into two or three subareas.

g C/m² per day	Upwelling areas
1	Peru (1), Canary (1), Benguela (2), S. Arabia, Vietnam, Gulf of Thailand
0.3–1.0	Peru (2), Canary (3), Benguela (1), Somali, Guinea dome, Malagasy (Madagascar) wedge, Orissa coast, Java, Sri Lanka (Ceylon), Flores, Banda, E. Arafura
0.3	California, Costa Rica dome, Chile, Benguela (3), New Guinea, Andaman Is., NW. Australia
0.1	Open ocean

an order of magnitude, but the differences in total production are much greater, perhaps two or three orders of magnitude; so such differences are those in area and in length of season. The distribution of winds along the world coastlines and across the ocean itself govern the distribution of total production.

A comparison between the intensities of production in the coastal upwelling areas may be made with those in the temperate production cycles. Off the north-east coast of England, Cushing & Vućetič (1963) described the spring production cycle from March to May 1954. From Table 3, assuming that the carbon–wet weight ratio was 0.065, the peak production was 4.25 g C/m^2 per day and the average during the productive period was 1.25 g C/m^2 per day. This average value is close to that found in the upwelling areas; individual peak values in upwelling areas are sometimes higher than that recorded off the north-east coast of England. Thus, the production cycles in upwelling areas and in temperate waters are not only homologous in principle, but can be compared in quantity.

A further study with the same very rough method was made of the production in the Indian Ocean (Cushing, 1973*a*). During the International Indian Ocean Expedition, a large number of radiocarbon observations were made and many catches of zooplankton were hauled from the sea. Indeed, it was possible to class both groups of observation in 5° squares across considerable areas of the Indian Ocean during both the south-west monsoon (April to September) and the north-east monsoon (October to March). There was enough information to calculate transfer coefficients from primary production to the underestimate of secondary production; the transfer coefficient is simply the ratio of the low secondary production to primary production. Figure 13 shows the dependence of transfer coefficient upon the intensity of primary production in g C/m^2 per day. At a middle level of primary production, the transfer coefficient is 10%, whereas at high primary production it is only 5% and at low production it is 15%. For an increase in production of about one order of magnitude, the transfer coefficient is reduced by a factor of three. Whatever the true value of secondary production, the average transfer coefficient of 10% suggests that the underestimate is not too great.

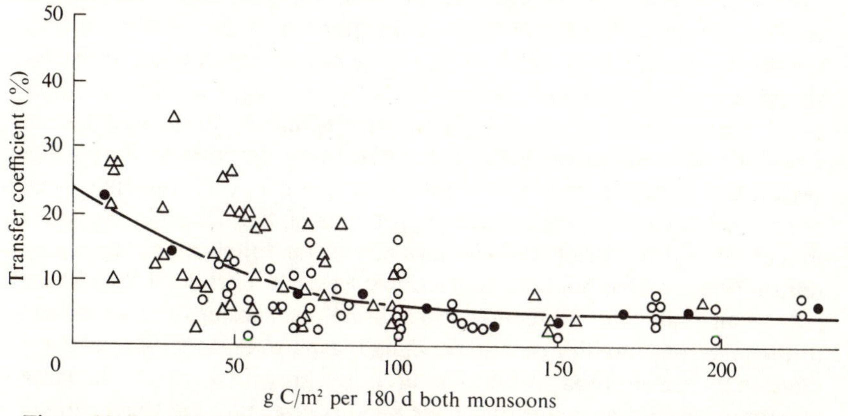

Figure 13. Dependence of transfer coefficients from primary production to an underestimate of secondary production, as function of primary production (Cushing, 1973b); SW. monsoon (○); NE. monsoon (△); means (●).

The information was taken from the upwelling areas in the Indian Ocean and from the areas in the centre of the ocean. Consequently, the transfer of living material is three times as efficient in the open ocean as in the upwelling areas. In the open ocean, with a quasi-steady state production cycle, the communities are diverse and the transfer of material from one trophic level to the next is efficient. By contrast, in the upwelling areas, the production cycle is disbalanced, the communities are simple and the transfer of material is inefficient. It is so inefficient that diatoms and other algae are spilt on the sea floor in productive processes and form a sediment of green mud (Sewell, 1935).

From the point of view of the commercial fisheries, the upwelling areas are much more important than the open ocean because (a) the intensity of production in g C/m^2 per day even with reduced transfer coefficients, will yield three times as much material to the herbivores (b) herring-like fishes exploit the herbivores directly in the upwelling area, whereas, in the open ocean, the herbivores are taken by small fishes, for example, myctophids of a few centimetres in length, which are eaten by tuna. The yield of tuna-like fishes in the world ocean is about 5 million tons (Gulland, 1970b), which implies a production of 10–15 million tons of tuna and perhaps ten times as much of myctophids each year. The difference between the upwelling areas and the open ocean in fisheries is that the herring-like fishes are caught in dense concentrations within perhaps twenty and certainly fifty miles of the coast, whereas the myctophids and their analogues are spread across the whole deep ocean in a dispersed phase and are almost impossible to catch in sufficient quantity to make any boat pay.

There is possibly another factor which affects the production of herring-like fishes. The sardines off California (Ahlstrom, 1966) and the anchoveta off Peru (Flores & Elias, 1967) appear to spawn near the coastline as if at the very points of upwelling. Indeed, the anchoveta exploits the herbivores so effectively that the distributions of anchoveta eggs exclude those of zooplankton almost completely. There are mechanisms in the structure of upwelling area which carry the fish away from the equator during the season and return them out of season. For example, off California, hake are carried north from their spawning grounds off San Diego in the counter-current during spring and summer and in the following autumn they return in the eastern boundary current (Alverson & Larkins, 1969). There may be an analogous mechanism which brings the sardines and anchoveta to the point of upwelling at the spawning season so that their larvae have more than enough food to eat. The little fish grow and exploit the large quantities of food available and their parents dominate the third trophic level: indeed there is reason to suppose that the anchoveta may compete with zooplankton for algal food on some occasions (de Mendiola, 1971). In a similar way, as will be shown in a later chapter, the pilchards in the English Channel dominated the third trophic level for a period of about forty years. If the temperate production cycle corresponded point by point with that in an upwelling area, such domination might be expected at the third trophic level.

Conclusion

The production cycle in an upwelling area is of the same character as that in temperate waters. The quantities produced in the upwelling areas are about the same as those of the cycle peaks in higher latitudes. The two forms of production cycle resemble each other in that there is a delay in both between the start of production and the appearance of effective grazing. As a consequence, each cycle is of high amplitude and much living material is produced both in the upwelling areas and in high latitudes. However, that in an upwelling area may not be described by the simple bell-shaped curve characteristic of the temperate cycle. The physical structures in an upwelling are complex, composed of large vertical and horizontal swirls, with divergences at the surface far out to sea. The mixing processes will tend to sustain production as the upwelled water drifts offshore. Off the coast of southern Arabia, areas of high production extend seaward for perhaps a thousand kilometres.

The water in an upwelling area, before it starts to rise, is clear and the solar radiation is high. If the average rate of upwelling over a period of weeks is about one metre per day, the quantity produced is high; if the rate of upwelling is higher; the quantity produced is less. The total quantity produced (as opposed to the intensity of production) is a consequence of the duration of the season of wind stress over an extensive area. The four main upwelling areas, California, Peru, Canary and Benguela are richer in primary and secondary production than the middle range of areas and in their turn, these are richer than the minor areas. The differences between the three ranges are physically in the strength of wind and in the time for which it blows. Hence, in a very rough way the longer the wind endures, the greater the production; so it follows from the equations given above that the rate of upwelling must, in general, be low. One of the differences between upwelling areas which may be important is the persistence of the wind. Off California, the winds tend to be intermittent whereas those off southern Arabia and Peru are persistent. If the wind stops blowing, production stops and the herbivores may die.

There are two conclusions of more general significance: (1) the transfer coefficients are three times higher in the 'poor' areas than in the 'rich' areas, so the difference in production intensity of one order of magnitude is reduced by a factor of three in the second trophic level in the production of herbivores; (2) the well known correlation between zooplankton and phosphorus implies that the decline in phosphorus in temperate waters with time and that in space from the edge of the subtropical anticyclone to its centre is a consequence of productive processes and does not cause them. The two conclusions are related in that each is a consequence of the part played by grazing in the production cycle. The difference between the inefficient rich areas and the efficient poor ones depends on the delay period in the cycle, which depends on the onset of grazing. In the open tropical ocean production is restrained by grazing, possibly at all stages of the cycles, whereas in the upwelling areas and in temperate waters production is only restrained by grazing at a late stage in the cycle, by which time a considerable quantity of living material has been produced.

The final interesting point in this chapter is the parallelism in two forms of stability. In descriptive terms it has long been suggested that a stable community is diverse, that is, there are many species, each few in numbers, such as that in the deep tropical ocean where the quasi-steady state cycle of production is found. Because the cycle is of low amplitude, the numbers are more or less stable. In contrast, in the high amplitude cycle of high latitudes and upwelling areas, there are few species each of high abundance, that is the community is not stable, nor is it diverse. So instability in numbers is associated with low diversity or instability in community structure and vice versa. Hence, the processes which determine the number of niches available are perhaps also those which regulate numbers between species.

3

The part played by nutrients in the sea

Introduction

Towards the end of the last century, and in the first decade of the present one, work began on marine production in Germany. Hensen & Apstein applied quantitative methods to plankton investigations in the sea and in freshwater. Brandt and Raben analysed chemically both seawater and freshwater.

Apstein (1896) noticed that different lakes varied considerably in their quantities of living material, and Brandt (1899) showed that these variations were correlated with differences in nitrogen content in the lakes. Similar observations with similar results were made in the sea. Raben (1905) showed that nitrogen, silicon and phosphorus were present in the sea in very small quantities and concluded that the density of marine algae would be expected to fluctuate according to the proportions of those nutrients.

Historically, these observations were of considerable importance because of their contribution to the formulation of the agricultural model of marine production. These ideas were summarized by Johnstone (1908). Justus von Liebig had stated in his famous Law of the Minimum that the growth of a plant depended on the amount of that nutrient which was present in the least quantity. The agricultural model assumed that the quantity of algae produced depended solely on the quantity of available nutrient, as simply as the growth of wheat in a field depends on the quantity of nutrients in the soil. No interaction with the grazing animals was postulated. Thus, algae were produced to the nutrient limit, and then were eaten by the herbivores. It is not surprising that the predator–prey relationship was not exploited to explain the processes, because Ross did not formulate the malaria equations until 1910 and the more general relationship was established by Lotka in 1925 and by Volterra in 1926.

The next stage in the historical development took place at Plymouth in the twenties. Atkins (1923,1925*a*), adopting the agricultural model, measured the quantity of phosphorus per unit wet weight of algae (as exemplified by *Phaeodactylum*). He also measured the seasonal decline of nutrients (phosphorus, nitrate nitrogen, silica and alkalinity) at the surface and at the bottom at International Station E1 near the Eddystone

Lighthouse in the western English Channel. From the decline in phosphorus during the summer and the quantity of phosphorus per wet weight of algae, an estimate of production was obtained. For example, a reduction of 40 mg P_2O_5 (which were the units used by Atkins) was found to be equivalent to 2.68 10^{10} cells of *Phaeodactylum*. From such estimates the annual production was estimated to be 1400 tons/km^2 per year of wet weight of algae; the same result was obtained in estimates of nitrate nitrogen and of alkalinity, but with silica the annual production amounted to 200 tons/km^2 per year.

Figure 14 (adapted from Cooper, 1938) shows a nearly continuous series of annual cycles in phosphorus for fifteen years at the surface and at the bottom. In each year the phosphorus was exhausted, or nearly so, at the surface in June, July or August. However, as shown in Figure 4, the main spring outburst took place in late March or early April and the decline in phosphorus did not start until the peak had been reached in the production cycle. Atkins (1925*b*) believed that production was stopped by lack of nutrients in the euphotic layer by the formation of the thermocline, but not until May or June. Production continues during the summer and, because of the higher rates, may be greater than appears from the magnitude of the stock, but the major quantity in temperate waters is produced during the spring outburst.

Also at Plymouth, Harvey, Cooper, Lebour & Russell (1935) measured the algal quantities from net samples as chlorophyll (in plant pigment units, PPU, or units of greenness) in a seasonal cycle, which is shown in Figure 4. An equivalent was established: 10^3 PPU = 0.08 mg P, so a reduction of 6–8 mg P/m^3 during the second half of the spring outburst was equivalent to 0.75–1.00 10^5 PPU, or thirty to forty times the stock. The difference between the production and the stock was taken to be the quantity eaten, so practically all the production was eaten by the herbivores. The rate of production was 0.11 mg P per day and as the herbivores amounted on average to 0.29 mg P, so the animals must have eaten 40% of their body weight per day. Today it must be pointed out that the stock of algae from the net was underestimated because many of the cells passed through the meshes. The quantity eaten was overestimated but the percentage of body weight per day taken by the herbivores was properly estimated (in phosphorus). The major discovery was that production was many times greater than the stock and that as it was reduced by grazing, it was transferred to animal flesh.

In the thirties, at Plymouth, there were apparently alternative views of the mechanisms of marine production. In the first view, based on the agricultural model, nutrients were considered to limit production as they were used up. In the second, the quantity produced was represented by the difference between the winter maximum of phosphorus and the

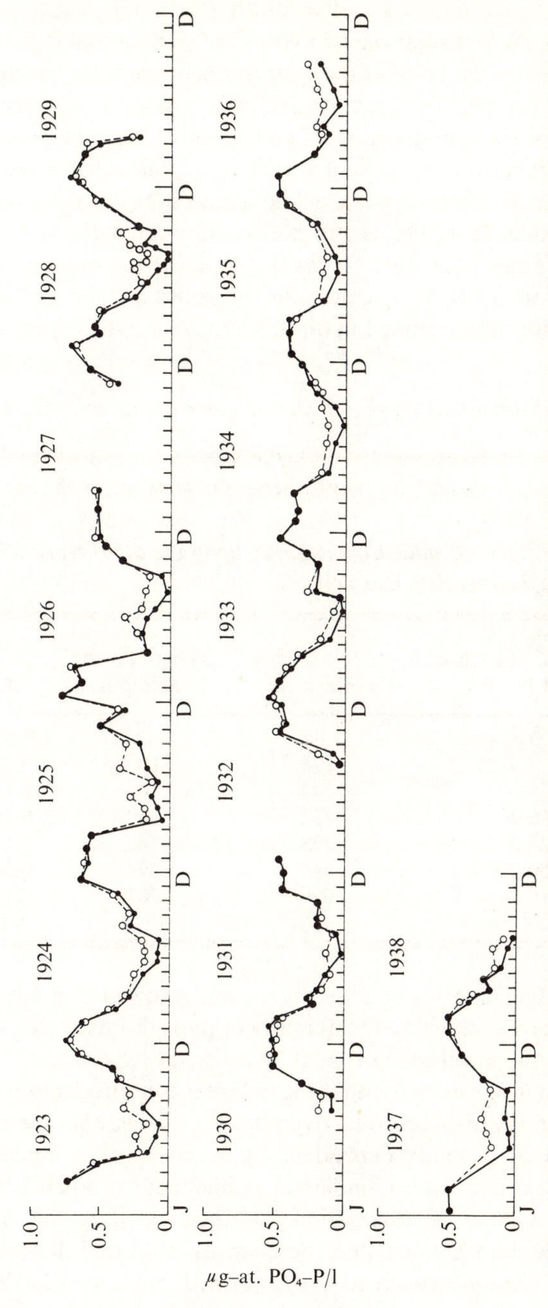

Figure 14. The concentrations of phosphate–phosphorus in μg-at./l at the surface (solid line) and near the bottom (broken line) at International Station E1 in the western English Channel from 1923 to 1938 (from Cooper, 1938).

summer minimum and the outburst was considered to be controlled by grazing and the quantity produced was only fractionally greater than the quantity grazed. At first sight the two views are distinct, but if the quantity grazed is practically the same as the quantity produced, the production of 1400 tons/km^2 per year of algae is effectively a grazed quantity and the quantity of herbivores produced is, of course, much less. The two ways of estimating production are reconciled as follows. Shown in Figure 4, the decline in nutrients after the peak of the algal outburst is the transfer of nutrients to animal flesh, faeces and particulate matter through the transient medium of the algal stock. After the transfer, the nutrient levels are low and in summertime production can be restrained by grazing or by nutrient lack until it is resumed in an autumn outburst (if it takes place).

Field observations on the effects of nutrients in production

Before the field observations are described, an important experiment by Ketchum (1939*a,b*) should be mentioned. He grew phosphorus-deficient

Table 5 *Effects of added phosphorus upon the algal reproductive rate of phosphorus-deficient cells*

Quantity of phosphorus added (μg P/l)	Increase (n cells × 10^6)	Days to reach maximum	R
0.5	86	6	0.07
2.5	198	10	0.14
5.0	344	14	0.14
10.0	432	14	0.14
25.0	493	16	0.14
50.0	744	19	0.14
110.0	520	20	0.14

cells in different quantities of phosphorus and showed that the quantity produced was proportional to the quantity of phosphorus in the water (but not to the quantity of nitrate nitrogen). Because the experiment lasted only for thirty hours it was assumed that the differences in production were due to differences in the algal reproductive rate. It followed that the reproductive rate might be directly dependent upon phosphorus concentration. Cushing (1955) carried out a similar experiment for a much longer time period with a number of observations between the start and end of the experiment. Ketchum's result that the quantity produced depended upon the quantity of phosphorus added was confirmed, but it was shown that the algal reproductive rate was constant at all levels of added phosphate except the lowest. Table 5 shows the results. The experiment was carried out at rather low light intensity, to show division rate and obtain a number of observations on algal numbers and hence on division rate. The two experiments can be reconciled by assuming that the nutrient limit of the reproductive rate is low (<2.5 μg P/l), but that differences in production arose because the phosphorus was exhausted more quickly in the lower concentrations.

It was the plastic-bag experiments (McAllister, Parsons, Stephens & Strickland, 1961; Antia *et al.*, 1963) that showed the real effect of possible nutrient lack at sea. Seawater, from which the planktonic animals had been

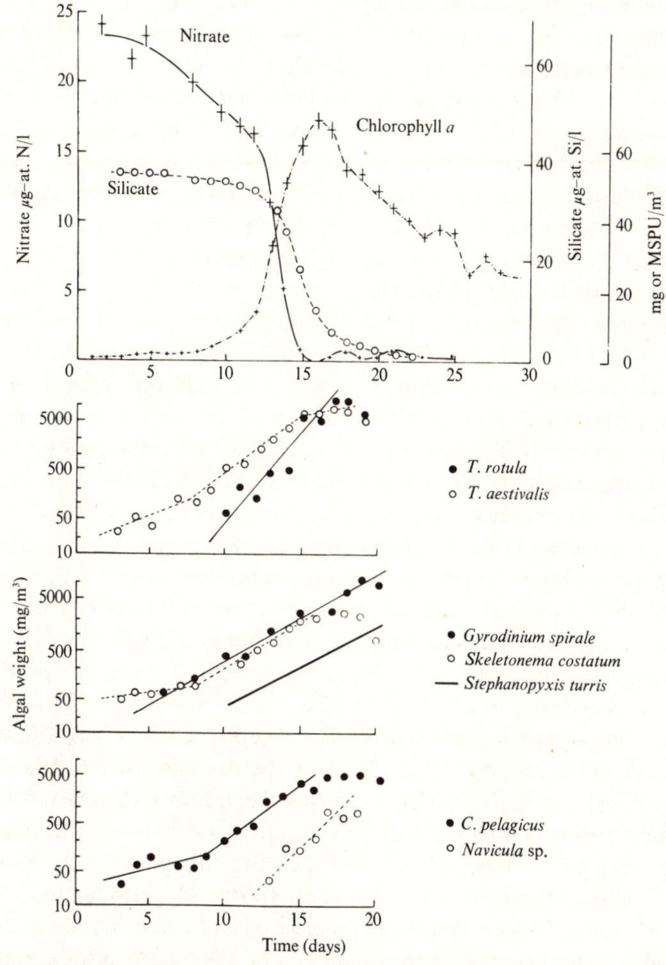

Figure 15. The results of the plastic-bag experiment (McAllister *et al.*, 1961); the decline in nitrate and silicate is shown together with the increase in chlorophyll *a*. The logarithmic increases of seven algal species in the bag are also shown.

filtered, was suspended in a large plastic bag in the ocean for a considerable period. The algae were counted frequently and numerous measurements were made of the nutrients in the water. The bag was large so that the surface of living algae was much greater than the surface of the bag itself. Then the bacterial populations on the bag surface must have been small as compared with the algal ones (Strickland & Terhune, 1961). There were two important results shown in Figure 15. The first was that the algal numbers increased by many times, which indicates that the algal stock in the sea is restrained by the grazing activity of the animals that were filtered off as the plastic bag was filled. The second important result is that there was no diminution of the algal reproductive rate (as shown by the logarithmic increase in numbers of a few algal species) until the nitrate nitrogen had reached very low levels. After the nitrate had run out, half the crop in the bag was produced, showing that enough nitrate remained in cell storage after it had been taken from the water. It has been shown recently that nitrate limitation occurs only at very low levels (see following section). Nutrients are stored in algal cells (Goldberg, Walker & Whisenand, 1951; Mackereth, 1953; Cushing, 1955); for example *Asterionella formosa* in Windermere stores enough phosphorus before the spring outburst to last the population right through it.

The limited effect of nutrient lack on the production cycle was shown in another way in the study of a *Calanus* patch (Cushing *et al.*, 1963*c*). A patch of *Calanus finmarchicus* Gunner was followed with two ships between early March and early June as it drifted south from the northern sea to the central North Sea. As the patch was more than 50 km across, the patch was easily found and delimited and it was shown that the losses due to diffusion were not too great. The production cycle was measured in mm^3/l of algae and the nutrients, phosphate, nitrate and silicate were sampled throughout the period. The algal reproductive rate was estimated in two ways: (1) from the dependence of algal reproductive rate upon light intensity, a quasi-theoretical estimate of reproductive rate can be derived from observed measures of irradiance by months during the productive season; this estimate was raised by the production ratio (D_c/D_m where D_c is the compensation depth and D_m is the depth of mixing) as it developed in the season (Cushing, 1959); (2) from the decrease in the cell sizes of diatoms (Cushing, 1955), which estimates the number of divisions which take place during a time interval; that is, all the effects of light intensity, sinking, vertical mixing and nutrient lack are taken up in the estimate. Figure 1 shows the results of the patch study. The production cycle is described in mm^3/l as a bell-shaped curve of stock and it was reduced to low levels of numbers before the quantities of phosphate, nitrate or silicate were reduced to any significant degree, although they started to decrease subsequently. The figure also shows how the algal reproductive rate increased during the period examined: the algal division rates were not reduced during the period of the spring outburst and the observed values support the theoretical trend. If the algal reproductive rate is not reduced by nutrient lack, the algal stock must be reduced by some other factor after the peak. It was shown that a large part of the algal mortality during the production cycle was due to grazing. This result is really the same as that found by Harvey, Cooper, Lebour & Russell (1935).

The spring outburst in temperate waters, which is the major component of production there, may take place independently of the nutrient content of the water. In earlier chapters a distinction was drawn between the discontinuous production cycle of high latitude and temperate waters and the continuous one of the deep tropical ocean. The two forms of cycle appear to be linked in the spatial correlation between zooplankton and phosphorus which was explained by the regeneration of phosphorus by the herbivores. Perhaps both forms of cycle are sustained by continuous regeneration. In the deep tropical ocean the nutrient quantities are often very low indeed, often below the levels of proper detection. One of the characteristics of the continuous production cycle is that the opposed parameters must be approximately equal; for example, the algal reproductive rates are equalled by the grazing mortality generated by the herbivores. Similarly, the uptake of nutrients by the productive algae may be balanced by their regeneration by the herbivores. Hence, the effect of nutrient lack in the continuous tropical production cycle may be mitigated by nutrient regeneration.

Gardiner (1937) was the first to suggest that plankton animals might regenerate phosphorus into the sea. The crucial observations were those of Menzel & Ryther (1960). In the surface layer (0–100 m) off Bermuda, the quantities of phosphorus ranged from 0.02–0.16 μg-at PO_4–P/l all the year round; the quantities of nitrogen were between 'undetectable' to 1.8 μg-at. NO_3-N/l and those of silica ranged from 0.3 to 1.8 μg-at. SiO_4-Si/l. Down to about 400 m in the Sargasso Sea, there is a lens of water, characterized by a temperature of 18 °C and a salinity of 36.5‰ and it contains on average 2–3 μg-at. NO_3-N/l and 0.10–0.15 μg at. PO_4-P/l. Temperature and salinity are nearly constant in depth down to 400 m over an extensive area and it is unlikely that there is any large-scale mixture of nutrients from below into the euphotic layer. Between January and May, 46 g C/m^2 were produced ($\equiv$570 mg-at. NO_3-N/m^2 and 38 mg-at. PO_4-P/m^2) and in the water, there was a loss of 250 mg-at. NO_3-N/m^2 and a gain of 35 mg-at. PO_4-P/m^2. The production was more than twice the loss of nitrate nitrogen and, in phosphorus, the quantity in the water increased by nearly the amount produced. The nutrients therefore were being regenerated rapidly and they must have been used by the algae immediately they were made available.

Ryther *et al.* (1966) examined the diurnal variation in nutrient levels down to 300 m in the Sargasso Sea and the average values of nitrate down to 150 m throughout the twenty-four hours were:

h	1330	1600	2100	0300	0400	0800
μg-at. NO_3-N/l	0.20	0.18	0.19	0.18	0.25	0.37

Thus, there is a diurnal variation by a factor of two, and the peak values were found in the early morning when the vertically migrating animals might be expected to excrete in the upper layers where they had been feeding all night although no correlation was established with zooplankton. If the nutrient limit is low, production can continue so long as regeneration is immediate.

Harris (1959), in his studies in Long Island Sound in 1952–3, showed that the algae used 0.61 μg-at N/cm^2 per day and that the zooplankton animals excreted 0.40 μg-at. N/cm^2 per day. Rigler (1961) compared the excretion of phosphorus by *Daphnia magna* Strauss with that reported by Cushing (1954*a*) by *Calanus finmarchicus*; the first excreted 0.032 μg/mg dry weight per hour and the second, 0.048 μg/mg dry weight per hour. Pomeroy, Matthews & Min (1963) showed that in Gulf Stream water, the daily excretion of inorganic phosphate and soluble organic phosphate was equivalent to the quantity needed for photosynthesis. In Shelf water, the quantity regenerated amounted to about one-third of that needed for photosynthesis. In the Gulf Stream water production may be continuous, whereas that in the Shelf water is discontinuous. The delay in regeneration (in days or weeks rather than months) must be the same in both forms of cycle, but in the discontinuous one there is a greater quantity to be regenerated.

Johannes (1964) showed that phosphorus (dissolved inorganic, dissolved organic and particulate) in *Lembos intermedius* Schellenberg (a gammarid amplipod) was turned over in 6.6 h; Barlow & Bishop (1965) have found the same thing in freshwater. Butler, Corner & Marshall (1969) have shown that stage V *Calanus finmarchicus* excreted 2.27 μg N/mg dry weight per day, whereas 'small copepods' excreted 3.98 μg N/mg dry weight per day. Johannes (1964) established a body equivalent excretion time in hours, which is the time taken to release a quantity equivalent to the body content of phosphorus. For a range of animal sizes (1 ng to 1 g) the release rates were calculated. An animal weighing 1 g dry weight needs 700 hours to release its body content of phosphorus; an animal weighing 1 mg needs 50 hours and one weighing 1 μg needs 2 hours. In a hypothetical population of 10% by weight of ciliates and 90% of *Calanus*, the ciliates contribute 70% of the total dissolved phosphorus. Thus, any consideration of nutrient regeneration in the sea should take into account the size distribution of the zooplankton.

The plastic bag experiments showed that the reproductive rate of the algae was modified at very low nutrient levels and that the low algal stock in the open ocean was restrained by grazing. The *Calanus* patch observations showed that the spring outburst in temperate waters can be completed with no decline in nutrients and with a continuous increase in algal reproductive rate. The measurements in the Sargasso Sea demonstrated that the rate of nutrient regeneration matched that taken up in production. The correlation between zooplankton and phosphorus in the surface waters of the Pacific suggests that the observed values of phosphorus were regenerated. The herbivorous animals in the plankton can regenerate nutrients at a high rate. Thus there may be mechanisms by which the apparent nutrient lack of the deep tropical ocean is mitigated.

However, there must be a nutrient level at which algal reproductive rates become reduced, as will be shown in the next section. The point here is that production can continue in the sea at low nutrient levels because of the high rate of regeneration. In the deep ocean animals migrate to the surface from 1000 m or more. It is possible that nutrients are regenerated in depth, for finally most of the seabed of the deep ocean is bare red clay. The regeneration in depth takes place by the reduction of faecal pellets and particulate material to solution. Because the animals migrate vertically, they can feed at depth on faecal pellets and particulate matter, amongst other foods, and release nutrients at the surface. The delay observed between absorption and regeneration in the temperate production cycle may be reproduced in depth in the deep tropical ocean. Consequently, the regeneration of nutrients in the photic layer might be greater than expected. It is this dynamic structure of nutrient pathways in the production cycle which provides the basis for the survival of production mechanisms in the still ocean.

The nature of nutrient lack

When nutrients have been exhausted, either in the water or in storage, the algae must stop dividing. They can divide once a day and at high stock levels, the last stocks of nutrients could be exhausted quickly. The most important question is to determine experimentally the quantity of nutrients in the water at which the algal division rates are reduced. Dugdale (1967) has applied the Michaelis–Menten equation of enzyme kinetics to the problem:

$$R = R_{max}N'/(K_s+N'), \qquad (14)$$

where R_{max} is the maximum reproductive rate or specific growth rate of the algae; R is the reproductive rate or specific growth rate of the algae; N' is the nutrient (or substrate concentration); K_s is the half saturation constant,

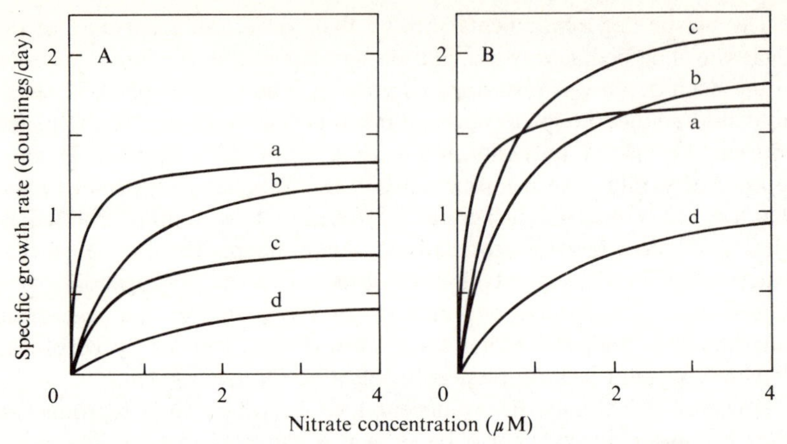

Figure 16. The effect of nutrient lack upon the algal reproductive rate for four common algal species (Eppley *et al.*, 1969) at two light intensities A and B. (*a*) *Coccolithus huxleyi*; (*b*) *Ditylum brightwellii*; (*c*) *Skeletonoma costatum*; (*d*) *Dunaliella tertiolecta.*

i.e. the quantity of nutrient at which $R = R_{max}/2$. The linear form of the equation is

$$N' = R_{max}\,(N'/R) - K_s, \tag{15}$$

which describes the uptake of nutrient by the algae. Droop (1968) used the method, in a chemostat, to describe the uptake and possible limiting effect of the micronutrient, vitamin B12, in the sea; he showed that lack of vitamin probably played little part in restraining the growth of algae. Eppley, Rogers & McCarthy (1969) have measured the half saturation constants of four common algal species (*Coccolithus huxleyi* (Lohm) Kampt; *Ditylum brightwellii* (West); *Skeletonema costatum* (Gréville); *Dunaliella tertiolecta* (Butcher)) in nitrate and ammonium. The half-saturation constants were estimated in μM-NO_3-N/l at 0.05 and 0.2 of the

Table 6 *Half saturation constants*, K_s, *and nitrate values for* R_{max}, *i.e.* N'_{max} *in* μM

	K_s		N'_{max}	
	$0.05I_0$	$0.20I_0$	$0.05I_0$	$0.20I_0$
C. huxleyi	0.1	0.1	1.0	1.5
D. brightwellii	0.5	0.5	3.5	4.0
S. costatum	0.4	0.4	3.0	4.0
D. tertiolecta	1.0	0.2	4.0	1.0

I_0 is the surface light intensity.

surface of light intensity (Figure 16); values of K_s and N_{max} are given in Table 6.

Eppley & Thomas (1969) have extended the method to many algal species and it can be shown that the half-saturation constant increases as function of algal cell size.

The values of N'_{max} for nitrate are within the highest values found in the Sargasso Sea studies of Menzel & Ryther (1960) and the half-saturation constants are in general less than the average values in the water ('undetectable' to 1.8 μg-at. NO_3-N/l). If there is a nightly regeneration of nutrient in the surface layers, then the four algal species, cited in Table 6, would have continued production at some level between $(R_{max})/2$ and R_{max}.

The loss of nutrients in the production cycle

Cushing (1959) formulated the loss of nutrients in the following way. The uptake of nutrient U_N, was assumed proportional to production, P_R, i.e. $U_N = P_R N_c'$, where N_c' is the quantity of nutrient per cell. Regeneration R_N was assumed proportional to the quantity grazed (P_G), i.e. $R_N = N_c' P_G$. Because $N_c'(P_R - P_G) = N_c'(P_1 - P_0)$, the loss to the system L, is given by:

$$L' = (N_0' - N_1') + N_c'(P_0 - P_1).$$

The uptake of nutrient in production is immediate but regeneration may be delayed, so at any point in time $L' > N_c'(P_0 - P_1)$ and this delay is accounted for in the term $(N_0' - N_1')$.

In Windermere, Lund (1950) made the most careful measurements of the production of *Asterionella formosa*, together with measurements of the algal reproductive rate and of the silica content of the water. The quantity

Table 7 *The loss of silica during the production cycle in Lake Windermere*

	N'_0–N'_1 (mg/l)	$(P_0$–$P_1)10^3$ (n/l)	$N'_c(P_0$–$P_1)$	L'(mg/l)
1	0.1	−660	−0.09	0.01
2	0.3	−1400	−0.20	0.10
3	0.0	−1500	−0.21	−0.21
4	0.5	−1000	−0.14	0.36
5	0.5	+2500	0.35	0.85
6	0.3	+1400	0.20	0.50
7	0.0	+820	0.11	0.11
				Σ1.83

of silica per cell has been well established (0.14 ng SiO_2; Lund, 1950). Table 7 shows how the loss was calculated.

The initial quantity of silica in the water was 2.50 mg/l SiO_2. Stock increased in the first four weeks to 0.70 mg and as it declined it was transferred to the loss account. If the regeneration of silica was immediate, production would have been seven times greater than the stock; if it were delayed, the ratio of production to stock would have been less. The difference between the initial quantity and the summed quantity in Table 7 is 0.67 mg silica. The least production–stock ratio is given by 1.83/0.70 = 2.61. Silica is not transferred to animal flesh like phosphorus or nitrogen, but may be retained in particulate matter and a proportion may sink to the bottom of the lake. Lund (1964) interprets the results quite differently, assuming that *Asterionella* is not eaten and as the cells do not sink quickly enough, their loss is accounted for by being washed down the river (which implies that the reproduction of the algae stops at the peak of the production cycle).

In the work on the *Calanus* patch (Cushing & Nicholson, 1963) the loss of nutrients was estimated by relating the change in nutrient on the change in algal volume. Figure 17*a* shows the relationship in phosphorus; the slope estimates N_c and the intercept on the ordinate,

$$L' \text{ [i.e. } N_0' - N_i' = L' + N_c'(P_1 - P_0)].$$

The value of N_c' (0.616 μg-at. P/mm^3 of algal volume) is higher than might be expected from the C/P ratio, but as the phosphate values were high (>0.5 μg-at. PO_4-P/l) the quantity in store might have been high. The quantity of phosphorus used in production is (P_R/N_c') and if this quantity is divided into the winter maximum phosphorus value, the rate of turnover is estimated. If $N_c = 0.616$ and the winter maximum is 0.75 μg-at.P/l (P_R/N_c') is turned over in four days; if $N_c = 0.094$ (from the C/P ratio), (P_R/N_c) is turned over every twenty-four days. Hence the rapid regeneration sustains much heavier production than the simple ‘agricultural’ model. Figure 17 (*b*) shows a similar relationship in silica, where the change in algal volume is related to the change in nutrient one week later. The loss is very low and $N_c \simeq 10.0$ μg-at./mm^3, and this relationship must mean that silica is regenerated quite quickly to the sea. The loss in phosphorus was 0.037 μg-at.PO_4-P/l, which is close to the quantity in animal flesh (0.031 μg-at. PO_4-P/l). The production of 1400 tons/km^2 per year off Plymouth (i.e. Atkins’ estimate) represents the quantity of nutrient which ends up in animal flesh, whereas the 200 tons/km^2 per year in silica represents algal stock and some delayed regeneration.

In the previous chapter, the correlation between zooplankton and phosphorus was discussed and it was concluded that the observations of phosphorus were comprised mainly of regenerated material. In temperate

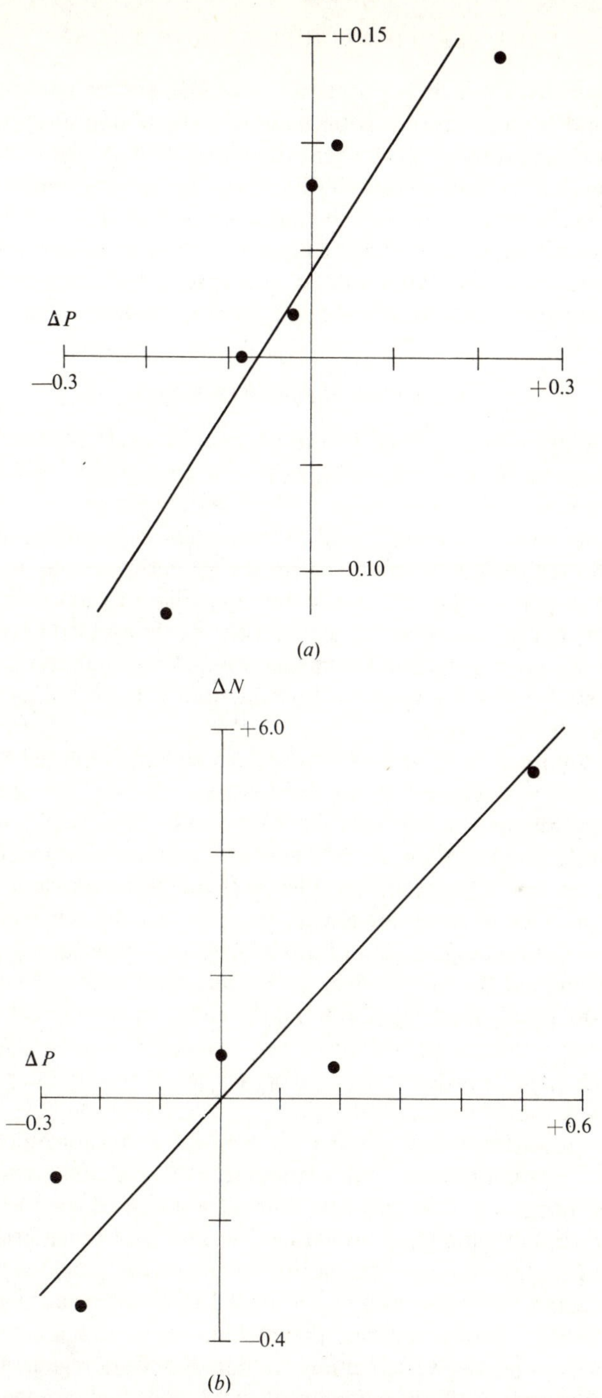

Figure 17. (*a*) The dependence of the decrement of phosphorus (ΔN) upon the increment or decrement of algal volume (ΔP) (Cushing & Nicholson, 1963); (*b*) the dependence of the decrement of silica (ΔN) one week after the increment or decrement of algal volume (ΔP) (Cushing & Nicholson, 1963).

waters, there is an inverse correlation in time between the decline in nutrient and the increase in living material. The temporal correlation is generated by the transfer of nutrient to animal flesh. Much of the spatial correlation in high latitudes and in the upwelling areas is generated by the temporal processes; where there is much phosphorus it is the product of many animals and where it is absent it may well be in animal flesh. Then, in a general sense, the magnitude of nutrients in the sea is the effect of productive processes in the sea and not necessarily their cause.

Nutrients in the production cycle

The part played by nutrients in the production cycle is more complex than is implied by the original agricultural analogy. In the temperate cycle, nutrients are turned over as algae are eaten and the decline of nutrients observed in the sea is really a measure of their storage in animals. During the temperate summer, the predatory herbivores and algal prey remain in a quasi-steady state with minor oscillations until the autumn. The overwintering copepods sink into cold deep layers (Ostvedt, 1955) and, with the grazing restraint removed, an autumn outburst may occur. It only lasts for a short time until production stops in failing light and increasing vertical mixture.

In high summer, in temperate waters, production is limited by its own processes at low stock and at low transfer. At this time, the quantities of nutrient are low and if nutrient lack halts production it does so in July, August or September when the nutrients reach their lowest values (Figure 14). Because the half saturation constants are less than was originally thought, production need not always be restrained by nutrient lack but perhaps by too much grazing and too little food. To decide whether production is limited by nutrient lack or not, measurements of algal reproductive rate, uptake and regeneration are needed throughout the day and night during a critical period. However, the point may be unimportant because the major transfer of material occurs before the limit to production is reached, however it is caused.

In the quasi-steady state system in tropical and subtropical waters, outside the upwelling areas, the amplitude of the seasonal cycle is low. The algae might use the nutrient as it is regenerated by the vertically migrating animals which feed upon them. The transfer of material must be continuous, uptake must be balanced by regeneration, algal reproductive rate by grazing mortality, and so on up the food chain, as suggested by Cushing (1959). Links between the trophic levels are secured by the dependence of uptake on production, the dependence of regeneration upon the quantity grazed, the dependence of algal reproductive rate upon nutrient and the dependence of grazing mortality upon the quantity of herbivores. Production continues in tropical waters throughout the year and the problem of establishing any effect of nutrient lack is as difficult as that in the temperate summer; within 10 or 20 m of the surface, nitrate is so low that production must be restrained, where photosynthesis might be inhibited in any case.

Cushing (1961) examined the failure of the Plymouth herring in the thirties, together with other biological changes, and suggested that the decline of winter phosphorus in the English Channel by about one-third after 1930 was the consequence of a profound change in the ecosystem (which will be described in detail in a later chapter). An ecosystem of herring and macroplankton changed to one of pilchards and very little macroplankton. If more phosphorus ends in animals in the summer, less will remain in the water in the following winter. It was suggested that the decrement in phosphorus after the change in 1930–1 represented an increment in larval pilchards and their predators. The observation of winter phosphorus is thus an effect of the production in the previous year and it may not limit production in the subsequent year. The long-established correlation between production and nutrients is not spurious, but the nutrient levels observed may be a consequence of productive processes.

Two models of nutrient transfer in the sea

Riley (1965) has developed a series of equations to simulate regional variation in the production cycle (Fladen ground, North Pacific and eastern tropical Pacific). In a two-layered system:

$$\frac{\mathrm{d}P}{\mathrm{d}t} = P(P_r - gh - v/L - m), \tag{16}$$

where P is the quantity of algae in g C/m^3; $P_r[= Ph/P\ (= 0.3)]$ where Ph is the daily quantity produced by photosynthesis; g is a grazing coefficient ($= 3.4$ l/d per g C per m^3); h is the quantity of herbivores in g C/m^3; v is the sinking rate of algae in m/d ($= 3$) and L is the depth of the upper or photic layer in m; m is the coefficient of exchange between the two layers. There is a respiratory coefficient ($= 0.035$) and a nutrient limit (at 0.40 μg-at. P/l, $P_r = 0.265$) and it decreases linearly with decreasing phosphorus. The nutrient limit is high and the reproductive rate (P_r) is perhaps too low. Changes in phosphorus in the water were formulated:

$$\frac{\mathrm{d}P}{\mathrm{d}t} = cP(egh - P_r) + m(p'_0 - p'), \tag{17}$$

where p' is the quantity of phosphorus in the photic layer and p'_0 that in the lower layer; $c = 0.774$, which converts μg-at P/l to g C/m^3; $e = 0.85$, the proportion regenerated (assuming 15% going to growth). The factor ghP

represents the total carbon consumed by herbivores and $cghP$, the phosphorus content of the food. The first part of the exponent is $(ecghP-cPP_r)$ or regeneration by grazing less uptake by the algal stock, i.e. $N_c(P_G-P_R)$. The model requires that phosphorus mixes from the lower layer to the photic layer, which may be transferred by vertically migrating animals. Riley's equation (eqn 15) is well suited to estimate the changes of nutrient in time, whether the vertical exchange is physical or biological.

Dugdale (1967) has expressed the transfer of nutrients to algal stock in terms of enzyme kinetics (see eqn 14); then

$$N' = K_t/[(R_{max}/V_L')-1], \tag{18}$$

where K_t is a transport coefficient ($= N_c'$); V_L is the loss rate of algae/unit algal stock [$= G$, etc.]. This equation resembles Riley's (eqn 15) in that the changes in nutrients are expressed in biological terms. Further,

$$P_N = (m/V_L')\{N_0' - K_t/[(R_{max}/V_L')-1]\}, \tag{19}$$

where P_N is the phytoplankton in units of nutrient; m is the mixing coefficient between the layer which is nutrient limited and that below; N_0 is the nutrient in the lower layer. It follows that the size of the algal population depends on the term

$$K_t/[(R_{max}/V_L')-1].$$

Dugdale's model is more extensive than presented here because there is a network of transport coefficients between all the components of nutrient in the system.

Conclusion

Twenty years ago any study of marine production was executed largely by observations of the nutrient concentrations in the water. From time to time it was observed that production would stop when phosphorus or nitrogen was still present in fair quantities and this led to the search for trace elements and other micro-nutrients. Only Vitamin B12 has been examined in any detail and Droop (1968) believes that only rarely can it limit production in the sea. However, if the major nutrients do not always affect the course of the production cycle, it is unlikely that the minor ones do so, because the ratios of the different nutrients in the sea and in the living organisms remain roughly constant. There is also a procedural difficulty in that the examination of successive nutrients has the character of an infinite regress.

The steps of argument were visible in the work of Harvey, Cooper, Lebour & Russell (1935), showing that the production cycle was essentially a dual process, a predator–prey relationship traced in nutrient concentrations. It was the first step in the replacement of the agricultural model by the predator–prey model. Three sets of observations were made at sea, using the plastic bag experiments, the *Calanus* patch observations and the Sargasso Sea observations on nutrient regeneration. Three conclusions emerged: (1) that production cycles, both continuous and discontinuous, are limited more by grazing and less by nutrient lack; (2) that the nutrient level at which the algal reproductive rate is reduced is lower than expected; (3) that rapid and immediate regeneration of nutrients is probably the rule in both forms of production cycle.

The spatial correlation between zooplankton and phosphorus and the temporal one between production cycle and nutrient concentration are both generated in the same way, i.e. because nutrient observations are the result of productive processes and not the other way round. The description of the production cycle as a general form of predator–prey relationship has been established for a long time, as will be described in the next chapter. It has replaced the agricultural model.

However, if a production cycle breaks down, for example, by the failure of grazers to appear by an accident in timing, nutrients are taken up until they run out and production stops. The agricultural model finds its use as a fail-safe device. Indeed, there may be special cases in which the agricultural model is the rule.

The original use and attraction of nutrient observations was that production could be described quickly with fairly simple measurements. Perhaps such measurements could be used in the form of the Dugdale model to describe the predator–prey relationship directly.

4

Models of production in the sea

Introduction

The production of living material in the sea is based on the growth of algae in the euphotic layer and the transfer of the grazed material to the herbivores. At higher trophic levels, production is harder to estimate, although a number of attempts have been made. In temperate and high latitude waters, production depends largely upon the temporal variation in the production of algae and the spring outburst is its most important component. In the surface waters of the deep tropical ocean, production is in a quasi-steady state and the temporal variation is of less importance. As indicated in the last chapter the relationship between the algae and the grazing zooplankton has been treated as a predator–prey process and the quantities of both are often expressed in carbon. It is a considerable simplification, for there are large numbers of algal species and nearly every phylum of the animal kingdom is represented in the zooplankton. The relatively uniform volume of the ocean may provide an environmental stability absent from the variable surface on which land animals live. Such a stable environment may justify the simplification.

However simplified the processes are they remain sufficiently complex to be analysed by model techniques. By such means a number of equations are put together to describe a series of linked events. In principle, there is no difference between the procedures to establish a model and those used to derive a law or to design an experiment. In other words, it represents a hypothesis, if a complicated one, and the means of testing it must be more extensive than those for simpler hypotheses. There are three forms of test and the first is in the fitness of the parameters. For example, the dependence of photosynthesis upon light intensity is a simple relationship that can be well established with ordinary statistical methods. The second test involves the general scientific framework; for example, is the photosynthetic process adequately described in physiological terms? that is, in an examination of one parameter. The relation between parameters requires a different question, for example, do the animals get enough to eat in the full model structure? Such questions are not answered by test procedures, but in the architecture of the model, which must stand on its own like any building. The third test is the overall fit of the model to observed data, which are often too few in the face of all the degrees of freedom taken in the complexities of the model. The three forms of test are common to model techniques and other forms of scientific procedure.

The first attempt at a quantitative description of the production cycle in model terms was made by Lohmann (1908), who counted plants and animals in centrifuged samples of seawater from Kiel Bay and he expressed the results in mm^3/l as the organisms were measured. He found that the seasonal cycle from August to August varied as shown in Figure 18; algae comprised 58% of the living material and the remainder comprised animals. In order to create such a high proportion of animals, Lohmann suggested that the algae divided at a rate of 0.3 per day and that the metazoa needed one-tenth body weight per day as food and that the

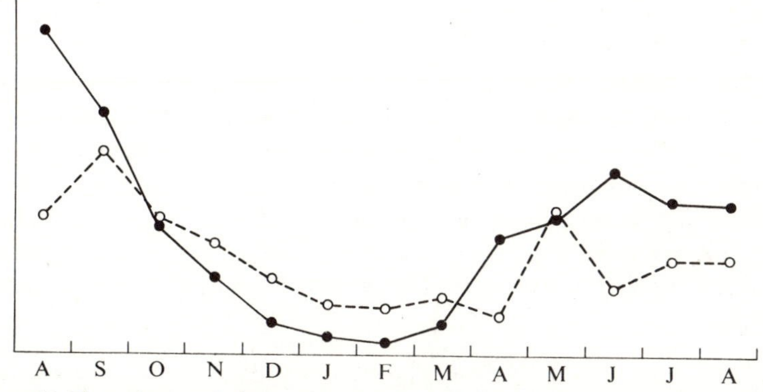

Figure 18. The volumes of plants and animals in Kiel Bay from one August to the next (mm^3/l); the quantities of algae were estimated from centrifuged water samples and those of herbivores from net hauls (Lohmann, 1908).

protozoa needed one-half body weight per day. Today, we can show that the reproductive rates of the algae are higher, perhaps as much as one division per day. One of Lohmann's contemporaries, Pütter (1909), studied the oxygen consumption of some pelagic and benthic animals and suggested that the algal reproductive rates used by Lohmann were not high enough to feed the animals; he concluded that the animals must subsist on organic material, as suspended detritus or in solution in the water. Pütter's famous hypothesis was denied by Krogh (1931) because he thought that the animals could not, or would not, subsist on such materials, but it would be nullified today by the evidence of somewhat higher algal reproductive rates. The idea, however, is currently relevant because when algae are scarce in the sea, the zooplankton animals must live on their fat (Conover, 1962) or upon detritus. Algae are rich in lipids, which are transferred to the zooplankton and provide a source of energy; detritus is always available, not as a rich source of energy, but as a means of subsistence. It has been shown that organic material may be brought out of solution near the surface by its adsorption on air bubbles and so provide a source of food (Baylor & Sutcliffe, 1963; Riley, 1963).

The arguments used in the first decade of the century by Lohmann and Pütter are echoed today in a number of ways. In model terms, zooplankton starve because the algal reproductive rates are sometimes set too low. Consequently, the question is asked again whether animals can subsist on sources of food other than algae. It is interesting, however, that the agricultural model of production processes, which was being propagated at that time by Brandt & Raben (1919), was virtually ignored by Lohmann and Pütter. Perhaps Lohmann realized that the production cycle should be described in terms of a predator–prey relation.

The first models

The first equations to describe the changes in numbers in an animal population were written by Ross (1910) on the parasite–host relationship in the malaria problem. The parameters were measured independently (for example, the number of mosquito bites per man) of the numbers in the population. The later more general models of Lotka (1925) and Volterra (1926) did not have this analytic character. The production cycle was first described formally in such an analytic way by Fleming (1939) in the following form:

$$\mathrm{d}P/\mathrm{d}t = P[R-(G_0+c_g t)], \tag{20}$$

where P is the quantity of algae in the sea (in numbers, carbon or chlorophyll per unit volume); R is the algal reproductive rate (as an instantaneous coefficient); G_0 is the initial grazing mortality (as an instantaneous coefficient); c_g is the rate at which the grazing mortality increases with time; and t is time, usually in days.

At the maximum, at time T, $\mathrm{d}P/\mathrm{d}t = 0$ and $R-G_0 = c_g T$: also

$$\ln (Pt/P_0) = (R-G_0)t-c_g t^2/2 = c_g t(T-t/2)$$

$$c_g = 2 \ln(Pt/P_0)/t(2T-t). \tag{21}$$

Thus, given the magnitude of the production cycle at its peak in measurable units, the time it took to develop, and an estimate of the algal reproductive rate, the total production can be estimated for any period of time (see Figure 19, fitted to the data of Harvey, Cooper, Lebour & Russell, 1935, which were expressed in PPU, an early unit of chlorophyll). The cycle was described by two exponents, R, the algal reproductive rate, and G, the grazing rate (which itself increases with time). When $R>G$ the curve of production rises in time and when $R<G$ it falls; $R = G$ when production reaches a maximum. This bell-shaped curve describes the

production cycle, but the exponents R and G_0 could not be estimated accurately when the model was designed. The model is unrealistic in two respects (1) the reproductive rate is assumed to be constant, whereas it probably increases during the productive season and (2) the grazing rate is assumed to increase linearly, which may overestimate the grazing mortality at the end of the cycle because the zooplankton does not increase linearly, but rises to a maximum.

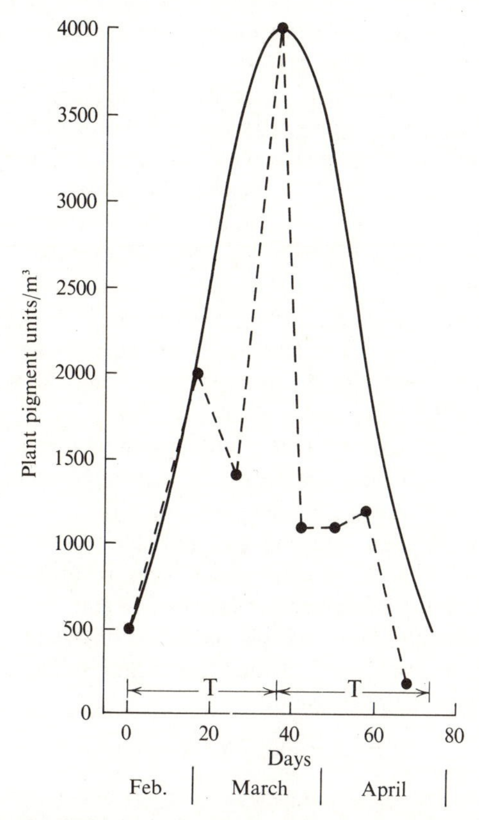

Figure 19. The fit of Fleming's (1939) curve to the data of Harvey *et al.* (1935).

The next step was made by Riley (1946), who expressed the algal reproductive rate in terms of photosynthesis and respiration and their environmental determinants. Riley's model is:

$$\mathrm{d}P/\mathrm{d}t = P\,(Ph - Re - G), \tag{22}$$

where Ph is the instantaneous rate of production of organic material by photosynthesis; Re is the instantaneous rate of loss of organic material by respiration. The model is a variant of Fleming's, in which $R = (Ph - Re)$.

It was fitted stepwise to the observations and so the increment in G in Fleming's model with time is expressed by the increase in the observed zooplankton. Fleming's model was designed to describe the bell-shaped curve and Riley's to describe its modification by environmental factors.

In more detail,

$$P_h = p''I_0\{[1-\exp(-kz_1)]/kz_1\}(1-N'')V, \tag{23}$$

where p'' is a coefficient that describes the dependence of photosynthesis on light intensity; I_0 is the surface irradiance in the photosynthetic range (400–700 nm) in ly/min; K is the extinction coefficient of the photosynthetic radiant energy in the sea; z_1 is the depth of the photic layer in m (approximately that of 1% penetration of photosynthetic irradiance), $(1-N'')$ is a nutrient limit factor which modifies the reproductive rate linearly from 0.55 μg-at. PO_4-P/1 to zero, i.e. $N'' = (0.55-x)/0.55$; where V is a vertical turbulence factor (as ratio of photic layer to depth of mixing).

The respiratory rate is given by $R_e = 0.0175 \exp 0.069\ T$, where T is absolute temperature.

The grazing rate is given by $Gh = g_h$, where h is the weight of herbivores per unit volume; g is a coefficient based on experiments on respiration made by Marshall, Nicholls & Orr (1935); in effect the herbivores took about 7% body weight per day in food.

It is a more complex model than Fleming's and its great virtue is in the detailed environmental modulation of the algal reproductive rate by the average irradiance expressed by $\{[1-\exp(-kz_1)]/kz_1\}$, of respiration by temperature, of the algal reproductive rate by vertical mixing and by nutrient limitation. The latter expressed the results of Ketchum's (1939*b*) experiments which were discussed in the last chapter.

Figure 20 shows the fit of Riley's model to observations on Georges Bank off the eastern seaboard of the United States. The calculation was made in a stepwise manner, using:

$$P_1 = P_0\exp(Ph-Re-G)t, \tag{24}$$

where P_0 and P_1 are algal quantities at time t_0 and t_1, respectively. The exponents were varied at each step in time between observations according to changes in the environmental conditions and the fit to data is really a fit of theoretical increments to those observed.

If the first two models are compared, the essential exponent in Fleming's is $(R-G)$, whereas that in Riley's is $\{R(1-N'')-G\}$. If N'' is near zero, the models are similar in principle. However, it has now been shown that at 0.55 μg-at. PO_4-P/l, N'' was probably too high (McAllister, Parsons, Stephens & Strickland, 1961; Antia *et al.*, 1963; Eppley, Rogers & McCarthy, 1969); The fit to data in Figure 20 is quite good, which means that the ratio R/G was properly estimated. However, with N'' too high, R and

G were both underestimated and the animals in the model did not get enough to eat; the ration of about 7% body weight per day is low compared with recent estimates.

Riley, Stommel & Bumpus (1949) examined the production cycle in a vertical column in the sea. The basic biological model is the same as the

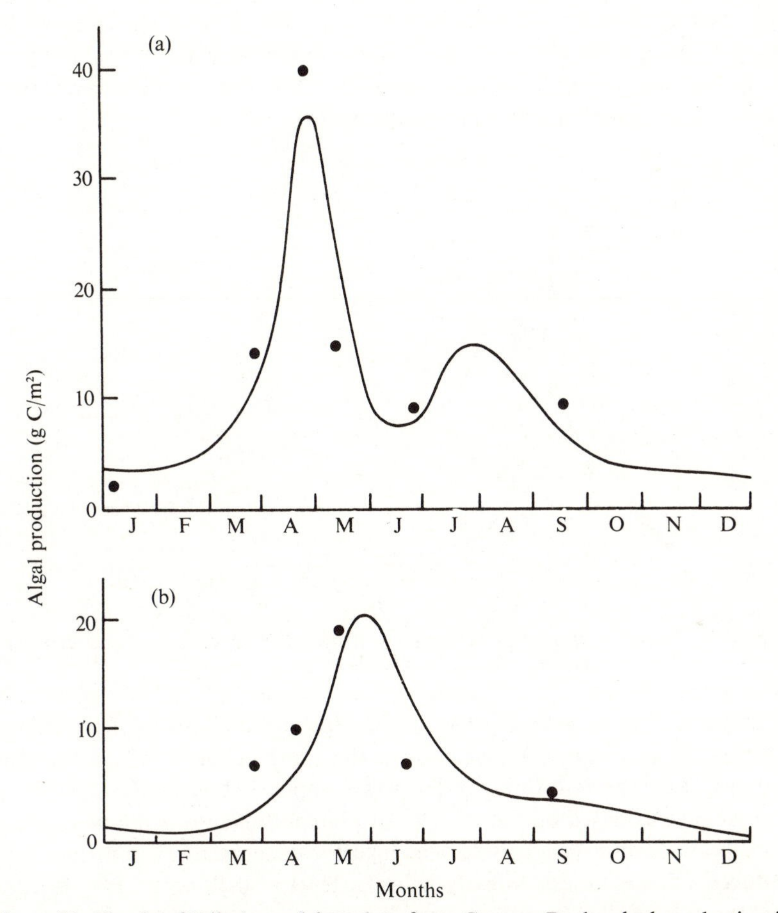

Figure 20. The fit of Riley's model to data from Georges Bank; algal production is shown in (*a*) and the production of herbivores in (*b*) (Riley, 1946).

previous one, but the losses of algae from the euphotic layer by sinking and by eddy diffusivity are described. The model may be written as follows:

$$dP/dt = P(P_h - R_e - G) + (\delta/\delta z)(A/\rho)(\delta P/\delta z) - v(\delta P/\delta z), \qquad (25)$$

where *A* is the coefficient of vertical eddy diffusivity (in g/cm per s); *v* is the sinking rate of the algae at 0 °C; a temperature correction is needed; ρ is density.

The equation was solved by finite differences and the changes were considered across a standard depth. The animals were considered as being spread over the algal populations in depth and so the effects of their vertical migration were averaged. The coefficients of eddy diffusivity were perhaps not well estimated, but more important is the general structure of the model. Because the algae sink and diffuse and because the animals migrate to the surface at night, future models must take into account the vertical structure of the production mechanisms.

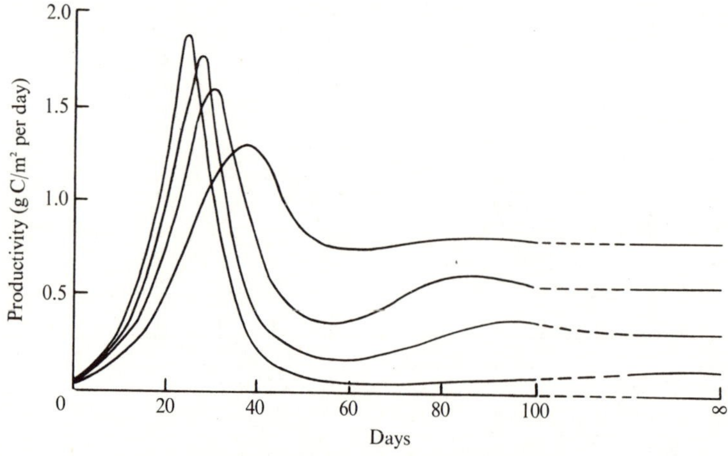

Figure 21. The effect of vertical mixing upon the production cycle, in theoretical terms (Steele, 1958) at four levels of vertical mixing.

Steele (1958) developed a model of algal production in the northern North Sea based on the layer above the thermocline at 40 m and that below it. Like the description of Harvey, Cooper, Lebour & Russell (1935); the model was based on the transfer of phosphorus into living material. A constant rate of mixing between the layers was postulated; a nutrient limit reduced photosynthesis linearly below a level of 0.40 μg-at. PO_4-P/l; the sinking rate of the algae was set at 3 m per day and a grazing rate of 3.4 m^3 of algae swept clear/g zooplankton carbon was used; Riley's respiration coefficient was put into the model. It is a development of Riley's models, both that on Georges Bank and that on the vertical structure of the ocean. As with Riley's model, some of the parameters would be altered in magnitude today. Perhaps the most important point is that the immediate regeneration of nutrients like phosphorus should be explicitly incorporated in the model. The effect of different mixing rates on algal stock is shown in Figure 21. It was the first demonstration that the production cycle should vary in timing, amplitude and spread under different rates of mixing or different wind strengths.

A model of quite different character was described by Cushing (1959). An empirical curve was constructed relating the potential algal reproductive rate to light intensity. The potential rate was raised by the production ratio (D_c/D_m) (where D_c is the compensation depth and D_m is the depth of mixing) and the product, the actual estimated rate, increased from 0.1 to 0.5 during the season. From observations made at sea by Cushing *et al.* (1963), such rates were underestimated (Figure 1 shows an increase from 0.2 to 1.0), although the upward trend was found. An encounter theory of grazing was developed in which the quantity eaten depended upon the volume searched and the handling time of a copepod; in detail, the volume searched was improperly estimated and no threshold in algal density was used. It was then thought unnecessary because grazing would only start after herbivore spawning occurred at a given food level during the early stages of the production cycle. However, John Steele pointed out to me that in the late stages grazing would be overestimated without such a threshold. The timing and magnitude of the North Sea production cycle were matched quite well, as was the increase in numbers of herbivores. The model was constructed without a nutrient limit to the algal reproductive rate during the production cycle; in Cushing & Vućetić (1963), it was shown that the spring outburst occurred in the central North Sea before the nutrients declined to any degree.

Fleming's achievement was to put the results of Harvey's work into a more rigorous form and to start the construction of a predictive model. Riley's model was successful in that a fair fit to data was obtained and the ratio (R/G) was correctly estimated, but both R and G were underestimated. Steele's model showed how variations in the production cycle are governed by wind mixing. Cushing's model suggested that the production cycle could be simulated in numbers in terms of algal reproductive rate and grazing mortality only. The argument that production is not limited by nutrient lack during the spring outburst need not be extended to the productive processes in later summer, July or August. The state of the science during the twenties and thirties may be recalled: Harvey, Cooper, Lebour & Russell (1935) showed that production was controlled by grazing (in spring and early summer) and Atkins (1923) suggested that it was limited by nutrient lack (in July and August after the thermocline had been established).

The fitness of the parameters

Each of the early series of models failed in some particular way, but each established an important point despite failure. For example, Riley showed that (R/G) could be correctly estimated and Steele demonstrated that variability in the production cycle was generated by variation in the depth of mixing. Science progresses in this way and in this spirit the fitness of the present parameters is examined.

The ecological description of photosynthesis

The description of photosynthesis in physiological or in biochemical terms is complex and extensive today. In the sea, or in any aquatic system, the ecologist is interested in the dependence of photosynthesis upon the light intensity which is attenuated in the water column. A number of empirical equations have been developed (Smith, 1936; van Oorschot, 1953) and they describe various relationships reaching an asymptote in light intensity. However, they ignore the inhibition of photosynthesis at high irradiance close to the surface of the sea after midday (Doty & Oguri, 1957). Steele (1962) developed an equation which describes the dependence of photosynthesis on all intensities of radiation including the high inhibitory levels; a later form in Steele (1965) is

$$p = p_m(I/I_m)\exp[1-(I/I_m)], \tag{26}$$

where p is the quantity of living material produced by photosynthesis, for example, in g C/g chlorophyll per day or in g C/g C per day; p_m is the maximum value of p, I is irradiance within the photosynthetic range (390–720 nm), in ly/min; very often it will be the average light intensity in the photic layer during the day; I_m is the irradiance at the depth at which p_m occurs. This equation can be used with any measure of photosynthesis. As noted earlier, Nielsen & Hansen (1959) estimated respiration directly from the dependence of radiocarbon produced upon light intensity; indeed, Steele (1965) suggested that respiration, $r = 0.2\,p_m$ on a daily basis, because Nielsen & Hansen's results showed that $r = 0.1\,p_m$ during daylight. The radiocarbon method is well suited to the estimation of photosynthesis at a series of depths in a vertical section. Wolfe & Schelske (1967) have shown that scintillation counting is an effective and absolute method of measuring the uptake of radiocarbon. The original method of calibration based on a self-absorption curve underestimated the true production by nearly 50% (Nielsen, 1965; Goldman, 1968). The recent improvements have made the method considerably more flexible. The most important point is that the radiocarbon method, Steele's equation and Nielsen and Hansen's method of estimating respiration rate can be combined to give estimates of the algal reproductive rate.

Lanskaya (1963) showed that the average division rate of some common algae could reach 1.0–1.5 per day. Cushing (1963) produced some evidence that the division rates of the algal community, weighted by the abundance of the common species increased during the spring to reach values of between 0.5 and 1.0 division per day. Jitts, McAllister, Stephens & Strickland (1964) investigated the variation of the division rates of a number of algal species with light and temperature in a large aluminium block; it was heated at one end and each tube had a different attenuator disc because there was a common light source. Results were expressed as a three-dimensional diagram of temperature (from 8 °C to 28 °C) on radiation (from 0.05 to 40 ly per minute) with division rate per hour in the vertical dimension. Figure 22 shows such a 'response diagram' for three algal species (similar work has been carried out by Thomas, 1966). Figure

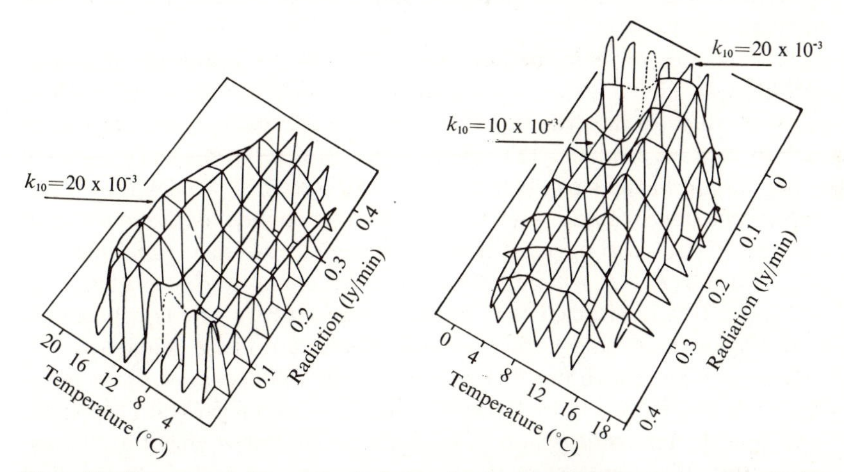

Figure 22. The response diagram in algal reproductive rate of *Thalassiosira nordenskioldii* in terms of light intensity (ly/min) and temperature (°C) (Jitts *et al.*, 1964); the hourly reproductive rate is k.

22 shows that the maximum division rate (in the third dimension rising from the surface) of *Thalassiosira nordenskioldii* Cleve is found at rather low values of irradiance, about 0.05 ly per minute; on average, this is a value which will tend to occur in the lower half of the euphotic layer.

Eppley & Sloan (1966) have developed an equation relating division rate to the absorption of light by chlorophyll *a*, to light intensity and to temperature; for *Dunaliella tertiolecta* Butcher the observed division rate was closely correlated with the calculated one. Cushing, Nicholson & Fox (1968) developed a method by which the living and dead particles in the sea could be counted with a Coulter counter: sea water in a bottle, from which the animals had been filtered off, was exposed to daylight in a tank of running seawater. If no detritus is present, then a linear regression of the ratios of increments of algal volume ($= \exp(Rt)$) on time, at equal time intervals, has zero slope. If detritus is present, the slope is positively biased; then by subtracting model estimates of detritus, the true value of detritus emerges when the slope becomes zero. From such an estimate, values of detritus, stock and algal division rate can be derived from one experimental series. Sheldon & Parsons (1967) have used the Coulter counter to estimate the algal reproductive rate in a different way, by comparing the size distributions at the start and end of the experiment. The methods described above yield their results in terms of algal volume because the Coulter counter records the volumes of particles. Strathmann (1967) has devised methods for the estimation of the carbon content of algal cells from cell volume or from plasma volume. He found a fair correlation between carbon/cell and cell volume, which differed to some degree between species.

There are three main methods of estimating the quantities of phytoplankton, in cell volume, in carbon and in chlorophyll: the Coulter counter method estimates the division rate of algae in cell volume and the radiocarbon method estimates that of the algae in carbon, provided that the living carbon in the stock can be measured. Recently, Eppley (1968) has applied the same methods used with the Coulter counter to the radiocarbon method: if radiocarbon measurements are made at equal time intervals $(P_1 - P_0)$ and $(P_2 - P_1)$, preferably a day apart, the ratio $(P_2 - P_1)/(P_1 - P_0) = \exp(Rt)$. It is assumed that the algal reproductive rate is constant in both time intervals and it must be realized that errors are geometric.

It is possible that chlorophyll measurements will become more reliable. The methods developed by Yentsch (1965) and Lorenzen (1967) separate chlorophyll from its degradation products by the measurement of fluorescence. It will be seen that methods are becoming available for measuring algal production in carbon, chlorophyll, particle volume and cell volume. A comprehensive photosynthetic equation is available which describes the dependence of photosynthesis on irradiance and hence on the depth of water. But perhaps the most important conclusion is that the algal reproductive rates reach values of 1.0–1.5 per day at relatively low irradiances. Consequently, the average algal reproductive rate probably lies between 0.5 per day and 1.0 per day, rather than 0.3 per day and 0.5 per day, as was believed about a decade ago.

The sinking of algal cells in sea water

The sinking of algal cells is a most important component of any dynamic study. As a population of algae grows old, it tends to sink in the water column. The early models of Riley (1946) and Steele (1958) assumed that the cells sank at an average rate of 3 metres per day. Munk & Riley (1952) arrived at much the same conclusion based on purely theoretical grounds, developed from Stokes' Law, which describes the resistance of spherical bodies to the movement of a fluid. Smayda & Boleyn (1965) have shown that some very small diatoms sank in still seawaters at 0.10–2.10 metres per day. Eppley, Holmes & Strickland (1967*b*) have measured the sinking rates in still water at 20 °C with a fluorometer; Figure 23 shows average sinking rates in metres per day as a function of cell diameter in μm. The average sinking rate is about 1 metre per day which can increase by nearly

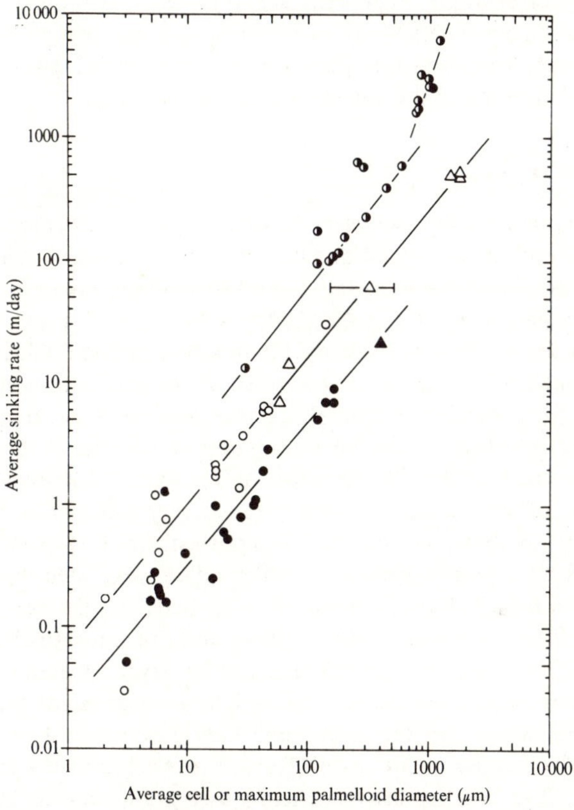

Figure 23. The average sinking rates of some algae as function of cell diameter (Smayda, 1970). (●) Actively growing cells; (○) senescent cells; (◑) palmelloid stages of *Cyclococcolithus fragilis*; (△) empty frustules; (▲) calculated rate. The numbers refer to different species.

five times in a senescent culture, if such cultures occur in the sea or in lakes (Marshall & Orr, 1930; Lund, Mackereth & Mortimer, 1963). From Figure 23 it may be shown that the largest cells of about 200 μm in diameter sink at about 6 metres per day and the smallest, of about 5 μm in diameter, at about 0.1 metre per day. In most waters the average cell diameter is less than 30 μm, and so the average sinking rate is probably less than 0.6 metre per day. As the populations of *Asterionella formosa* grow old they sink (Lund, Mackereth & Mortimer, 1963) and during the spring outburst, dead frustules accumulate as quickly as 0.7 mg/m^2 per week (Tutin, 1955). Smayda (1970) has shown that the descent of faecal pellets follows Stokes' Law. The density of algal cells may change from time to time and the spines may brake the sinking, as Ostwald (1903) suggested long ago. In the sea, the most important effect is probably that of vertical turbulence; for example, algal cells are not found on the sea bed in the southern North Sea where the tides are swift. The problem of sinking is an important one, if the loss rate from the photic layer is high, but it needs formulating in terms of the vertical mixing of the water.

The analysis of grazing

There has been a conflict between the quantities of food eaten in grazing experiments and the estimated quantities from field observations. Harvey, Cooper, Lebour & Russell (1935) showed from the uptake of phosphorus that the animals were eating 40% body weight per day as an average throughout the production cycle off Plymouth. Cushing & Vućetiċ (1963) obtained much higher values at the peak of the production cycle in the North Sea for a particular animal, *Calanus finmarchicus* (approx. 300% body weight per day); at the beginning and at the end of the cycle, the values were much lower. The average rations for all animals throughout the duration of the production cycle were not very different from Harvey's value of 40% body weight per day. In contrast, from Marshall, Nicholls & Orr's (1935) oxygen experiments, Riley (1946) suggested that the daily ration of the zooplankton amounted to 7% body weight per day. Later experiments by Marshall & Orr (1958) and others supported Riley's conclusion. Yet Winberg's (1956) data on the oxygen metabolism of fish and crustacea suggest that an animal weighing 1 mg might take a ration of 60% body weight per day; Zeuthen's (1947) extensive series of experiments on the oxygen metabolism of small invertebrates led to the same conclusion. The conflict between field observation and experiment has not yet been fully resolved, but some progress has been made, particularly by Corner, Head & Kilvington (1972).

Petipa (1964), working on *Calanus helgolandicus* Claus in the Black Sea, found a high daily ration during the period of algal abundance. For a period of about twenty days, she estimated the diurnal variation in feeding at various depths for all stages of development of the animal. There was a diurnal rhythm in feeding intensity and in fat deposition, which was most marked in the older animals which migrated further in a vertical sense; however, Hutchinson (1967) showed, from estimates of drag, that vertical migration used little excess energy. The gut contents and the fat droplets along the gut wall were measured microscopically and expressed in volume.

The fat is absorbed directly from the algae when the animals feed heavily and excrete faecal pellets every five to fifteen minutes. The fat is said to pass through the gut wall by a series of fine canals each about 0.5 μm in diameter. The daily ration of *Calanus* was calculated first from the gut contents and estimated digestion rates and secondly from the diurnal variation in fat content. Table 8 shows the daily rations as per cent body weight per day for the developmental stages by both methods. Perhaps the point about vertical migration is not the upward movement, but the distance travelled.

The daily ration estimated from diurnal changes in fat content is greater than that from gut contents. The former method has not yet been repeated, but the latter is a well-tried one in other fields of fisheries biology. The main conclusion is that the daily ration for such small animals is high, as it is for small ones that live on the land.

Table 8 *Daily ration as per cent body weight per day of* Calanus helgolandicus *in the Black Sea* (*Petipa, 1964*)

		Copepodite stage						
	Nauplii	I	II	III	IV	V	Adults	
From gut contents	97	47	23	27	55	56	33	97
From diurnal fat differences	128	68	120	183	60	121	313	93

Cushing (1968*a*) developed a theory of grazing, based on an earlier encounter theory (Cushing, 1959). The herbivore is assumed to search a volume which is the product of the area swept by the sensors and the distance swum. As the animal stops to eat, the volume searched becomes reduced. Algal mortality may be expressed in the volume searched as a proportion of one litre. It is increased by the number of herbivores per litre and decreased by failure to capture cells in the volume searched. As algal density is increased, the time taken in eating is proportionately greater, so mortality decreases in a density-dependent way. The time to capture and eat one cell was shown to be proportional to the reciprocal of the volume searched, so the handling time might be only a small proportion of the searching time. The maximum volume searched varied with the weight of the grazer which was also shown for *Daphnia rosea* feeding on suspensions of yeast (Burns & Rigler, 1967). Like fish, copepods may swim at an average speed of three lengths per second. They have high escape speeds, but there is no evidence yet that the perceptive range is greater than one or two body lengths (Petipa, 1965); there is a little evidence that it is a rough function of length. All these processes were combined (Cushing, 1972*b*) in a single formulation of grazing mortality, G:

$$G = \frac{N[(3\pi\ 0.066l)^2 t]}{[1 + a'3\pi(0.066l)^2\ 7.10^{-6} w_g^{0\cdot 8} c'']10^{-3}}, \qquad (27)$$

where l is the length of the grazer in cm; w_g is the volume of the grazer in mm^3; c'' is the volume of the algal cells in μ^3; a' is the algal density in n/ml, greater than a threshold; N is the number of herbivores/l. Figure 24 shows how grazing mortality might eventually be estimated from algal densities and the size distributions of grazers and algae.

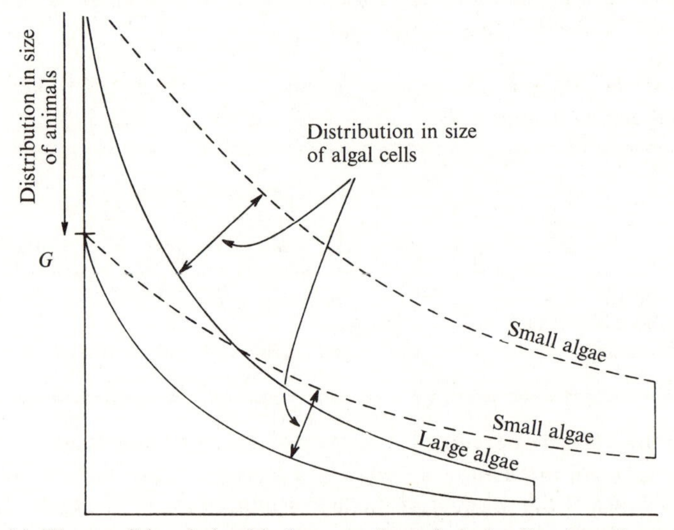

Figure 24. The possible relationship between sizes of algal cells, sizes of grazers and mortality (Cushing, 1968*a*).

Harris (1968) showed that the estimate of volume searched is biased upwards if the animal searches on an irregular course rather than on a straight course. He developed three equations: for the searching time in low algal densities;

$$t_s = 1/(a'\pi r^2 V_c) \qquad (28)$$

in medium algal densities;

$$t_s = (1/V_c)[(1/\pi\ r^2\ a') + r/3] \qquad (29)$$

and if the animal accelerates in attack;

$$t_s = (1/V_c)[(1/(\pi\ r^2\ a' - 2r/3) + r/V_a] \qquad (30)$$

in very high algal densities;

$$t_s = 0.698/V_a\,a'^{1/3}, \tag{31}$$

where V_c is the cruising speed of the animal in cm/s; r is the perceptive range in cm; V_a is the attack speed of the herbivore. A step forward in technique was made by Parsons *et al.* (1967) in applying the Coulter counter to grazing experiments. Ivlev's (1961) equation (eqn 35) was used to describe the results successfully. A threshold level of food density at which grazing starts was determined (as Adams & Steele (1966) had earlier suggested) and feeding was expressed in μ^3/animal per unit time on food density in μ^3. Such methods would make it possible to detect the selection of particles by size by the grazing animal; the work of Mullin (1963) indicated that a large herbivore like *Calanus* tends to take larger diatoms.

The herbivores rise towards the surface at night in their diurnal vertical migration from well below the photic layer. Grazing activity would be expected to vary by day and night. McAllister (1970) has devised a model in which the herbivores only graze at night. Petipa & Makarova (1969) assumed that the grazing rate varies as a sine curve in time, with one wavelength per day and the maximum amplitude at night. It was shown that the quantity grazed during one night in the Black Sea did vary in such a way. As indicated in an earlier chapter, such processes may play an important part in nutrient regeneration, particularly in the lower half of the photic layer at night.

The theory of predation assumes that it is a process that can be split into separate periods of time, search, capture, handling, digestion, etc. (Rashevsky, 1959; Holling, 1959; Watt, 1959; Cushing, 1959). Moreover, it depends on numbers and sizes of algae and herbivores. The density-dependence inherent in the formulation has been detected in many grazing experiments (Ryther, 1954; Mullin 1963; Haq, 1967; Frost, 1972). Any model in numerical terms can express such biological factors very easily.

Secondary production

Models of secondary production have been constructed, but, in general, not enough is known of the ordinary biology of the plankton animals, of which there are many groups and many species. A convenient method is the Allen curve (Allen, 1951), in which numbers are plotted on weight; as weight increases and numbers decrease in time, the curve of biomass moves to the right. It is admirably adapted to the study of a cohort in time; Neess & Dugdale (1959) applied it to an insect population and showed that there was a constant ratio of (M/K) in that cohort. The production of a single species has been calculated in rather a similar way by Shushkina (1968). The specific growth rate $((1/w)(\mathrm{d}w/\mathrm{d}t)$ was estimated for the copepod *Haloptilus longicornis* Claus., south of the island of Fiji. Production Q, is given by:

$$Q = \Sigma_i (C_w)_i w_i n_i, \tag{32}$$

where n_i is the number of animals in an age group; w_i is the average weight of animals in the age group; $(C_w)_i$ is the specific growth rate. B is biomass and the most valuable ratio is P_r/B, which is an estimate of the increment of production per unit time, or productivity. Shushkina estimated the productivity of this copepod in the following way:

Stage		*Weight* (μg)	*B* (%)	C_w/B (= P/B)
Copepodite	II	20	1.1	0.094
	III	30	2.4	0.187
	IV	60	3.3	0.224
	V	110	16.5	0.990
Adult ♀		250	76.5	3.825

The productivity is estimated on a daily basis, but can be equally well estimated weekly or monthly. This is an example of the Vinberg method of estimating secondary production and it depends upon a proper measure of specific growth rate.

An extension of the method is that used by fisheries biologists to estimate the production of eggs over a spawning ground. The numbers of eggs/m^2 in each developmental stage are divided by the coefficient of development rate in temperature to give the production of eggs/m^2 per day. The distribution of daily production in time may be integrated during the period under examination. It is an improved method because information is used from the whole distribution of production in time, whereas the other methods only use the mean of the distribution. However, information is needed on the temperature coefficients of development of the immature stages of the animals and for many animals in the plankton such information is lacking.

The present status of production cycle models

Common to the models described early is the predator–prey relationship between the herbivores and algae. Until about 1960 or so the intention was to model the relationship, to explain it and little more. During the following decade, much under the influence of John Strickland, the parameters were re-examined in some detail and at the present time there is a desire to build models of the production cycle. Common to all of them is Steele's (1965) equation that describes photosynthesis as a function of light intensity and Dugdale's (1967) equation that describes the effect of nutrient limitation. Recently, Parsons *et al.* (1967) have applied Ivlev's feeding equation (eqn 35) successfully to the grazing of zooplankton and it can be used to model the predation in the system. Some of the models constructed are published (e.g. Steele, 1974), and others are being published.

Steele's model is a comprehensive one in that three trophic levels are used primarily in order to describe the death rate of the grazing zooplankton. Like others of the present series it is designed to run for long periods, for example, the annual cycle in the northern North Sea and it is constructed in the form of a computer model; consequently, the variations under different conditions can be analysed in some detail. The predator–prey system is run at a somewhat high level of nutrient limitation with relatively few degrees of freedom. Hence, it should be possible to test it in a statistical manner with relatively few observations, but of course many more than in earlier series.

Many of the arguments that arose from the earlier models on the magnitude of parameters and on the significance of grazing or of nutrient limitation have largely dissolved in new information. However, the major simplifications remain that carbon or chlorophyll adequately represent the succession of algal species, that copepods are the main agents of grazing and that environmental factors can be fed directly into the system in much the same way as Riley suggested. The present need for models are in studies of eutrophication, the effects of pollutants and in the description of the production cycle in fisheries work. It is hoped that such a model can be developed that will describe the variability in timing of the production cycle and the need for one will emerge in later chapters. It will be based on the observations in the daily weather report, so that the onset of production and its development can be described quite fully with the use of equations 7 to 9. The functions developed by Steele and Dugdale will be employed together with one that estimates nutrient regeneration with an adaptation of Riley's formulation (eqn 17). The development of the copepod population will be based on the egg production, the mortality through the developmental stages and their growth, much of which is temperature- and food-dependent. It is possible, but not yet established, that the development of algal stock in time after the spring outburst is governed to some degree by the variation in grazing capacity with age. Very roughly, copepods die quickly during the naupliar stages and slowly during the copepodid stages II–IV and sharply in stages V and VI; the generations tend to be more or less distinct and the greatest grazing mortality is generated in the middle copepodid stages.

It is likely that models of the production cycle are now an accepted and general part of scientific practice rather than the local descriptions that were characteristic of the forties and fifties. In the future we might expect that they be exploited considerably across the whole field of production studies.

The links between trophic levels

At the centre of the deep ocean in the tropical anticyclone, the production cycle may be in a quasi-steady state. The euphotic layer is deep, up to 120 m (Clarke & Denton, 1962); during periods of moderate algal abundance, it shallows to 60 m due to the absorption of light by algal cells (Riley & Schurr, 1959). The herbivores live below the euphotic layer in daytime and they migrate towards the surface at night to feed. These animals range in size from small copepods, that live at about 100 m in daytime, to euphausids, that live at perhaps 400–600 m in the daytime. An array of animal sizes climbs from considerable depths to the photic layer at night; it is part of Vinogradov's (1970) 'ladder of migrations' by which material is transferred from the photic layer to much deeper water. The total density of herbivores at night in the photic layer may be enough to consume the daily increment of algal production. One of the consequences of the grazing in darkness is that the nutrient levels in the photic layer may be restored after a day's production (Ryther *et al.*, 1961). There is a diurnal variation in the photic layer in numbers of algae, in numbers of herbivores and in nutrient levels. In recent years it has been shown that many marine algae divide in darkness (Eppley *et al.*, 1967*a*). In the deep ocean the algae tend to be distributed in the lower half of the photic layer. The herbivores in the nightly vertical migration rise to the same depths and eat the algae and excrete there. The addition of excreted nutrients might well bring the algal reproductive rate nearer the maximum at the time when the daily increment of photosynthesis is started. At the end of the day the nutrients in the water may be reduced at a time when the algal reproductive rate is not much reduced. The work of Cahet, Fiala, Jacques & Panouse (1972) in the Mediterranean describes the distribution of photosynthesis in the photic layer and provides a little support for this thesis.

In the photic layer algae sink, die and are eaten which means that part of them is transferred to faecal pellets which also sink. Further, herbivores which migrate downwards in the daytime defaecate and are eaten in deeper water, so there is a downward transfer of material. In depth there are two processes that take place in parallel: (1) the transfer of material up through the trophic levels and ending in the bathypelagic community of ultimate carnivores; but it must be remembered that there is also a surface transfer at night through the trophic levels to the myctophid-like fish of the Deep Scattering Layer and to the oceanic tunas; (2) the transfer of material through the scavengers and decomposers into migratory fish which take it back to the surface. The whole system in the deep ocean is self-contained and efficient in that there is little loss of material to the sea bed. In contrast, in the production cycles of temperate and high latitudes and of coastal upwelling areas, considerable quantities of material are deposited on the sea bed. Indeed the presence of diatomaceous ooze on the continental shelves and elsewhere indicates the inefficiency of transfer of material between trophic levels. But, most important, this inefficient transfer nourishes the benthos and to some degree the commercial fisheries.

At the primary level, the algal reproductive rate depends upon irradiance and the nutrient concentration (when it is low) and the grazing rate is a function of the number of herbivores and of the algal density. The numbers of algae are modulated by the trophic levels above and below. In a similar way, the reproduction of the herbivores depends upon the quantity of food eaten and their mortality may be a function of available food and predation. The numbers of herbivores are determined by the quantity of food eaten. The lower trophic level of nutrients represents a balance between uptake and regeneration, both of which depend upon production and grazing in the 'primary' trophic level above.

Between each trophic level there are links, which may be classed as transfer links or modulation links. Nutrients are transferred to production and the algal reproductive rate is modulated by the nutrient concentrations at low nutrient levels. Algae are transferred to herbivores in the quantity grazed, whereas the quantity of herbivores modulates algal stock, algal production and hence the quantity grazed itself.

If the daily differences in algal numbers are low as in the tropical ocean and if their reproductive rate is only mildly modified by the nutrient concentration, then the balancing exponents must be nearly equal; uptake and regeneration of nutrients, algal reproductive rate and grazing mortality, herbivore reproduction and loss are in balance. The balance is achieved not by the transfer links but by the modulation links, nutrient concentration and herbivore stock. It has already been suggested that variance in nutrient concentration is minimized by the processes of vertical migration. The variation in algal stock is due to the quantity grazed, which should be enough to support the herbivores for a day. It is not unexpected that the relatively long-lived herbivores should exceed the algal stock in quantity by many times and that the nutrients should be exploited to the lowest level possible. These processes form the bases of the continuous or quasi-steady state production cycle.

The difference between such a cycle and that in temperate waters and in upwelling areas is only that the latter has to start from very low overwintering or initial levels. The start of production each spring generates a delay in the development of grazing capacity and hence a high amplitude production cycle. After the peak of such a cycle, material is transferred to the herbivores and the nutrients decline. However, the principles outlined for the continuous cycle also apply. The delay is of the greatest importance and so the processes occur in sequence. The links between trophic levels are the same in both forms of cycle, but occur one after the other in the high latitude cycle.

Production and the fisheries

In Chapter 3 the contrasting production cycles of the deep ocean and of high latitudes were compared. That of the deep ocean in the centre of the tropical anticyclone is a low-amplitude continuous cycle with many species each of which is not very abundant and so there is a diversity of animals. The discontinuous cycle in high latitudes and in upwelling areas is of high amplitude with few species each of which is abundant, i.e. there is little diversity. The first is numerically and structurally stable and the second is numerically unstable and perhaps structurally unstable. A stable structure is a network of food links which retains its identity from generation to generation. However, the structural stability must depend on the stability of numbers. Many temperate fishes use a variety of foods as they become available and as they grow through the life cycle. There is a difference of about an order of magnitude in production intensity (mg C/m^2 per day) between tropical and temperate cycles and a difference of three times in conversion efficiency, so the effective difference in intensity at the secondary trophic level is of the order of three times.

The oceanic ecosystem was described as an efficient one partly because most of the material is recycled in depth. The ocean is very deep and on the shallow continental shelves, because of less efficient transfer, relatively more material is spilled on the bottom where it supports the benthos. The links between the benthos and the rest of the oceanic food chain are somewhat indeterminate because they depend upon the corpses, faecal pellets, dead algae and detritus which sink to the sea bed. The efficiency of transfer by means of such residuals of productive processes is difficult to establish. Cod and hake, which support the principal demersal fisheries in the world, feed on benthos, on euphausids (which may be herbivores or scavengers) or on fish. The ecosystem which rests on the temperate production cycle (including those in upwelling areas) appears to be an inefficient one, but more living material may be produced within such systems than in the oceanic ones.

However, the efficiency of the system in the tropical ocean has considerable disadvantages for fishermen. Because the photic layer is deep and because the daily increments of algal material are relatively small, the animals tend to be rather dispersed. Consequently, the dense aggregations of primary and secondary carnivores characteristic of higher latitudes are absent in most offshore tropical waters: the animals that comprise and are associated with the Deep Scattering Layer are dispersed. It is no accident that when the Japanese fish for salmon south of the Aleutians, or for tuna in the subtropical ocean, they use gill nets and long lines (respectively) of up to 80 km in length. A reasonable catch of tuna or tuna-like fishes is one to five fish per 100 hooks or one to five fish on a line two and a half miles long. It is technically rather difficult to make the large catches in the tropical ocean that are common in higher latitudes.

In the subtropical oceans, there is a trophic level which is absent from the pelagic zone of temperate waters. The myctophid-like fishes of the Deep Scattering Layer feed on zooplankton and are eaten by tuna. They are about 5–10 cm in length and correspond to the yearling herring and sprats of higher latitudes. In temperate waters cod feed on herring, but also on euphausids and on benthic animals; perhaps it would be truer to say not that a trophic level was absent, but that the links between them were differently arranged. The myctophid-like fishes are dispersed; from studies on the Deep Scattering Layer, their density is of the order of $1/1000\ m^3$ (Johnson, Backus, Hersey & Owen, 1956). A mid-water trawl with a mouth of $20\ m \times 20\ m$ would capture about 4000 per hour, hauled at about 5 knots. At 15 g each, the total catch would weigh 60 kg, which would pay for only a few minutes' steaming by a trawler capable of towing such a mid-water trawl. In contrast purse seines working in the north-east Atlantic could catch hundreds of tons of herring in an hour, under favourable conditions of aggregation. There are three factors which counter the development of an oceanic fishery for these fishes (1) their small size; (2) their low density, and (3) their lack of aggregation.

The quantity of tuna available for capture in the tropical and subtropical ocean is not very great. In the Indian Ocean the four main species caught by the Japanese and others are already exploited at the maximum sustainable yield; the maximum sustained yield of the four species amounts to not more than three or four hundred thousand tons (Suda, 1971). The fishery is sustained by the high value of tuna in the American market, but the low yield is a direct consequence of the low amplitude production cycle, the extended food chain and the dispersal of the animals.

The main fisheries of the world are found in the high latitudes and in the upwelling areas. The greater proportion of the world catch of nearly 70 million tons is comprised of herring-like or cod-like fishes; ten million tons alone were taken from the anchoveta fishery in the Peruvian upwelling area until recently. Recently, Gulland (1970*b*) has estimated that the total potential world catch of fish and shellfish taken by conventional means will not amount to more than 120 or 140 million tons; for economic reasons the attained maximum catch might well be less. The additional catches will come from traditional grounds, from new ones, for example, in the Indian Ocean and from grounds over-exploited at the present time. Most of the catches are and will be taken from production cycles in temperate waters and in upwelling areas.

Later in this book it will be shown that stocks of fish in temperate waters are linked in their recruitment rather closely to the nature of the production cycle. Climatic changes modulate the abundances of fish stocks through this link. In order to understand its nature the dynamics of fish populations must be comprehended in considerable detail. The next section of this book is concerned with this subject.

5

The circuits of migration and the unity of fish stocks

Introduction

There are fisheries for spawning fish, for feeding fish, for overwintering fish, for fish gathered at oceanic boundaries and in the upwelling areas found in tropical and subtropical seas. Common to all of them, at least in temperate waters and in the upwelling areas, is some degree of regularity in their seasonal appearance. The cod (*Gadus morhua* L.) of the arctonorwegian stock spawn in the Vestfjord in northern Norway in February and March and they have been fished there at that season since the twelfth century (Rollefsen, 1955). The same regularity is observed in the fisheries for the maturing Pacific sockeye salmon (*Oncorhynchus nerka* (Walbaum)) in the Strait of San Juan de Fuca and the Strait of Georgia off British Columbia in Canada, where they gather before migrating up the Fraser River. The stocklets of the river system are identified in the Straits by the patterns on the scales of each fish (Henry, 1961) as they pass through the fishery during the same week, or thereabouts, each year. Each stocklet returns to its native stream, and a large number of them have been identified in the Fraser river system.

The timing of feeding fisheries is less precise, as might be expected; for example, immature albacore (*Thunnus alalunga* (Bonaterre)) appear regularly in the Kuroshio extension in summertime, when the young fish feed (van Campen, 1960; Suda, 1963). The Icelanders used to catch the Norwegian herring off the north coast of Iceland in July and August where the fish fed on aggregations of the copepod *Calanus finmarchicus* (Gunner). Whereas the date of peak spawning in temperate waters may vary by about a week (Cushing, 1970; see below), the season of a feeding fishery in the same waters might vary by about a month. Fisheries at an oceanic boundary depend upon its position and upon the arrival of the fish there; for example, the arctonorwegian cod are caught in June at the boundary between the West Spitzbergen current and the arctic water mass on the Svalbard shelf in the Barents Sea. In some years the fish are caught on the shelf quite close to Bear Island, but in other years they are found in deeper water off the shelf edge because the water mass boundary occurs in a different position (Lee, 1952). Uda (1952*a*) showed that the position of the Japanese bluefin tuna fishery in the early months of the year shifted considerably in two particular years (1939–40) as the axis of the Kuroshio current south of Honshu was displaced. The regularity of such fisheries is less precise than that of temperate spawning ones. Fisheries in upwelling areas owe their regularity partly to the dependence of the local stocks upon the physical and meteorological structures and partly to the appearance of oceanic fishes at the boundaries beyond 100 km offshore (Cushing, 1971*a*). The regularity in the timing of the different forms of fishery depends upon a number of factors, but the dominant and controlling factor is the regularity based on the circuits of migration of the fishes.

Harden Jones (1968) epitomized the migration of fish as movement from spawning ground to nursery ground, from nursery ground to feeding

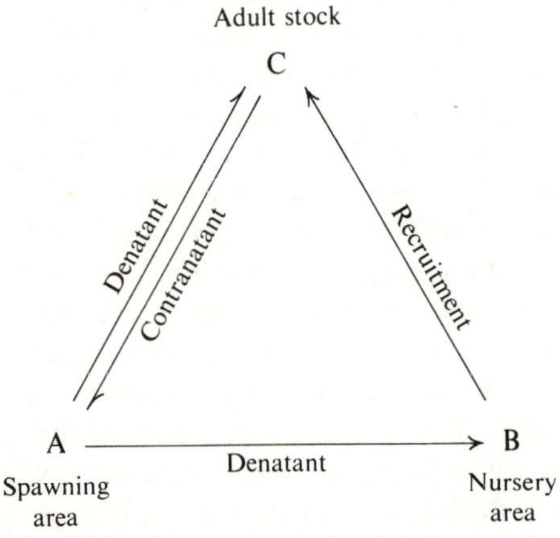

Figure 25. The triangle of fish migration; maturing fish move against a current, contranatantly to the spawning ground. Spent fish drift, denatantly, from spawning ground to feeding ground; larvae drift denatantly to the nursery ground in the same current. Recruits migrate from the nursery ground to join the adult stock on the feeding ground. The terms *contranatant* and *denatant* describe the nature of migration and carry no connotation of orientation.

ground and from feeding ground back to the spawning ground, in the form of a triangle during the life cycle (Figure 25). The spawning grounds, in temperate waters at least, tend to be fixed in time and space rather precisely and the larvae drift away from them in a regular current system towards the nursery ground. The nursery ground is often inshore and the little fish tend to reach it at or just after metamorphosis. Although metamorphosis is often a profound structural change, it is also an ecological one, the departure of fish larvae from the plankton. The larval drift, or the migration of larvae from spawning ground to nursery ground, is illustrated in Figure 26. The centre of the plaice spawning ground in the southern Bight of the North Sea has been established from a number of egg surveys which have been executed in the area since 1911. The point where the larvae sink towards the bottom lies off Texel Island (off northern Holland) and it has been described by larval surveys made in the Southern Bight of the North Sea since 1962 (Harding & Talbot 1973; Cushing, 1972*a*). In this region there is a water mass boundary between the homothermal channel water and the Dutch coastal water which is stratified; Dietrich (1954*a*) has shown that where the tides run parallel to this boundary the water near

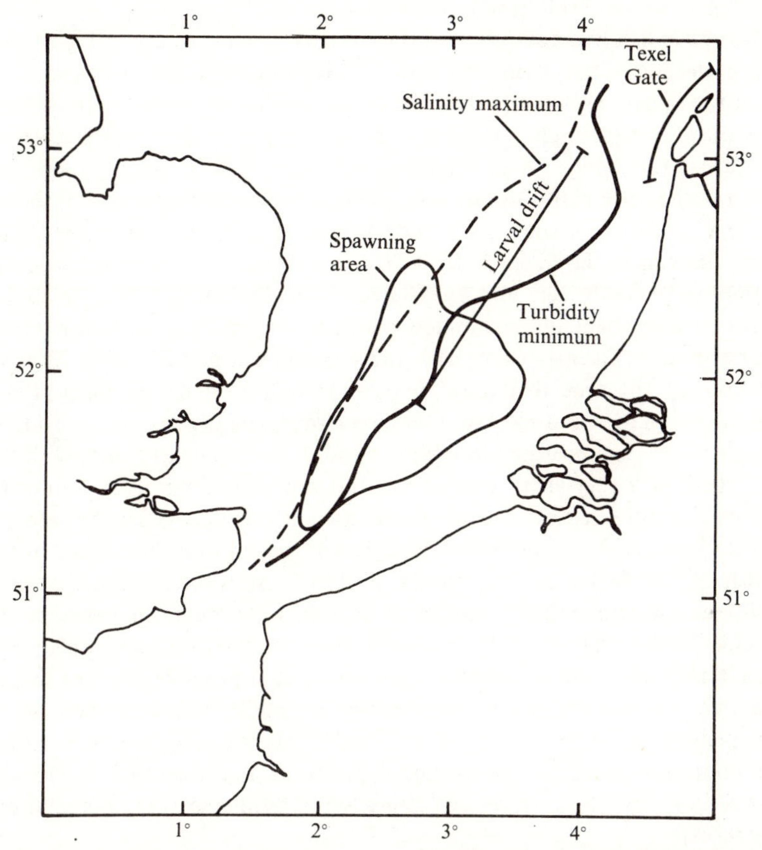

Figure 26. The larval drift of the plaice in the Southern Bight of the North Sea. The spawning area between the Thames and the Rhine was described by egg surveys since the first decade of the century. The larvae drift in the most saline and clear water towards the Texel Gate where they move in towards the Waddensee (Cushing, 1972*a*).

the bottom moves inshore at a particular state of tide. The little plaice are drifted inshore by such a residual current near the bottom and are found predominantly in the Dutch Waddensee, inside the western Friesian Islands close to Texel (Zijlstra, 1972). Such a hydrographic gate is found only off Texel Island and off the east Friesian Islands, south of Esbjerg, and the latter would be a convenient entry for larvae from the German Bight spawning ground. Hence, hydrographic mechanisms may exist which ensure the isolation of the nursery grounds of stocklets adjacent to each other. Ramster (1973) has shown that bottom sea-bed drifters moved inshore particularly in winter, to those beaches in the Irish Sea where metamorphosed plaice are found. The larval drift of the southern Bight plaice from the spawning ground between the Thames and the Rhine to the Texel gate is well established.

The little fish live and grow on the nursery ground where, if bigger fish live in deeper water (Heincke's Law, 1913), they might find some security from predators. However, when the metamorphosed plaice move inshore they must pass through a band of avid I group plaice and other predators, which eat the smaller ones (Steele & Edwards, 1969). During their first summer the little plaice live in very shallow water, between three fathoms and the tide line. As they grow they spread into deeper water and away from the coast; Beverton & Holt (1957) analysed the stock densities of immature plaice by age and by distance from the coast and calculated a diffusion coefficient that estimated the rate of spreading. In linear terms, over a period of nearly four years, the juvenile fish moved about 70 miles offshore by the time that they joined the adult animals. de Veen (1962) has tagged plaice (and sole) on their spawning grounds and, in the following years, the fish returned to their ground of first spawning and, with the exception of a very small proportion, did not stray from it. It cannot be shown that they return to their parent grounds as do the Pacific salmon, but de Veen & Boerema (1959) have demonstrated that differences in the width of the first-year ring on the otolith exists between the Southern Bight and German Bight stocklets. Hence, differentiation between the stocklets must originate early in the life cycle, on the two nursery grounds, which are more than a hundred miles apart. It is possible that the fish of the two stocklets return to their native grounds, although there is no independent evidence of this point. Harden Jones' diagram summarizes the migration circuit in the sea and it presumes that the stock or stocklet has achieved its isolation by exploiting the current structure, in evolutionary terms.

In temperate waters, spawning grounds and nursery grounds are found on relatively fixed positions and so the larval drift between them in a regular current is also found in the same position from year to year. The nursery grounds of the two plaice stocklets in the southern North Sea are much more isolated from each other than was expected. The spawning season, although it peaks in very nearly the same week each year, lasts for about three months, so the area of the larval drift and the region of the nursery ground are pre-empted by that spawning group for that period. Plaice stocklets spawn at a characteristic season and deny their larval drift, spawning ground and nursery ground to any competing potential stocklet, merely because the spawning season is long. The larval drift which is the base of Harden Jones' triangle provides a geographical basis for the stock and stability is maintained as it returns to the same spawning ground, or group of spawning grounds, each year. Frequently the animals must appear to migrate contranatantly, that is against the current which carries the larvae, but equally they could have drifted in a

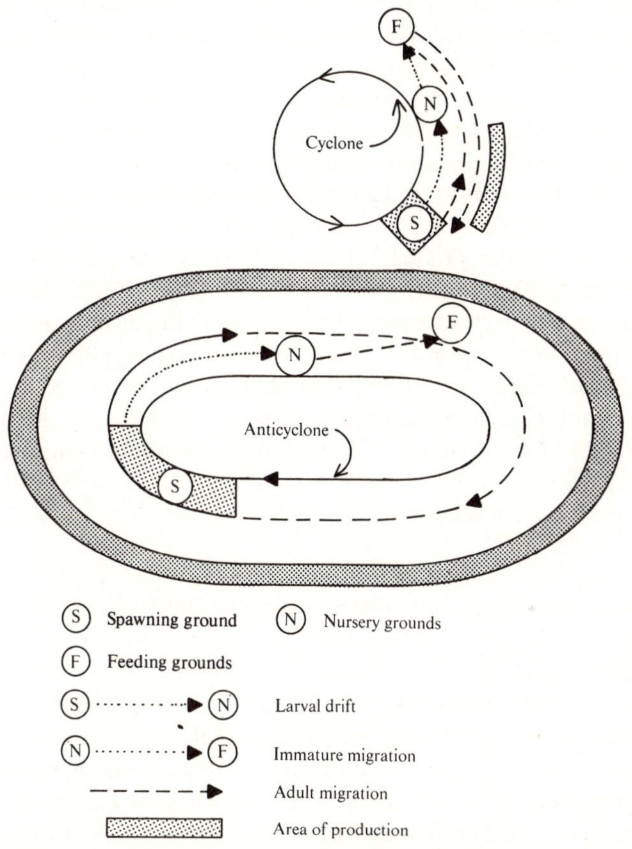

Figure 27. Migration and production in cyclone and in anticyclone; it is assumed that production in the first is discontinuous and in the second it is continuous. In the anticyclone the spawning ground spreads over thousands of miles; the production cycle is continuous and the whole system moves in drift. In the cyclone production is discontinuous and adult spawners have to move contranatantly on their spawning migrations.

counter-current. The important point is that the fish return each year to the same spawning ground, but the mechanisms of migration remain mysterious.

Harden Jones (1968) links migration, oceanic circulation and production cycles in a single diagram; he distinguished the continuous production cycles in the seas of the anticyclones and the discontinuous ones of higher latitudes, as indicated in an earlier chapter. In each diagram the inner circle represents the direction of water circulation. The outer represents the migration of the adult fish (- - - -→) and the middle one represents the larval drift (· · · · · ·→) and the migration of the immatures to the point at which they recruit to the adult stock. The period of production is indicated by a shaded segment and each diagram represents an annual circuit of migration. In the tropical ocean spawning is spread through long seasons and recruitment occurs at all times and places. I have adapted his diagram to the conditions in a temperate cyclone and a subtropical anticyclone (Figure 27). In temperate and high latitudes there is a larval drift restricted to the productive season, and recruitment probably occurs at a particular time of year. The circuit may be broken by a coastline, a water mass or the need to reach the bottom. In general terms migration is linked to current structure and to the production cycle. If stocks are isolated and have become genetically distinct, then stock unity is maintained in the circuit of migration within the oceanic structures. Individuals may be lost from a stock for a variety of causes, but the link between migration circuit and stock identity makes it unlikely that they would be gained by another with any evolutionary persistence; it is more likely that such individuals would establish a new stocklet. The study of migration is fundamental in fisheries biology because it provides evidence of stock unity independent of the difference between stocks.

Spawning in space and time

Positions of spawning

Marine fishes have many different forms of spawning ground. A small one of limited extent visited each year is the Eagle Bank in the Blackwater estuary in southern England, and the Blackwater herring spawn there; they are a local race of small estuarine herring such as are found in many English rivers which used to be known as Leach's herring (Yarrell, 1836). An extensive spawning ground is that of the albacore (Yoshida & Otsu, 1963) in the North Pacific Equatorial current, and it extends imprecisely from the region south of the Hawaiian Islands to the Philippines; not only is the area thousands of miles long, but it is hundreds of miles wide. Between the two extremes there is every variety of spawning ground. But in the definition of a spawning ground in geographical terms lurks the presumption that the fish make use of it exclusively; it is, however, rarely known whether they return to their native grounds.

The parent stream hypothesis as applied to the Pacific salmon (and steelhead trout) is thus of some importance. The hypothesis originated amongst the early workers in the field (Gilbert, 1916; Rich, 1937) but some, at least, of the tests are comparatively recent. In Cultus Lake, off the Fraser River in British Columbia, over six hundred thousand sockeye smolts were fin-clipped and at the outlet of the lake all smolts leaving it and salmon returning to it years later were counted; about 1% returned to Cultus Lake and none were reported elsewhere (Foerster, 1934,1936,1937). In McClinton Creek on Graham Island (one of the Queen Charlotte Islands off British Columbia) just over a hundred thousand pink salmon (*Oncorhynchus gorbuscha* (Walbaum)) smolts were fin-clipped and 3265 returned to the creek or to Masset Inlet (where the river reaches the sea); only 0.3% of the returns were recovered elsewhere (Pritchard, 1939). On the coast of California, Waddell and Scott creeks reach the sea at points five miles apart. Steelhead trout (*Salmo gairdneri* Richardson) were fin-clipped as smolts in each creek; 3% of the Waddell migrants returned and 0.07% strayed to Scott creek, whereas 2.1% of the Scott migrants returned and 0.03% strayed to Waddell creek (Taft & Shapovalov, 1938; Shapovalov, 1937). Clutter & Whitesel (1956) have shown that the sockeye smolts of the Cultus, Francois, Shuswap and Stuart stocklets differed markedly in the number or circuli (or non-annual rings (10–19) within an annual ring, or annulus) and that the returning smolts can be identified. This technique allowed Henry (1961) to identify the Fraser stocklets in the fishery in the Straits of Georgia and of San Juan de Fuca. This paragraph has summarized the evidence that the Pacific salmon (and steelhead trout) return to their parent streams to spawn and that the percentage of strays is surprisingly small.

From the point of view of stock unity, the degree of stray from generation to generation is of considerable importance. In purely marine fishes, it cannot be established by fin-clipping the very young fish, but in a multi-age stock, the degree of stray within a year class has been estimated. Harden Jones (1968) has analysed the tagging results on the British Columbian herring. Hundreds of thousands of metal tags were fixed with little guns into the peritoneal cavities of the herring and were recovered by magnets in the fish meal plants. The recoveries were classified by statistical areas. The proportion returning to given statistical areas increases with age until finally all are returned to the original area, so the percentage of strays decreases with age. Further, the interchange between statistical areas decreases with distance. There are three areas between which there is no interchange, West, Middle East and Lower East. But within any such group or stocklet, there may be as much as 20% stray from the ground of first spawning and this proportion decreases with age. Li (1955) suggested that a population consisted of a number of groups within which interchange occurs, to maximize the variability.

The spawning grounds of the herring (*Clupea harengus harengus* L.) in the north-east Atlantic are very small, are used each year and in each stock there appears to be a group of them. During a period of about fifteen years (1950–65), the spawning grounds of the Downs herring stock in the southern North Sea and eastern English Channel were under close observation by trawlers, by echo survey and by larval survey. Trawlers were

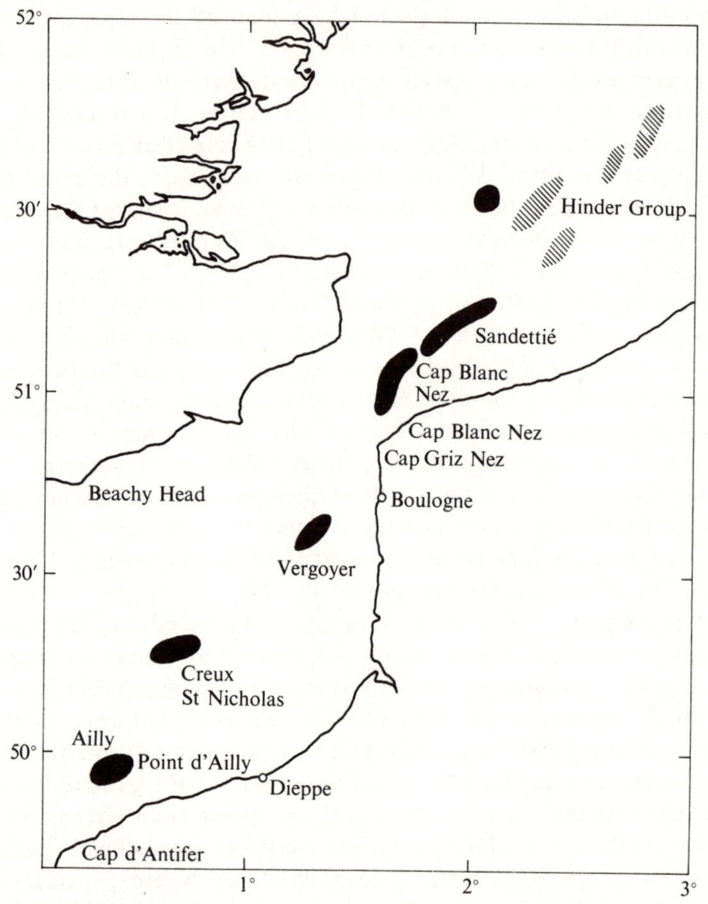

Figure 28. Herring spawning grounds in the southern North Sea and eastern English Channel (adapted from Cushing, 1968); the shaded areas are assembly grounds.

rarely found outside the patches shown in Figure 28. The densest echo traces were observed on them or near them. They were rarely indicated by larval survey because they are very small and the larvae drift and diffuse away from them quickly. Only one was examined in detail; that near the Sandettié LV was 500 yards wide by about 2000 yards long (Bolster & Bridger, 1957). Like other herring spawning grounds in the north-east

Atlantic, it was situated on gravel beds in a little less than 40 m of water. A spawning ground of spring spawning herring has been described (Parrish *et al.*, 1959; Stubbs & Lawrie, 1962) on the Ballantrae Bank in the Firth of Clyde; it is also of small extent and spawning occurs at the same position each year. Runnstrøm (1936) explored the spawning grounds of the Norwegian spring spawning herring for seven years with a Petersen grab. The grounds were of limited extent on gravel around the island of Utsira, not far from Haugesund on the south-west coast of Norway. Some were characterized as 'early February', 'late February' or 'early March' grounds. Runnstrøm (1934, 1936) had shown that there were three main groups of spawning herring, off Utsira, off Stadt (between Bergen and Trondheim) and off the Lofoten Islands, each separated from the other by hundreds of miles. Runnstrøm also showed that there were small but consistent meristic differences between them. Within each stocklet there are a number of spawning grounds, just as there appears to be in the Downs stock of autumn spawning herring.

Cushing & Bridger (1966) distinguished the Downs stock from the Dogger stock, which spawns in the neighbourhood of the Dogger Bank in October. Differences between the stocks in meristic characters had been known for a long time, but Cushing & Bridger distinguished them by their vital parameters, growth, recruitment and mortality. Within three fisheries on the Downs stock, mortality was common for a period of thirty years and distinct from that in the adjacent Dogger stock. Year classes were highly correlated within the three Downs fisheries, but there was no correlation between Downs recruitment and that to the Dogger and Buchan stocks. Length for age differed persistently for twenty years or more between the three stocks, although the differences were small (Anon, 1965). Within the Downs stock, however, there are two groups, Southern Bight and Eastern Channel, originally designated as Sandettié and Somme by le Gall (1935). There was no difference between the groups on vital parameters, but there has always been a small and persistent difference in vertebral count (about 0.15). When the Downs stock was reduced by fishing during the late fifties and early sixties, both components were reduced. But when the fishing effort had been reduced, only the Eastern Channel component recovered. The spawning in the Southern Bight did not recover, so there must have been a difference in the recruitment mechanism which was not detected by other means.

The distinction between the stocklets of plaice in the Southern North Sea has already been made. But a difference in recruitment had also been detected. Bückmann (1961) showed that, in the late fifties, the egg production in the German Bight increased by nearly an order of magnitude. Present surveys (Bannister, *et al.*, 1974) show that the German Bight stocklet is much the largest in the southern North Sea, because that in the Southern

Bight itself did not increase during the same period. Hence the two stocklets are distinct in recruitment and in otolith character.

The arctonorwegian cod stock spawns in the Vestfjord in northern Norway, but also at some distance to the south (3% in tags returned from the Barents Sea to the Vestfjord or south of it (Trout, 1957)). A stock is isolated in the Faroe Islands. Cod spawn on the south coast of Iceland and on both the east and west coasts of Greenland. Throughout a long period tagging experiments show that some Greenland fish have been recovered on the spawning ground south of Iceland (see Harden Jones, 1968, for a summary of the results). In a later section, the genetic evidence on the identity of cod stocks in the north Atlantic will be examined. Except for the Iceland ground, the spawning grounds of the cod are distinct and their identity from year to year appears to be maintained.

There is a trend in the precision of information from the Pacific salmon, which spawns on its parent or natal ground, to the oceanic tuna which may do the same thing if it is admitted that the natal ground is of enormous extent. There is a small proportion of stray in the Pacific salmon from its native ground and a downward trend of stray with age in the British Columbian herring. In the Atlantic herring, the structure of discrete spawning groups within a stock has been discovered, but the measure of stray is unknown. There is a superficial resemblance between herring and salmon in that both lay their eggs on the sea bed, but the eggs of cod and plaice are pelagic and the cod, at least, spawns in mid-water. Under such circumstances, separate spawning groups could not be maintained on a single spawning ground. There might be some stray between cod stocks which we now consider as genetically distinct and indeed it has been measured by the departure of some tagged cod from their traditional grounds. In upwelling areas, sardines appear to spawn at the points of upwelling (Ahlstrom, 1966; Flores, 1968). Because upwelling depends upon the longshore wind and the angle it makes to the coast, spawning grounds would be expected to shift from season to season and they do. Yet Sprague & Vrooman (1962) distinguished three groups of sardine in the Californian upwelling system. Even in a variable system the fish populations found ways of turning their migration circuits to achieve the needed isolation. The common ground to this array of facts about disparate stocks is that fish stocks have a unity which has not yet always been established properly.

The conclusion from this section is that fish stocks in temperate or high latitude waters tend to spawn at fixed positions within a current system. There may be a number of sites between which some interchange of spawners might be expected from year to year. The sites may be small and local, as in the herring which lays its eggs on the sea bed, or they may be less well defined as in cod or plaice, which spawn in mid-water. The boundary to a stock considered as a group of stocklets is exclusive. The reproductive isolation is maintained by the circuit of migration and the larval drift provides a geographical base between fixed spawning ground and fixed nursery ground and during the period of drift and perhaps also on the nursery grounds, the numbers of the stock are regulated. The larval drift provides a geographical base to the stock and during the period of drift, and perhaps also on the nursery grounds, the numbers in the stock are regulated. The whole circuit of migration is based on the larval drift; the stock has evolved in such a way that the larvae exploit the production cycle at the most profitable geographical position. The spawning season is long and so the nursery ground is pre-empted from competitors of the same species. In theory larvae might reach a particular nursery ground from a distant spawning if the currents were favourable; in fact they would have metamorphosed and sunk out of the current system long before they reached it. With such mechanisms is the reproductive isolation of a stock maintained. However, any such stock must sustain small losses in the current systems; for example, strays from the Icelandic cod stock established that at west Greenland during the thirties. This effect might well be achieved by a lengthy spawning at the right time to exclude immigrants and to deny a haven to emigrants elsewhere. Losses must take place and the appearance of a haven for emigrants may be a mechanism of the colonization of environments, like west Greenland for cod that live at Iceland.

Times of spawning

The time of spawning in temperate and high latitude waters is quite precise. There is an extensive spawning period, lasting perhaps two or three months in plaice and cod, but lasting only two or three weeks in herring (Runnstrøm, 1936; Ancellin & Nédelèc, 1959). Buchanan-Wollaston (1915,1922,1926), and Simpson (1959), have published the results of plaice egg surveys in the Southern Bight of the North Sea. Each series of cruises started before the spawning season and continued until it was over, so a mean date of spawning can be estimated. It is designated as the peak date of spawning and, with observations from many years, a mean date of peak spawning is estimated. For the plaice this date is 19 January and its standard deviation is about seven days. Runnstrøm (1936) made grab samples on the Utsira spawning grounds of the Norwegian herring for seven years. For more than a dozen spawning groups it was shown that the dates of spawning did not trend with time and that the standard deviation of the mean peak date was again about one week.

There is a considerable quantity of data on the peak spawning dates of the Pacific sockeye salmon, which is published in the Annual Reports of the

International Pacific Salmon Commission: for about fifty stocklets in the Fraser River system there are up to seventeen years of data available and only in one is a trend with time discerned. The standard deviation of the mean date of spawning for each of this large number of spawning groups is about a week. There is a long time series (of about seventy years) of observations on weekly catches of cod in the Vestfjord in northern Norway during the spawning season and the date of peak catch was assumed to be related to that of peak spawning. During the whole period, a delay in spawning of about ten days occurred in the twenties and thirties of the present century; it was a period of climatic change when the difference in pressure between the Icelandic low and the Azores high tended to decrease. The standard deviation about the trend in the peak date of arctic cod catches is about a week. For four well-known species, commercially dominant ones, the standard deviation of the peak date is low, which means that fish might spawn at fixed seasons. Even the trend with time, a possible adaptation to climatic change by the arctic cod, is low. Hence the generalization can be made that fish spawn at fixed seasons in temperate and high latitudes.

In tropical and subtropical waters the times of spawning are less precise. There is more than a decade of observations in egg surveys on the Californian sardine (Ahlstrom, 1966) and the standard deviation of the spawning date is nearer seven weeks than seven days, so peak spawning can be found in any of three or four months. Tuna larvae are found all over the Pacific during most seasons as if spawning were continuous both in space and time (Matsumoto, 1966). This account is a summary of that given in Cushing (1970). The four fish species – salmon, herring, plaice and cod – with relatively precise spawning times live in the high latitude cyclones, whereas the two species whose spawning time is variable live in the subtropical anticyclones. In the cyclones the production cycle is discontinuous and variable, whereas those in the anticyclones are probably continuous and relatively constant (see Figure 6). Food is always available for fish larvae in the anticyclones and so a precisely timed spawning season on a fixed ground is not needed. But, in the cyclones, food is available for only part of the year and the hatching of the larvae must be linked to production cycles (as will be shown later) and, as they are variable, the best chance for the larvae is obtained with a fixed spawning season on the same ground year after year.

From low latitude to high latitude, from the continuous production cycle to the discontinuous one, the timing of spawning seems to become more precise. The trend is the same as that in the structure of spawning grounds from the very extensive and possibly variable positions of sardines and albacores to the restricted and consistent ones of salmon and herring. Hence the constancy in season is linked to that in space by the nature of the production cycle. The migration circuit in high latitudes is more highly organized than the loose one of the tropical anticyclone; in the first, feeding and spawning are firmly separated in the seasonal cycle and in the second, the oceanic tuna appear to feed and spawn in the same season and at most seasons.

The nature of the larval drift

The migration of eggs and larvae from spawning ground to nursery ground is the larval drift. Eggs and larvae of the plaice of the southern North Sea drift from the Southern Bight to the Texel Gate (as described above), from the German Bight to the north Friesian Islands and from the Flamborough Off Ground to the English coast. The leptocephali (or eel larvae) drift from their spawning ground in the Sargasso Sea for perhaps two and a half years in the North Atlantic Drift before they reach European coastal waters (or three and a half years before they reach the Nile) (Schmidt, 1922) and metamorphose just before they swim up the rivers. The plaice stock is small in number and the larval drift extends over a limited distance, whereas the eel stock provides fish for all European and Mediterranean rivers (Harden Jones, 1968) and its larval drift ranges across 90 degrees of longitude. Because the larvae must spread by diffusion during the period of their drift, the nursery ground is larger, the longer the drift. Large stocks like the arctonorwegian cod or the Norwegian herring each have extensive nursery grounds at the end of the larval drifts which extend for 400 miles. The young cod live all over the shallows of the Barents Sea, on the Svalbard Shelf south of Spitzbergen, on Skolpen Bank in the central Barents Sea and on the banks off the coasts of Norway and Russia; the fat (juvenile) herring are caught on the shelf near the fjords along the Norwegian coastline north of Bergen for more than five hundred miles (Harden Jones, 1968).

Because the spawning ground in temperate waters is more or less fixed and because the larvae are carried from there to the nursery ground in a regular current system, the larval drift provides a fixed geographical base for the stock. When the fish leave the nursery ground, they diffuse away to deeper water until they join the adult stock on the feeding grounds. The latter are not very well defined because the fish range across them as far as they can until they are limited by depth or oceanic boundaries. For example, the Norwegian herring migrate right across the Norwegian Sea; in spring and early summer they tend to move northwards from the Norwegian coast across very rich feeding grounds in the centre of the area (Marty, 1959). In high summer they concentrate along the polar front between Jan Mayen and Iceland and move southwards along it until they join the East Icelandic current (Harden Jones, 1968). Similarly, the arctonorwegian cod spread across the Barents Sea and, as shown by the recovery of tagged fish liberated in the Vestfjord and by the spread of echo traces across the shelves in summer, the summer distribution is bounded only by the cooler arctic water (Figure 29) (Dannevig, 1953; Beverton & Lee, 1964). Thus, the feeding grounds of the adult cod and the Norwegian herring are limited

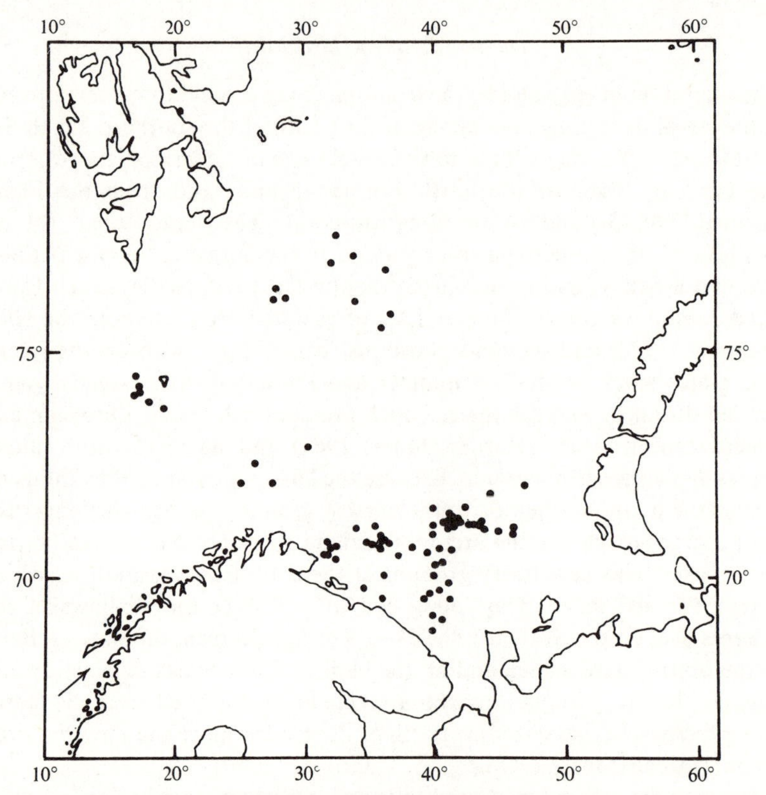

Figure 29. The position of tags recovered in the Barents Sea from a liberation in the Vestfjord (arrowed) (Dannevig, 1953)

by depth or oceanic boundaries and the fish feed there until they turn towards the spawning ground that may be as much as a thousand miles away. The furthest limit of such a feeding migration may be governed only by the distance which the fish can swim. For example, the Downs herring, which spawns between the Rhine and the Seine in winter, migrates northward to its feeding ground on the Fladen Ground in the northern North Sea in summer. They turn at 450 miles from the spawning ground and at 240 miles from the shelf edge (Wood, 1937; Cushing & Bridger, 1966). The circuit of migration in its extensive adult phase is based in temperate waters on the position of the larval drift.

The spawning season in temperate waters, despite the fixed season, lasts for a long time. The arctonorwegian cod are caught in the Vestfjord for about thirteen weeks and they spawn for most of the period (Hjort, 1914), so the spawning ground cannot be occupied by a competitor cod stock which spawns in the spring. There is a considerable spread in time of arrival of the larvae on the nursery ground and because of the diffusion along the larval drift, another exclusion operates there in space and time. The spread by diffusion adds to losses by death, and as the little fish settle from a volume to a surface, the nursery ground must be extensive to accommodate the numbers. The stock of cod is so large and the exclusion is so effective that there is no competing stock in the area offshore (although there are small stocks of fjord cod). From the siting of their spawning grounds, the nurseries of the plaice in the Southern Bight and the German Bight of the North Sea are quite far apart, so competition between stocklets cannot take place. In the Southern Bight in winter and early spring, a number of species coexist (herring, sandeel, plaice, dab, cod and whiting; Simpson, 1949*a*,*b*) all of which spawn at slightly different times on slightly different grounds. Although there are no competing plaice larvae from another stocklet, other species can compete with the plaice in the plankton. During the larval drift, because of the long spawning season and of diffusion, the stock occupies sections of time and space which ensures its reproductive isolation from stocks of the same species, yet at the same time competition between species can occur. The differences between stocks are generated and maintained during the periods of spawning and of larval drift and competition between species may be most intense merely because so many live together when they are about the same size and feed on the same food. At later stages in their life histories, fish like plaice lead rather exclusive lives and competition between species may be much reduced or effectively absent.

During the period of larval drift, when the opportunity for competition is greatest, the density of fish is greatest, so density-dependent mortality should then be maximal. Numbers of animals are regulated naturally by density-dependent factors (Solomon, 1949) such as fecundity in insects (Andrewartha & Birch, 1954) or in birds (Lack, 1954). In fish, however, fecundity is proportional to weight. If fecundity were to determine recruitment, the natural death rate of the adults would have to be much higher to modify the biomass sufficiently to generate the variation in recruitment needed to stabilize the population or density-dependent growth should be pronounced in adult fish and of course it decreases with age and tends to be absent in adults. Hence, it is likely that regulation is achieved by density-dependent mortality. There is a little evidence that density-dependent mortality occurs at least during the period of larval drift.

Le Cren (1973) estimated the mortality rates during the first two years of life (including the larval drift) of trout in Black Brows Beck in the English Lake District. In three-year classes the initial numbers varied by a factor of five and mortality varied by a factor of three in the same way, that is, mortality increased with increasing initial numbers. Further, the differences in mortality extended throughout the period of two years or so. As will be explained in a later chapter, it is possible that the density-dependent mortality is generated by the availability of food to the stock of larvae. Further, density-dependent mortality might endure throughout life, decreasing with age. Then the prime function of the regulation of numbers must occur during the larval drift, but the function might continue until recruitment.

The larval drift is the most important period in the life history of the fish stock. From the spawning ground to the nursery ground, it provides a geographical base for the distribution of the stock. Numbers are probably regulated during the period of larval drift, stock differences are established and competition between species is maintained. To understand the biology of a fish stock fully, the larval drift must be properly described.

The course of migration

From an extensive series of published tagging experiments, Russell (1937) showed that each cod stock in the North Atlantic was connected to its neighbour by one or more tagged fish which had strayed. More recently, Gulland & Williamson (1962) reported that a cod tagged in the North Sea was recovered on the Grand Banks off Newfoundland, a transatlantic migration. But the dramatic migration of strays is less important than the circuit maintained by the main body of the stock. The arctonorwegian cod travel up to 1600 miles a year from the Vestfjord in northern Norway to Spitzbergen or Novaya Zemlaya and back again (Cushing, 1966*a*). The Pacific hake (*Merluccius productus* Ayres) spawns off southern California in early spring and migrates northward to reach the coast of Oregon and Washington State by June or July, a distance of about 1200 miles. They travel 2400 miles per year and perhaps more (Alverson, 1967). The albacore makes transpacific migrations, sometimes in a period little short of a year (Otsu, 1960). As might be expected, bluefin tuna have been shown to cross the Atlantic from the eastern seaboard of the United States to the coast of Norway (Mather, 1962). Between March and May, 1954, a number of nuclear bombs were exploded at Bikini Island, in the North Equatorial Current at 165°E. During the following six months, radioactive fish were discarded at sea in the Kuroshio current and in the Kuroshio extension as far east as 176°W six months later, which is a distance of 3000 miles on a straight course but round the gyre it is nearer 7500 miles (Nakamura, 1969), at a minimum speed of thirty miles per day. The Pacific salmon migrate from their native rivers as smolts to grow up in the mid-Pacific south of the Aleutian Islands (Clemens, Foerster & Pritchard, 1939; Milne, 1957). Atlantic salmon spend the same period of their lives off West Greenland (Saunders, Kerswill & Elsdon, 1965). The larger fish of abundant species can migrate thousands of miles in a year.

However, a small fish like the Norwegian herring can migrate up to 1500 miles per year (from Devold's diagrams, 1963) and the North Sea herring will travel from 900 to 1200 miles on their annual migrations (from diagrams in Cushing & Bridger, 1966). In the southern North Sea plaice move as little as 200 miles from spawning ground to feeding ground and back again. Winter flounders (*Pseudopleuronectes americanus* Walbaum) spawn in brackish ponds on Rhode Island, USA and spread about fifteen miles to seaward in their feeding migration (Saila, 1961). A population of sprats winters in the Wash, an estuary in eastern England, and spawns offshore in deeper water and probably feeds in about the same area (Johnson, 1970); the total distance of their migration may be less than one hundred miles. The smaller fish do not travel as far as the bigger ones, as might be expected but nevertheless the distances are quite considerable. Of course, there are fish which do not migrate at all particularly in tropical waters but in temperate waters it is possible that most commercial fish make some migration during the year. This suggests that the migratory species are the abundant ones.

There are many ways in which a stock maintains itself within a current system, to put it in terms of evolutionary history. The Pacific salmon make use of an apparently simple mechanism to move up the Fraser; they swim up the river against the current at a speed which depends on the distance which they must travel to reach their native stream or lake (Killick, 1955). They may receive visual clues from the bottom of the river, as they swim against the current. In the sea such external referents to a progressive migration are hard to find, although Harden Jones (1968) has suggested that fish can see particles at an oceanic boundary, a rheocline; it is, however, a special case. The arctonorwegian cod drift away from their spawning grounds in the North Cape Current (Hjort, 1914) and in the West Spitzbergen Current. Eggvin (1963) suggested that they returned to the Vestfjord in a southbound counter-current in 300–700 m below the Atlantic stream. The speed of the current was put at 9 km per day, which is not very different from Maslov's (1944) observed speed of tagged fish at 7–8 km per day. Recently, Dickson & Baxter (1972) have shown that such a current exists off the Norwegian coast at about the right depth. The use of drift current and counter-current can possibly explain the migration of the Pacific hake; the counter-current beneath the southbound surface current in the upwelling system could carry the spent fish north from southern California to the Pacific north-west at about the right time, that is,

June and July. The California current at the surface could return the fish to the waters off southern California in the late autumn. The northern North Sea herring winter in the Skagerrak and were tagged at that season on the Bohuslån coast of Sweden. In the subsequent year, tags were recovered in a sequence in the seasons round the North Sea which took the same direction as the main North Sea swirl (Cushing, 1954*a*; Höglund, 1955). Again, the path of the Norwegian herring, as shown by tagging and by the sequence of the fisheries, is in the general track of water movements in the Norwegian Sea (Devold, 1963). From these examples, fish may drift in a current during part of their migration circuit, but when visual clues can be used (as by salmon in the Fraser Current) they can swim against a current.

However, migration is mediated by physiological mechanisms and this is expressed by the capacity to 'discard' or 'acquire' a current at an appropriate time. The larval plaice sink out of the northbound current at a particular length or age at which they can pass through the Texel Gate into the Waddensee. The Norwegian herring move south-eastwards from Iceland in the East Icelandic Current to a point north of the Faroe Islands where they start to cross the Atlantic stream at a depth of 140 m towards the Norwegian coast, where they spawn (Devold, 1966). Not enough is known of such mechanisms by which populations of fishes step from one current system to another because the information on the distribution of fish at all seasons and on the detailed structure of currents is much too general at the present time. But if stocks are to maintain themselves against loss by straying, quite precise mechanisms must exist to secure the migration circuits.

The course of migration has been described in the form of an ideal and rigid structure which may not account for the real variability of movement in the sea. The Beverton and Holt model of the diffusion of immature plaice away from the Dutch coast has been briefly described above. Jones (1959) has estimated the directional component of velocity, V', from a coastline and the dispersion coefficient, a^2, using the results of a tagging experiment on North Sea haddock:

$$V' = \Sigma r_n \cos\theta/\Sigma t \qquad (33)$$
$$a^2 = (1/n)[\Sigma(r_n^2/t) - (\Sigma r_n \cos\theta)^2/\Sigma t],$$

where r_n is the shortest distance travelled; θ is the direction measured in angle from the axis of the coastline; t is the number of days at liberty; n is the number of fish recovered. Jones showed that the haddock travelled at -0.076 to 0.82 miles per day and dispersed at a rate of 2.2 to 118 miles^2/per day. During summer and autumn, the tagged fish did not migrate at all, but in winter and spring they moved northwards at about 0.4–0.7 mile per day. The highest rates of dispersion are possibly biased upwards by a few fish which migrated for considerable distances. A similar method was applied by Saila (1961) to the results of a tagging experiment made on the winter flounder off Rhode Island. He showed that they moved 0.03 mile and dispersed at 0.67–1.10 mile^2/per day, indicating that there is hardly any directional component. Indeed, Saila estimated that 75% of the population would reach the coast after 90 days of random search. Errors of such estimates can be determined by transforming the polar co-ordinates into cartesian ones

$$\text{i.e. } x = (1/n)\sum_{i=1}^{n} \cos\theta; \qquad y = (1/n)\sum_{i=1}^{n} \sin\theta;$$

then $r_n = (x^2+y^2)^{1/2}$ and $\cos\alpha = x/r$, where α is the mean angle at which tags are recovered.

$$S' = [2(1-r_n)]^{1/2},$$

where S is the mean angular deviation in radians; Batschelet (1965) gives a fuller treatment. Thus, from a tagging experiment, the average directional velocity with a mean angular deviation can be derived, as well as a dispersion coefficient. Out in the open sea, away from a coastline, the method can be equally well applied if the direction travelled is measured from an arbitrary axis, for example true north, and the directions are estimated separately for easterly and westerly movement.

Saila & Shappy (1963) devised a model of Pacific salmon migration by assuming that the fish swam in straight lines of variable step length and that when they turned, they turned in any direction. The pattern of step lengths is described by a cardioid or heart-shaped curve, the radius of curvature of which (R_d) is given by $R_d = P'' + Q''\cos\theta$, where P'' and Q'' are constants. Let $A' = Q''/P''$, then $R_d = P''(1+A'\cos\theta)$. When $A' = 0$, R_d is equal at all angles and the search pattern is a constant; when A' is positive, it measures movement in a particular direction. For the Pacific salmon, when $A' = 0.3$, the return probability corresponds to the 10% recovery rate obtained. Hence, the possibility of estimating directed movement is raised and it can be measured with methods adapted to the ordinary tagging techniques of fisheries biologists.

Tagging experiments are easy to design and many have been executed with ends other than migration in view. So long as both the positions of liberation and of recapture and the time interval between them are known, then information on migration can be obtained. Thompson & Herrington (1930) studied the spread of the Pacific halibut (*Hippoglossus stenolepis* Schmidt) from their spawning grounds by plotting the frequency distribution of distance travelled. The spawning grounds of two groups were far apart, one off British Columbia and the other off Alaska (Thompson & van Cleve, 1936). One stocklet remained roughly in the same position but the other spread for considerable distances. From such distributions, a mean distance travelled can be estimated irrespective of direction. The different spread of tags in area was treated almost as a stock character. Recently, de Veen (1970) has examined the distributions of recaptured plaice in the North Sea, which were liberated at some distance from their original position of capture. The work was carried out in the first decade of this century as part of a study of transplantation. de Veen found a tendency amongst the recaptured fish to shift from the release position towards the original position of capture, but they rarely reached it. The shift tended to increase with distance up to about 150 miles.

There are three subjects in the study of migration; (1) the measurement of distance travelled in one current system; (2) the discovery of mechanisms by which fish enter and leave particular currents, and (3) the estimation of directional velocity, mean angular deviation, the spread of fish in area and the stray of fish between generations. By these methods, the dependence of a stock upon current systems which contain it can be elucidated. Further, the spread of characteristics can be readily estimated, by which the fit of a stock's migration to a current system can be measured. Finally, estimates of stock unity would be derived by measuring the stray from generation to generation.

The unity of a stock

Before a population can be fully studied, the spread of fish from one spawning ground to another within a year class should be estimated. Such estimates are hard to make and very often an estimate of death rate due to fishing is needed quickly. However, the problem of stock unity has been investigated from time to time by tagging.

The fishery for the arctonorwegian cod stock has persisted in the Vestfjord for centuries; Hjort (1914) charted the distribution of cod eggs there and showed that they drifted away north in the Atlantic Current offshore at about five miles per day. In the west Spitzbergen Current such a drift would take the little fish to the Svalbard shelf and the west coast of Spitzbergen. In the North Cape current a similar drift would carry them eastwards in the direction of Novaya Zemlya. After three to five months they settle on the bottom; Schmidt (1909) found them between 100 and 160 fathoms. Maslov (1944) charted the distribution of immature cod in the Barents Sea and found them on the shallow banks of the Svalbard Shelf, Skolpen and Goose Banks, and off Novaya Zemlya. The circuit of migration extends fully into the Barents Sea from the Vestfjord.

The first tagging experiments on this stock of cod were made by Hjort (1914) who pinned silver buttons to the gill covers of the fish with silver wire. Fish were tagged in the Vestfjord during the spawning season and were recovered later in the summer all along the northern coast of Norway.

Fish tagged in summer time at various points in the Barents Sea were recovered in the Vestfjord (Idelson, 1931). A later series of tagging experiments showed that in spring, fish moved eastwards towards Novaya Zemlya and in autumn, they moved back in a westerly direction (Maslov, 1944). Trout (1957) tagged fish in the Barents Sea; 349 were returned from the Lofoten Islands and about ten were recovered from the coast of Norway to the south. Although spawning was not completely confined to the Vestfjord, more than 97% of the fish tagged in the Barents Sea did return to it. Figure 31 shows the distribution of recoveries in the Barents Sea from taggings in the Vestfjord; it shows the distribution of the fish on their summer feeding grounds. Well-designed tagging experiments show the range of migration very clearly, so long as fishing vessels go wherever the fish live.

More detailed biological studies of the arctonorwegian cod have broadened our knowledge of the stock boundaries. Lee (1952) showed that fish catches were limited to the water warmer than 2 °C on the bottom and that the position of this isotherm in the Bear Island summer fishery shifted from year to year. Hylen *et al.* (1961) made echo surveys off the northern coast of Norway in spring and autumn; the distribution of large single fishes in mid-water, almost certainly cod, was limited by the 2 °C isotherm in 150 m. Richardson *et al.* (1959) charted the distribution of cod as echo signals close to the bottom and showed that, as the warmer Atlantic water pressed up the Svalbard Shelf, the cod gathered at the 2 °C isotherm on the bottom, which constitutes the boundary between the Atlantic water and the arctic water mass on top of the shelf. Woodhead & Woodhead (1959) have described a physiological mechanism by which cod are affected by water colder than 2 °C. The area in which the cod live in the Barents Sea, where they feed, can be charted as bounded by the 2 °C isotherms.

The most important tagging experiments on the nature of a fish stock are those executed by de Veen (1961,1962). Plaice were tagged on their spawning grounds in the German Bight and on the Flamborough Off Grounds. Tagged fish were recovered from the same grounds in the following season. There was no exchange between spawning grounds and no tags were returned from the Southern Bight spawning ground. The stray of tags from a year class appears to be low and it suggests that the three spawning groups may be separate stocklets.

Recently, stock structure has been analysed by geneticists. Blood serum proteins have been used for a long time to establish genetic differences, for example, in haemoglobins in man. In fish, various techniques have been used; J. E. Cushing (1964) gives a useful review. Sprague & Vrooman (1962), with the use of red cell antigen separated three groups of sardines off California, including the Gulf of California. Fujino & Kang (1968) have analysed differences in esterases in tuna in the Pacific. Sindermann &

Mairs (1959) established certain differences between groups of herring off the Gulf of Maine. de Ligny (1969) showed that there was a difference between the herring spawning in the Southern Bight of the North Sea and those spawning in the Eastern Channel.

Sick (1961) introduced the technique of zone electrophoresis into fisheries research. The most important application is that in transferrins and haemoglobins in cod. Figure 30 shows the relationships between the North Atlantic cod stocks (de Ligny, 1969). Figure 30(*a*) gives the distribution of the haemoglobin allele HbI. The stocks in the North Sea are quite distinct from the other North Atlantic stocks, in a genetic sense. Figure 30(*b*) shows the distribution of the transferrin allele Tf^c throughout the North Atlantic. The North Sea stock is genetically uniform and the Baltic, Faroe, West Greenland and the Newfoundland stocks are genetically distinct. Jamieson (1970) distinguishes seven transferrin alleles in the North Atlantic. Samples usually follow the Hardy–Weinberg law, i.e. that the homozygotes *a* and *b*, with their heterozygotes *ab*, are distributed as $(a^2+2\ ab+b^2)$. If the samples do so, the characters are assumed to be alleles. Then the proportions of alleles can be used in $(2\times n)$ tables to test the genetic distinctness of the stocks as sampled, with a χ^2 test, and the chance of mixture is as low as one in 10^4. A consequence is that most of the cod stocks in the North Atlantic are distinct. The groups within the North Sea are not so, but that on Faroe Bank was separated from that on Faroe Plateau in two successive years on the basis of a single transferrin allele (Jamieson & Jones, 1967).

The traditional way of establishing differences between fish stocks has been with the use of meristic and morphometric characters. Schmidt (1909) originally separated cod stocks on the basis of vertebral and fin ray counts. Much statistical work was carried out on the differences between such samples on the assumption that they were environmentally generated. Recently, Purdom & Wyatt (1969) have shown, with rearing experiments at a constant temperature, that there are genetic differences in the number of vertebrae between the Southern Bight plaice and the Irish Sea plaice. Le Gall (1935) distinguished between the races of herring in the north-east Atlantic on the basis of morphometric measurements. Royce (1964) showed with a form of discriminatory analysis that there were differences in a number of morphometric measurements on yellowfin tuna all along the equator in the Pacific. Similarly, the Asian and American salmon in the same ocean have been separated on the basis of their distinct parasites (Margolis, Cleaver, Fukuda & Godfrey, 1966) and by discriminatory analysis of morphometric characters (Fukuhara, Murai, Lalanne & Sribhibhadh, 1962). Such methods are of value today only when genetic differences cannot be discovered. For example, de Ligny (1969) has distinguished a number of loci in the southern North Sea plaice, but found no

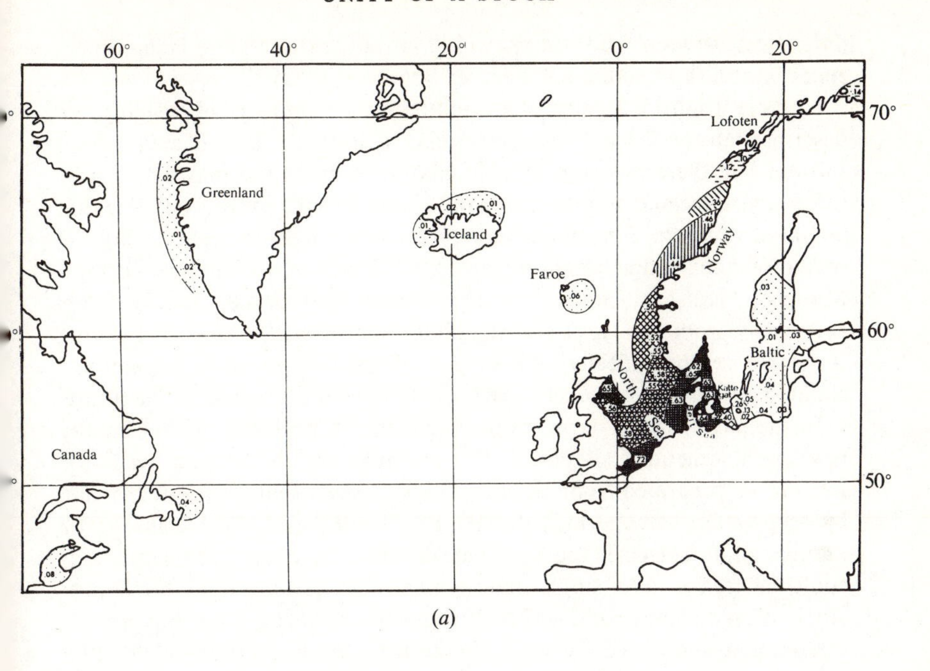

(*a*)

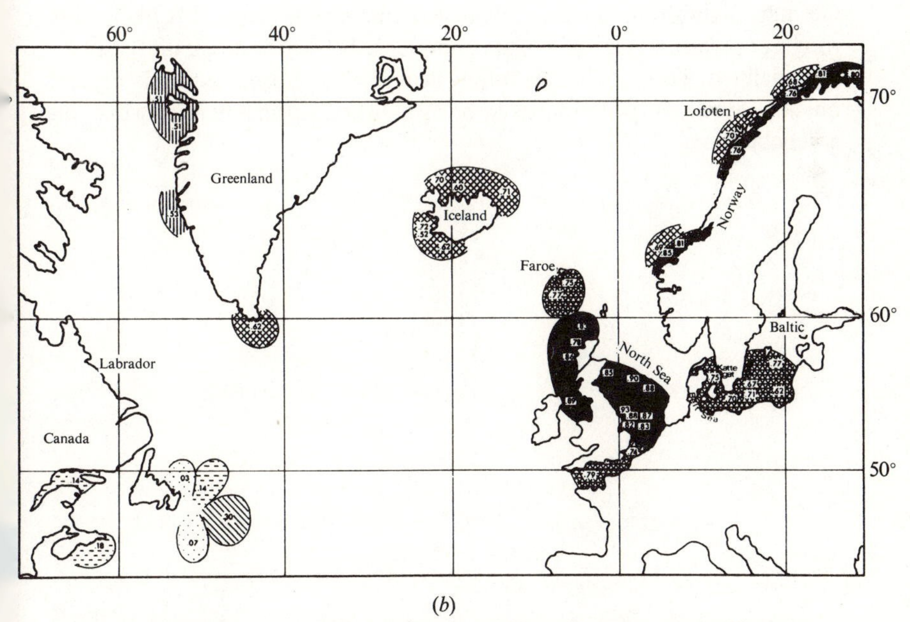

(*b*)

Figure 30. The distribution of blood groups in the North Atlantic cod; (*a*) The haemoglobin allele HbI; (*b*) the transferrin allele Tf^c (de Ligny, 1969).

differences between the three spawning groups, despite there being differences in otolith structure (de Veen & Boerema, 1959).

Another traditional way of examining stock unity is by tagging, as described above for the arctonorwegian cod stock. But Tåning (1937) showed that there was considerable mixture between the spawning stocks off southern Iceland and those off south-west Greenland. It is possible that coasts of southern Greenland were colonized by Icelandic fish during the period of climatic amelioration by a drift of larvae in the Irminger current. Møller (1968) has shown that the arctonorwegian cod stock differs genetically from the fjord cod, but both spawn in the Vestfjord during the same season. Ideally, the degree of mixture in such populations should be estimated by genetic methods, but this must remain in the future.

In the past, for the purposes of management, problems of stock unity have been sometimes simplified. For example, the cod around the Faroe Islands are separated from all other stocks, except that on Faroe plateau, by deep water, across which the fish might not travel. Again, amongst a number of stocklets, if the vital parameters were always the same, they could be treated as a single management unit. For example, the Downs stock of herring may consist of two genetically distinct groups, but growth, recruitment and mortality were common to the two groups. The three spawning groups of plaice in the southern North Sea were treated as a single management group until the late fifties, but then recruitment to the German Bight group increased sharply (Bückmann, 1961). It is likely that rough population assessments are completed before the stock unity can be fully defined. Today, when fishing is heavy, international action is needed quickly. Stock unity also needs to be established as quickly as possible, on a genetic basis.

6

Growth and death in the sea

Introduction

Animals grow to avoid death and, in the sea, the faster they grow, the more they avoid predation by becoming predators themselves. As the fish grows, its predators become larger and as they are more widely dispersed, mortality is reduced; so a small increment in growth leads to a small decrement in death rate. The cumulative difference between growth and death throughout the life history is expressed in the final size of the fish, be it large or small. The eggs of most marine fishes are about the same size, and large fish are more fecund than small ones and they tend to live longer. The larvae of the most fecund fishes probably suffer a greater mortality than those of the less fecund ones, and such mortality may well be density-dependent to a fair degree. It is such mortality that generates stability in the population. In an evolutionary sense, a comparison between species or groups shows that animals which grow slowly to considerable sizes achieve stability in numbers. Stabilization is then the result of a high cumulative difference between growth and mortality which ends in high fecundity. For an individual, however, growth is the competitive process by which death is to be avoided as long as possible.

Growth in the sea

The sizes of animals in the sea, from tintinnids to blue whales, range across fourteen to fifteen orders of magnitude. Even within their life cycles fishes may grow by a factor of 10^6 or 10^7 and this is because the eggs are small. The great range in size makes growth easy to describe and, indeed, because the increments in weight decline smoothly in time, the growth of animals seems to be a simple process. Gray (1926) showed that the equations used to graduate growth data in age have a descriptive value only and that the constants have little biological meaning. Medawar (1945) has summarized the form of information that might be derived from data with the use of a growth equation: weight increases with age in a multiplicative manner so growth rate is uniform and the specific growth rate declines with age, more slowly with increasing age, and the specific acceleration is negative. It is the simplest description of the complex processes of morphogenesis, development and synthesis in age to which is given the name of growth.

During their adult lives after maturation, cod or plaice may grow by a factor of twenty to thirty. For practical reasons, fisheries biologists need to describe the growth of fish as accurately as they can. The growth of plaice is well described by the von Bertalannfy equation (Beverton & Holt, 1957) for all ages after metamorphosis (Figure 31); the weights of larvae and of metamorphosing fishes are shown and they are not fitted by the curve. The equation is usually formulated as follows:

$$l_t = L_\infty[1-\exp(-Kt)] \tag{34}$$

$$W_t = W_\infty[1-\exp(-Kt)]^3, \tag{35}$$

where l_t is the length at time t; W_t is the weight at time t; L_∞ is the asymptotic length at infinite age; W_∞ is the asymptotic weight at infinite age; and K is the rate at which the asymptote is approached in time expressed either in length or weight.

The constants are readily evaluated from measurements in length or weight at age; Gulland (1969) gives methods of fitting the data. The constands find their greatest use as convenient expressions in the fish population models.

Parker & Larkin (1959) compared the growth of red salmon and steelhead trout both in the sea and in freshwater. Growth was expressed as a power function of age and the exponent increased significantly as the fish migrated from freshwater to the sea. Because both species grew faster in the sea, Parker and Larkin believed that growth should be expressed in 'stanzas', each corresponding to a different environment and each characterized by a different exponent. The growth pattern in the plaice (Figure 31) before metamorphosis is distinct from that in later life and the point was explained by Beverton & Holt (1957) in saying that the pattern of growth was not isometric with age. More particularly, when salmon fry move from lake to river and when smolts migrate from the river to the sea or when plaice larvae settle from the plankton on to sandy beaches, there are profound changes in the quality and quantity of food. Indeed the ratio of growth to mortality changes considerably after metamorphosis.

When the growth patterns of animals are compared, one of the most important factors is absolute size. Winberg (1956) tabulated daily ration and daily weight gain as percentage of body weight by sizes of fish. The smaller animals (of about 1 mg wet weight) eat as much as their own body weight in a day and they grow very quickly and gain perhaps as much as one-third of their body weight per day in growth. As fish grow, their growth rate declines as in all other animals and so does their daily ration as proportion of body weight. The quantity eaten increases from about 1

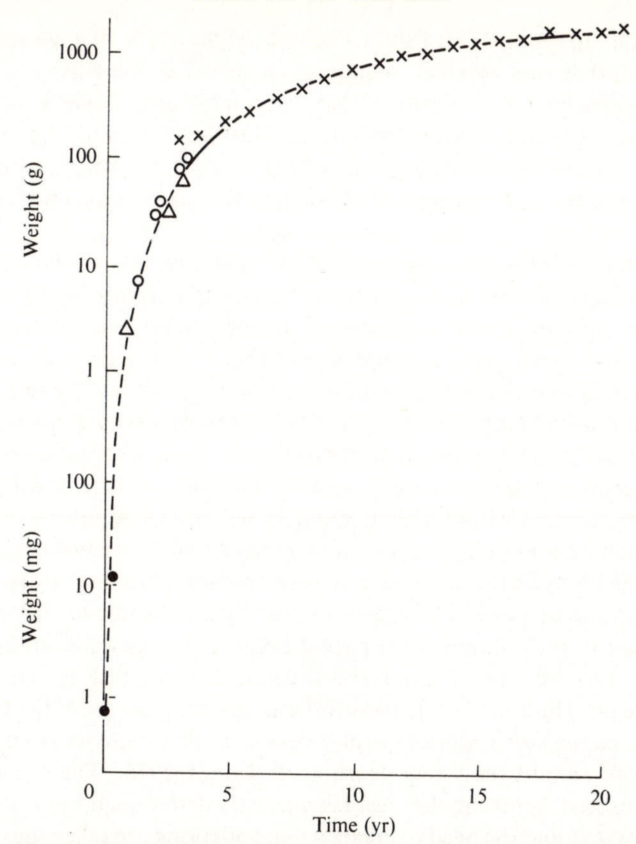

Figure 31. The growth of plaice from the age of metamorphosis to 27 years of age (×) fitted by the von Bertalannfy equation (Beverton & Holt, 1957); the weights of yolk-sac larvae and of late stage ones (●) are marked on the figure to show that they are not fitted by the curve; the weight of 0-group fish is represented by (△) and Wallace's observations on juvenile fish by (○).

Table 9 *The daily intake of food per unit volume in the sea*

Weight (g)	Daily ration (g)	n/m^3	Intake/unit vol. per day (g/m^3)
0.001	0.0005	5	0.0025
0.1	0.025	0.1	0.0025
10.0	1.0	0.002	0.0020
1000.0	40.0	50.10^{-6}	0.0020

Note: the estimates of larval density are taken from Cushing (1957), Cushing & Bridger (1966) and Harding & Talbot (1973) and the density estimates for larger fish are taken from Cushing (1967*b*) on the basis of acoustic observations.

mg per day in the smallest fish of 1 mg in weight to 25–50 g per day in the larger fish that may weigh as much as 1 kg. Thus, each big fish has a much greater capacity for destruction than each little one. Table 9 shows the daily intake per unit volume for animals between 1 mg and 1 kg in weight. Because weight and density are inversely related in a rough manner the daily intake per unit volume is very roughly the same irrespective of size of animal.

This generalization is based on a very small quantity of data on fishes and it is justified only by the dominant position occupied by fishes in the food web. Again it is a statement of average conditions over long time periods. If it were true we would expect that as fish grew their mortality would decrease with age as a function of age; the point will be reiterated in the next section. To put it another way, fish must sweep greater volumes for food and be vulnerable to larger but more thinly distributed predators as they grow through the trophic levels. This characteristic of fish populations may contain within it their capacity to stabilize numbers and at the same time exploit fully the carrying capacity of the environment. Yet within the life cycle the biomass of a fish population increases by two orders of magnitude or more. The eggs spawned by one female in the water are equivalent to the volume of the parent because the egg takes up five to six times its own volume of water and the gonad is one-fifth to one-sixth of body weight (Fulton, 1897). Before the larvae start to grow the eggs and yolk-sac larvae suffer mortality for a period so that biomass is reduced by a factor of about twenty-five (Harding & Talbot, 1973). The age at which growth exceeds mortality has not been exactly determined but if the factor of twenty-five and the need to create a male offspring are taken into account a deficit of up to two orders of magnitude in biomass is possible. During the later lives of the fishes this deficit between generations is made up as growth exceeds mortality until the cohort has replaced itself. Between generations one might expect growth and death to be so balanced that biomass remained about the same. The deficit within the early life cycle implies that growth must exceed mortality considerably in later life.

The analysis of growth in fishes received a great impetus from the experimental work of Ivlev (1961). He suggested that the daily ration, R_a, depended on food concentration, p_n, as a function of the maximum ration $R_{a_{max}}$ and feeding coefficient k_f:

$$R_a = R_{a_{max}}[1 - \exp(-k_f p_n)]. \qquad (36)$$

Figure 32 illustrates some of Ivlev's results for carp feeding on bream roe, for roach feeding on chironomid larvae and for bleak feeding on *Daphnia*. There is a level of ration and also of food density at which the maintenance requirements of the animal are fulfilled. Any food converted in excess of maintenance is laid down as growth and any ration that fails to satisfy maintenance is a starvation ration. It is a much simplified picture of the processes of growth but it provides a useful framework with which to start. A complication was added when Ivlev found that ration increased with the patchiness of food; when the coefficient of aggregation (or coefficient of variation of food density) was increased by a factor of five, the ration increased by three times. Patchiness is the rule in the sea largely because of variations in reproduction and in predation in space and time; indeed, Steele (1961*b*), has modelled such a generation of patchiness

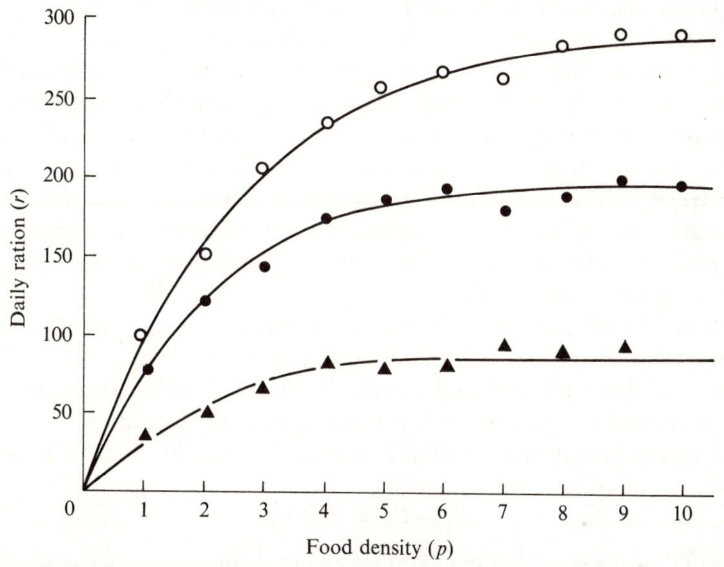

Figure 32. The dependence of ration on food density; carp feeding on bream roe, roach feeding on chironomid larvae and bleak feeding on *Daphnia*. The data are fitted by Ivlev's equation (Ivlev, 1961).

purely on the basis of the variation in vital population rates, rather than in the physical variability of the environment. Further, the population might be expected to be insulated from such physical variation to some degree. It is true that the distributions of algae are likely to be determined by physical factors to some extent (Platt, 1970), but they reproduce and suffer death by grazing. Ivlev's result can be readily explained because the distance between patches may be less than the summed distances between evenly distributed individuals. Growth is not only a simple conversion of food but is also a complex reflection of all the variances in the local dynamics of the food chain. Later it will be suggested that the distribution of such patchiness plays an important part in the generation of recruitment.

Not only does growth depend upon the density of food but the efficiency of growth does so as well. For example, if algal cells are thinly distributed an animal like *Daphnia* has to search too much and if they are densely distributed the cells may pass through the gut without being digested. Slobodkin (1966) showed that there is a middle density at which the growth efficiency (as growth per unit of food) is at its greatest. Such an efficiency factor decreases with increasing caloric content, that is, younger animals grow more efficiently than older ones. Cushing (1964) showed that the feeding efficiency of herring (as ml gut content per encounter) was greatest when feeding on *Calanus*, which may be a prey of optimal size for the herring; the ratio was less both for small animals such as *Pseudocalanus* and larger ones like euphausids. Ivlev (1961) demonstrated experimentally that there is an optimum size of prey for each predator. Then, as an animal grows it passes through the encounter fields of a succession of predators each of which is larger and less numerous than its predecessor. Therefore its death rate must decrease with age very roughly in proportion to the growth rate. A number of principles converge, that ration depends on density, on patchiness and on the relative sizes of predator and of prey. As an animal grows, it does so relatively more quickly when it is small and eats prey of moderate size at middle densities and it grows most at any age when its prey is patchily distributed.

Studies on the physiology of growth of fishes are very extensive. Ivlev (1961) partitioned the energy of a food ration into growth, excretion, heat and work (both internal and external); Winberg (1956) shortened the formulation to express the whole process in terms of respiration. A current form of the energy budget of fish growth is given by Warren & Davis (1967):

$$Q_{cc} - Q_w = Q_g + Q_r, \tag{37}$$

where Q_{cc} is the energy in the daily ration; Q_w is the energy excreted; Q_g is the energy in the increment of weight, that is, growth; Q_r is the energy of metabolism.

$$Q_r = Q_s + Q_d + Q_a, \tag{38}$$

where Q_s is the standard metabolism in units of oxygen consumption; Q_d is the specific dynamic activity; and Q_a is the energy expended in activity.

The standard metabolism is a measure which corresponds to the basal metabolic rate in human physiology. The specific dynamic activity is the energy of transfer from its sources, proteins, fats and carbohydrates. Warren and Davis give examples of the trend of such parameters in temperature for a number of fishes, expressed in units of energy.

The efficiency of growth was also formulated by Ivlev and is commonly expressed in one of two ratios: (1) growth per unit intake; (2) growth per unit of (intake less maintenance). Birkett (1969) expressed the efficiencies in the following form:

$$\Delta W = \xi^{+}(A_s - A_{s_n}), \tag{39}$$

where ΔW is the gain in weight in units of nitrogen; A_s is the absorption in units of nitrogen; A_{s_n} is the maintenance level of absorption in units of nitrogen; ξ^+ is the efficiency of nitrogen conversion. In such a relationship, excretion is shown as a negative intercept on the ordinate. Figure

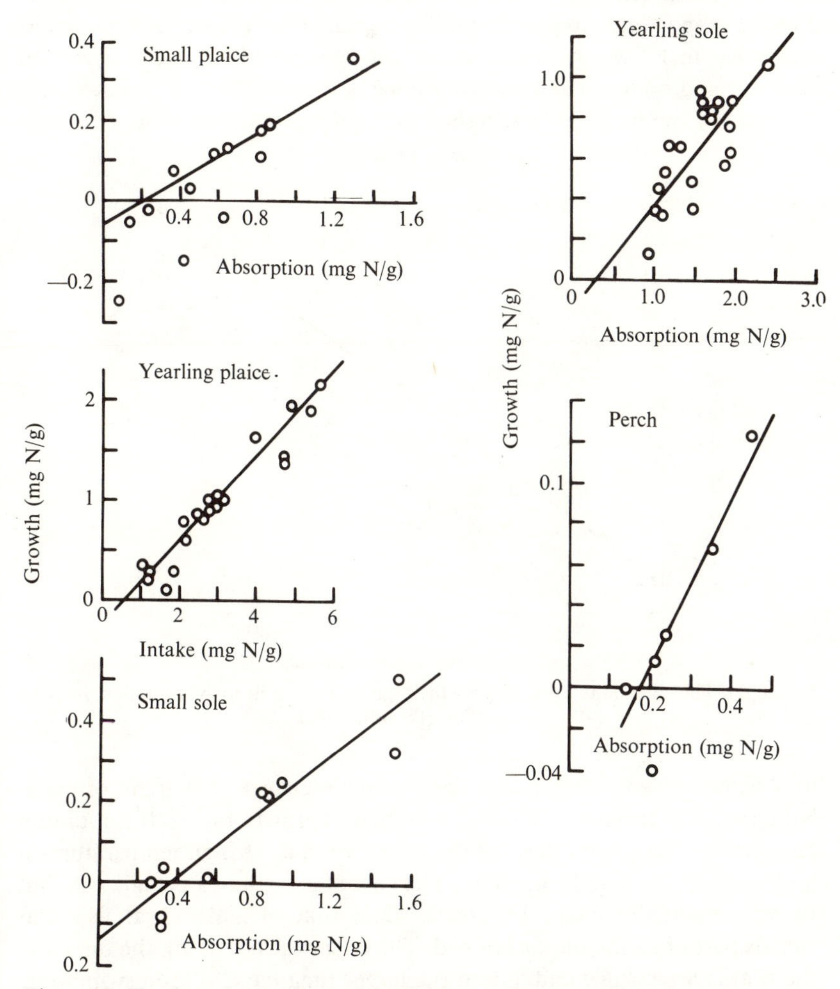

Figure 33. The dependence of growth on food eaten by the small flatfish and the perch (Birkett, 1969); the slope of the line is the efficiency of food conversion.

33 shows the experimental data for plaice where growth is related to absorption in a linear manner; the slope of the line is the efficiency of food conversion, that is, growth per unit of absorption or intake. There is a constant maintenance level of absorption and a constant rate of nitrogen

excretion, E_x, under the experimental conditions. From the constants another equation can be

$$E_x = [(1-\xi^+)/\xi^+]\Delta W + A_{s_n}, \tag{40}$$

where the expression $(1-\xi^+)/\xi^+$ is the metabolic cost of growth. It is directly proportional to growth and is estimated in the form of experiment illustrated in Figure 34 which shows the relationship between excretion and assimilation in small plaice; the slope of the line is $(1-\xi^+)$. The cost is the ratio of the proportion metabolized to the proportion retained and it

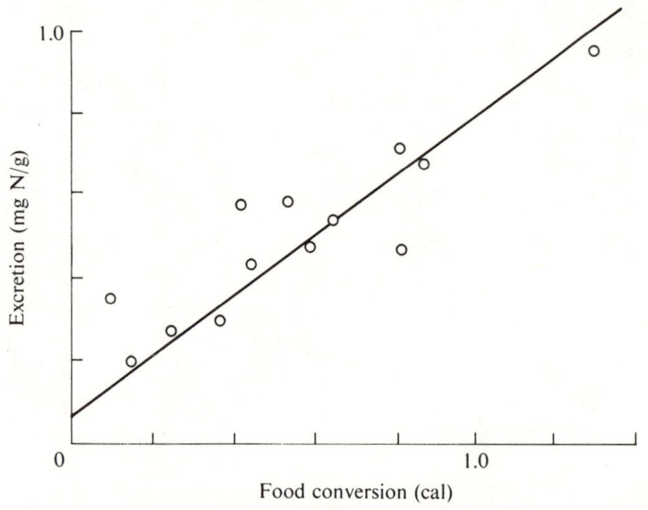

Figure 34. The metabolic cost of growth is given by the ratio of excretion to food conversion (Birkett, 1969).

may be conceived as a dynamic form of maintenance. The trend of metabolic cost with temperature is less than that of growth rate so the optimum temperature for growth lies between the two, and at high temperatures a high growth rate may be wasteful. The cost of growth in a pelagic fish that swims perpetually might be greater than that of a demersal fish that spends part of its life on the sea bed. The biggest fishes of all, the tuna and the sharks are pelagic and indeed the larger tuna have to keep swimming. They are both predators that can afford to swim idly between meals and perhaps just fast enough to maintain depth and conserve the cost of growth.

Paloheimo & Dickie (1965,1966*a*,*b*) expressed the growth of fishes in the following way. The first postulate is:

$$T' = \alpha W^{\gamma}, \tag{41}$$

where T' is ml O_2 per g body weight; W is weight; α and γ are constants.

The second is:

$$K_1 = (\Delta W/R_a)/\Delta t = [\exp(-a - bR_a)], \tag{42}$$

where K_1 is a coefficient of efficiency; R_a is the total ration; a and b are constants. Then

$$\Delta W/\Delta t = R_a[\exp(-a - bR_a)] \tag{43}$$

$$\frac{\Delta W}{\Delta t} = R_a - T',$$

$$T' = \alpha W^{\gamma} = R_a\{1 - [\exp(-a - bR_a)]\}. \tag{44}$$

Much experimental material was used to support the two basic postulates. The first is a well-established one and expresses the dependence of respiration on weight; it used to be thought that $\gamma = 0.67$ because respiration was thought to be a function of surface, but perhaps the exponent is best established empirically because the surface is hard to define. The second postulate really expresses the fact that the efficiency of food conversion declines with the amount of food, much as specific growth rate declines with age or weight. The final expression puts growth in physiological terms and such an expression is distinct from the purely descriptive equations used to graduate length or weight data. In the future such expressions may replace the descriptive ones.

In general, growth in fishes has been studied by fisheries biologists with limited ends in view and they have used exponential and descriptive equations such as that of von Bertalannfy. The expressions developed by Paloheimo & Dickie include physiological terms. However, the population effects like those associated with density are still not understood. During the thirties it was thought that density-dependent growth might vitiate any conservation of biomass (Bückmann, 1932), but Hempel (1955) showed that density-dependent growth could not be detected amongst adult plaice in the German Bight of the North Sea. On the other hand, Raitt (1939) found that the growth of immature haddock was clearly density-dependent and Raitt, D.F.S. (1968) demonstrated the same effect for small gadoids in the northern North Sea. Southward (1967) studied the yearly increments of growth of the Pacific halibut and found that they decreased with age, that is, with the decreasing density of the cohort, but not with the density of the adult stock. Brocksen, Davis & Warren (1970) make the same point with juvenile sockeye, that is, growth is less with greater biomass and they also related growth to the biomass of zooplankton. Such effects might well be expected if it is recalled that any stock must comprise a sequence of cohorts, each of which is independent of the others. With very little evidence, it is possible that the growth of adult fishes is not usually density-dependent; the density-dependence of growth appears to decline with age and so it is very difficult to detect in populations of adult fish.

Death in the sea

The forms of death in the sea are as diverse as those on land and predation is probably its major agent. The bacteria in the sea as numbers per unit volume are much less dense than the numbers per unit surface on the land; however, on a mud surface on the sea bed they may be equally high. Populations of fishes have been reduced by disease, for example, the loss of mackerel due to the fungus *Icthyosporidium* off the eastern seaboard of the United States (Sproston, 1944) and salmon populations suffer from recurring diseases like ulcerative necrosis. However, such events are exceptional. Very old fish have been caught in unexploited populations, for example, *Tilapia esculenta* Graham, in the Kavirondo Gulf of Lake Victoria, which were possibly senile (Graham, 1958). Biologists familiar with the problems of ageing quote the Gompertz Law, which states that mortality increases exponentially with age amongst older adult animals (Gompertz, 1825). In studies of fish populations, it is often convenient to assume a constant natural mortality in the abundant age groups, but if such an assumption is extended to the virgin stock, biomass might be overestimated because the animals might appear to grow to excessive ages. In the sea the effect of both disease and senility may make the victims more vulnerable to predation.

Any study of predation is one of attack and escape in units of time. Fish can cruise at three lengths per second for long periods, but the time spent in attack or escape is very short indeed. Acceleration builds up in a fraction of a second and the maximum speed of perhaps ten lengths per second is reached in perhaps 0.5 s, then the animal decelerates (Bainbridge, 1960). The visual range of a fish is short because light is scattered in the water, so a predator can only attack in a short distance and a prey has not far to go before it vanishes (provided that it has enough acceleration). A predator's acceleration must exceed that of a prey by a sufficient margin to catch it before it disappears. To avoid capture the prey should accelerate when the predator is decelerating. From Bainbridge's work, quoted above, Gray (1968) deduced that $V' \propto L^{0 \cdot 6}$, where V' is the cruising speed of the fish in cm/s and L is the length in cm. The resistance of a rigid model of a pelagic fish is R_s. Then

$$R_s = K'\rho A_r V'^2/2g \qquad (45)$$

where K' is a dimensionless coefficient; ρ is the density of water; A_r is the maximum cross-sectional area in m^3; V' is the velocity in m/s.

In water-tunnel experiments, Webb (1971) related power (g cal/h) to velocity (cm/s) in a young trout by estimating the drag and it was related also to metabolism for a range of temperatures. Greer-Walker (1970) has used Gray's formula to show that the resistance in kg increases with speed much more sharply for larger fish than for smaller ones. Hence, the relative cruising speed in lengths per second decreases with length and the attack or escape speed decreases relatively with length. Judged by the power expended, big fish are not so efficient in attack or escape as smaller ones. Greer-Walker has also shown that the diameters of myofibrils in the white muscle, that used for acceleration, decreases in the older animals; in cod more than 0.9 to 1.0 m in length the capacity for acceleration must be considerably reduced. Not only does the attack or escape speed decrease relatively with size but after a particular size or age it decreases absolutely. Consequently, as the fish grows older it becomes more vulnerable to predation and less able to catch its food. There is thus a physiological basis for senescence and old fish might even die of predation. Although it is a property of individual fishes, senescent mortality is generated in the network of predators.

The predatory mechanisms of insects were analysed by Holling (1965) and much of the theory that he developed is useful in marine ecology. The time to capture and eat a prey, t_c, comprises four periods: (1) the digestive pause t_α, during which no attack takes place; (2) the time spent searching, t_s, when the digestive pause is over; (3) the time spent in pursuit, t_p, and (4) the time spent in eating and handling. The essential step is to split the processes of predation into components of time. The metabolic, behavioural and ecological processes are expressed in the same units, which is a most important step in reducing biological complexity to mathematical terms. The principle appeared at the same time in four publications, Holling (1959); Rashevsky (1959); Watt (1959); and Cushing (1959), the last described an early version of the encounter theory of grazing, which was given in an earlier chapter.

The analysis of hunger proceeds:

$$H = H_k + (H_0 - H_k)[\exp(-\alpha t)], \qquad (46)$$

where H is g food needed to satiate, i.e. the quantity required to make the whole feeding cycle stop; H_0 is the degree of fullness of the gut at time t_0; H_k is the maximum capacity of the gut in ml; α is a constant, the specific rate at which the gut is filled.

The time to search is given by:

$$t_s = \frac{1}{2V_r\beta(H_0 - H_{t_0})N_0 S_s} - \frac{\pi\beta(H_0 - H_{t_0})}{2V_r}, \qquad (47)$$

where V_r is the predator speed less that of the prey in cm/s; β is a constant determined by the shape of the search field and the rate at which it expands with hunger; H_{t0} is the quantity of food in g needed to satiate when the search starts or the hunger threshold for attack; N_0 is the prey density in n/ml; S_s is the strike success, or the number of prey captured per unit of attack. The time to search is effectively the reciprocal of the difference between the speeds of predator and prey modified by environmental factors as expressed in the search field.

The time spent in pursuit is:

$$t_p = (\beta / V''_p)(H_t - H_{tt}),$$

where V''_p is the attack speed of the predator in cm/s; H_t is the quantity of food needed to satiate at the point in time of prey capture; H_{tt} is the attack threshold (the degree of fullness of the predator's gut needed to elicit an attack) at the point in time of prey capture.

In order to understand the constants more fully, Holling's development of these equations should be consulted in his original papers. The initial and important step was a measureable definition of hunger in terms of gut fullness, which, in turn, determined the search field and the time spent in pursuit. The application of such methods to marine ecology would require a closer examination of animal behaviour and physiology. Many of the animals do not live too readily in aquaria, and the little ones swim quickly out of the depth of focus of a microscope. Knox & Brooks (1969) have developed cinematic holography in an attempt to solve this problem.

Cushing (1968*a*) and Harris (1968) investigated a much simpler case, the grazing by copepods on algal populations, summarized in an earlier chapter. The time to capture and eat comprised mainly the time to search, which implies that plankton animals tend to eat continuously if they can and that the filter mechanisms do not clog at the densities examined. The same conclusion emerges from Beklemishev's (1957) concept of 'superfluous' feeding because, in high algal densities, faecal pellets are produced by copepods with little digestion. Corner, Head & Kilvington (1972) showed recently that *Calanus* destroys a large proportion of a population of *Biddulphia sinensis* (which is a large diatom), and ingests them at low efficiencies. Hence, there is some reason for supposing that feeding is a continuous process in copepods, perhaps above some threshold concentration (Adams & Steele, 1966; Parsons *et al.*, 1967). Cushing suggested that the grazing mortality was a function of the grazer's weight and that its attack range was a function of length, itself a function of attack speed. An expression was developed in which the grazing mortality of the algae was estimated in terms of the cruising speed of the grazer, its weight and attack range and the density and sizes of the algal cells. By purely mechanical filtration at one point the animals could only sweep a few millilitres per day; by moving most of the time and locating cells by encounter, even if the perceptive range is short, they can obtain enough to eat. Figure

24 shows an imaginary relationship between grazing mortality and algal density applied to the size distributions of copepods and algae. If it were true, estimates of grazing mortality might be made from size and density distributions. Such a possibility arises only from the simple form of predation envisaged.

A much more complex form of predation is found amongst the large fish that prey on shoals of little ones. From acoustic evidence, many small and medium-sized fish live in mid-water in shoals, at least in daytime. The habit is not only of advantage to the prey within the shoal but also to the predator attacking them. Brock & Riffenburgh (1960) defined two probabilities, that of a single prey being sighted by a predator and that of a predator sighting a shoal. Let there be r_a/c' prey along the radius, r_a, of a spherical shoal. Then the number of fish in the shoal, n_f, is given by:

$$n_f = (4\pi/3)(r_a/c')^3. \tag{48}$$

If the prey are to obtain any advantage in shoaling, the quantity eaten by the predator, n_e:

$$n_e \ll \frac{r_a{}^3 n_f}{[r_a + c'(3n_f/4\pi)^{\frac{1}{3}}]^3}, \tag{49}$$

where r is the sighting range of the predator. Brock and Riffenburgh tabulated numbers of prey eaten by prey densities and sighting ranges, needed to violate this inequality. When the prey numbers are high or when the sighting range is greater than the distance between shoals, the numbers eaten are very high. That is, the advantages of shoaling are nullified at short ranges, at night or in turbid water. The same conclusion emerges from the theory of search developed in antisubmarine warfare (Olson, 1964). Saila & Flowers (1969) have extended the argument: the number of prey, n_f, entering a circle of radius R_d about an observer is given by:

$$n_f = (4R_d N/\pi)(U + V_p'')E(\partial), \tag{50}$$

where N is the number of prey per unit area; U is the prey speed; V_p'' is the predator speed; $E(\partial)$ is the elliptic integral of the second kind. It follows from this formula that the greater the speed of the prey, more come in range than escape, so a wise predator works during periods of prey activity. Prey are active at dawn and dusk when it is rather harder for the predator to achieve its theoretical advantage. Further, in order to survive, the prey should move slowly. If the predator's search field were restricted to a broad sector ahead of it, it would only need cruise at about twice the speed of the prey. The underwater films of tuna and their prey taken from the bow chamber of *RV Thomas Cromwell* show that both prey and predator swim rather slowly and with mutual unconcern.

The theoretical basis of the study of death in the sea is in its infancy. It is mainly a study of predation, and the predatory mortality may be density-dependent because the time to capture, eat and digest is roughly constant whereas the time to search is inversely proportional to density. There is some indication of the methods by which an active study of death in the sea could be made, but they can only be investigated with a combination of model and experiment.

The age structure of a fish population

Deevey (1947) examined the trend of natural mortality with age for a number of animal species. During juvenile life mortality tends to decrease with age, but after middle age the death rate increases with age. The rate is least during early adult life when the reproductive activity is greatest. The decrease of mortality with age in immature life occurs as numbers decrease; in early adult life numbers remain nearly steady and decline with age very slowly, but in later life numbers decline because the death rate increases exponentially. This pattern of mortality in age is general and we would expect to find it amongst fishes.

Fishes are very fecund and lay as many as 10^4 to 10^7 eggs in each spawning. The death rate of eggs and early larvae is high, perhaps 5–10% per day and obviously such a rate must subsequently decline. Pearcy (1962) studied the larval and immature life of the winter flounder (*Pseudopleuronectes americanus* Walbaum) and found that the death rate declines with increasing age during the larval stages. Predation is density-dependent if the time to search predominates, and with a constant number of predators the time to search varies with the number of prey and thus death may then be density-dependent. As they grow, the animals pass through a number of trophic levels and usually end as rather large second-order carnivores. Because the range of growth is considerable (e.g. 1 mg to 1 kg), fish pass through a number of predatory fields as they grow. Each predator is larger and less numerous than its predecessor in the series. Each can kill more effectively per unit volume than its predecessors, but because they are much more thinly distributed the chance of death decreases in time. The sequence depends on growth, itself a function of age. Hence the component of density-dependent mortality due to predation in some fishes may be described as a function of age.

In the next section another mechanism is described; larval fish that eat well and swim well avoid predation and vice versa and so density-dependent mortality could again be described as a function of age. The two mechanisms are hypothetical, but not exclusive; indeed density-dependence might arise in the time to search or in the availability of food or both.

Death is due to density-independent as well as density-dependent causes and intuitively density-independent causes may predominate. By either of the mechanisms described above any mortality independent of density must alter the degree of density-dependence; in other words a density-independent effect is modulated by the density processes merely because they are persistent. Provided that the density-dependent proportion of mortality is not too low (say one-third), a density-dependent function of age can include density-independent effects – particularly if based on a larval mortality that must include both.

Suppose that mortality is a density-dependent function of age in the restricted way indicated above:

Let $N_t = N_{(t-\delta t)}[\exp(-kN_t\delta_t)] \simeq N_{(t-\delta t)}(1-kN_t\delta t+\ldots).$ (51)

$$N_t - N_{(t-\delta t)} = \delta N_t = -kN_t^2\delta t.$$

$$\int\frac{\delta N_t}{N_t^2} = -\int k\delta t.$$

$$\text{Therefore}\left[-\frac{1}{N}\right]_{N_0}^{N_t} = \left[-kt\right]_0^t.$$

$$\text{Therefore}\ -\frac{1}{N_t}+\frac{1}{N_0} = -kt.$$

Therefore let $kN = M$; $N_t = N_0/(1+M_0t)$; $M_0 = (N_0-N_t)/N_t,t$. (52)

The initial approximation is valid so long as $M<0.5$; N_o is the initial number, i.e. at hatching or at first feeding.

The mortality rate of larval plaice is 80% per month (Harding & Talbot, 1973); then $N_{30}/N_0 = 0.2 = 1/(1+30\ M_0)$. Therefore $M_0 = 0.133$. N_0 can be calculated assuming that the cohort replaces itself by the age of 16, i.e. $N_t = 2$ at the critical age when growth rate equals mortality rate. In the adult age groups, the trend in mortality is as calculated above for a wide range of initial values and so the critical age can be estimated. The estimate of larval mortality includes both density dependent and density-independent components and it is assumed that the proportion remains the same during the period. With this method differences in natural mortality between species would really depend on the estimated larval mortality. The mortality rate of 0 group fish is 40% per month at arrival on the beach in June or 10% per month during the following winter (Cushing, 1974). The mean date of spawning is 19 January (Cushing, 1970), so the time of first feeding is probably in early March. If the initial mortality of 80% per month lasts into early April, it must be reduced to 38% per month by early June when the little fish are swimming on to the beach. The adult natural mortality rate is taken from the loss rates of the transwartime year classes of plaice in the southern North Sea (Beverton & Holt, 1957), between the ages of five, six or seven before the Second World War, to thirteen, fourteen or fifteen after it. The mortality rate estimated in this way is 10% per year, although Beverton (1964) in a later analysis suggested that $M♀ = 0.08$ and that $M♂ = 0.12$ (on an annual basis). The average theoretical mortality rate estimated from the curve in Figure 35(*a*) is 9.7% per year. Thus, given the initial larval mortality rate, the trend of death rate with age from larval life to the critical age can be fairly well simulated.

The curve in Figure 35(*a*) was located by establishing the critical age.

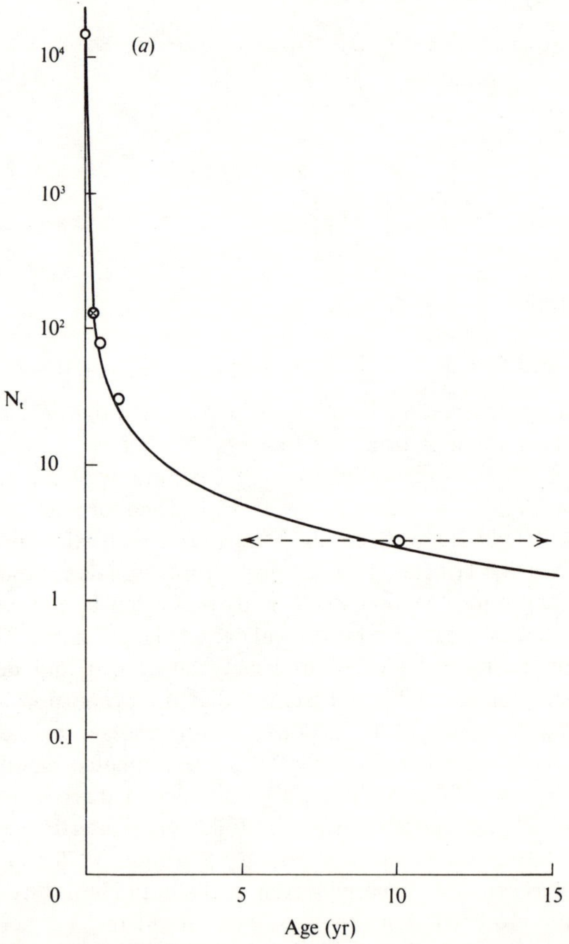

Figure 35. (*a*) The trend of natural mortality with age for the southern North Sea plaice; the curve is established from the observed larval mortality (●) and the critical age. See text for explanation. (*b*) The trend of natural mortality with age for the Gulf of Maine herring (from the data of Graham *et al.*, 1972). (*c*) The Baltic cod (from the data of Poulsen, 1931) (Cushing, 1974).

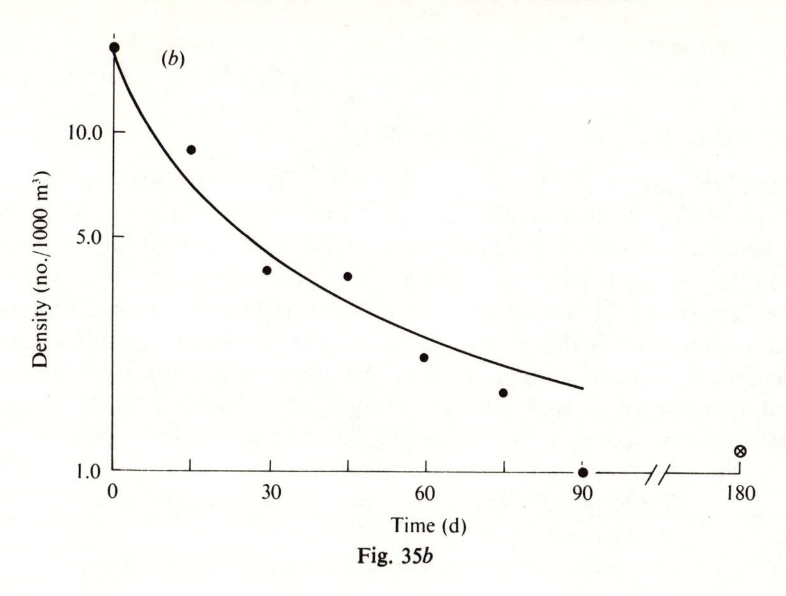

Fig. 35*b*

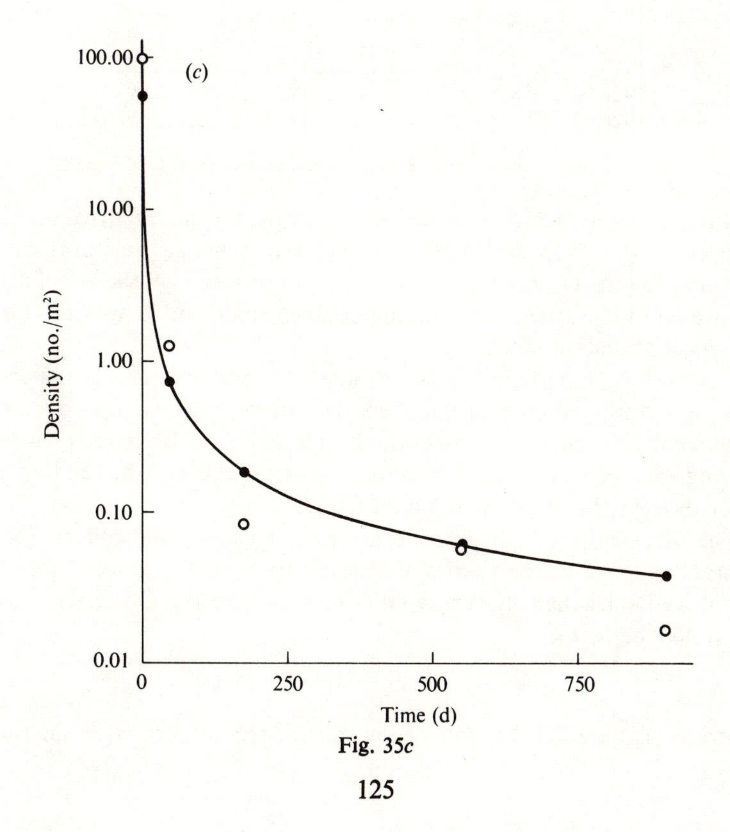

Fig. 35*c*

Another method is:

$$N_2/N_1 = (1+M_0t_1)/(1+M_0t_2)$$

Therefore $M_0 = (N_1-N_2)/(N_1t_2-N_1t_1)$.

Similarly, $N_0/M_0 = (t_2-t_1)\ \{N_1N_2/(N_1-N_2)\}$. Figure 35(*b*) shows the curve fitted to observations on the Gulf of Maine herring (Graham, Chenoweth & Davis, 1972) and Figure 35(*c*) that on the Baltic cod (Poulsen, 1931). The herring data were averaged for a period of years, samples being taken in the Boothbay area with high-speed nets. The cod observations were made with a Hensen net (for eggs), a Petersen young fish trawl and a tog, which is a small inshore trawl originally used for catching eels; information was available for the five-year classes 1923–7. A most interesting point emerges from the ranking of year classes as in Table 10. The

Table 10 *The ranking of year classes at different stages during the life cycle of the Baltic cod*

Stock	Larval drift	0-group	II-group
1924	1923	1923	1925
1927	1925	1925	1923
1923	1926	1926	1926
1925	1927	1927	1927
1926	1924	1924	1924

ranking of year classes is established during the larval drift with one exception; the 1923 and 1925 year classes exchange positions on the nursery ground. The compensatory nature of processes in the larval drift is illustrated by the reversal of ranking between stock and larval drift for the two most abundant stocks.

Not only is some light thrown on the nature of the density-dependent processes during immature life, but the curve that describes a density-dependent mortality as a function of age fits data for cod, plaice and herring. The generality that emerges reflects the life of fishes as they pass through food chains of common structure.

The suggestion that there is a critical age by which the cohorts replace themselves on average carries the implication that senescent mortality might well start then. Beverton (1964) formulated the Gompertz Law in the following form:

$$N_t/N_0 = \exp[-\exp(m_0+m_1t)], \tag{53}$$

where m_0 is the mortality rate at the start of senescence; m_1 is the rate of increase of mortality with age. However, no successful analysis has yet been made that demonstrates the existence of senescent mortality.

The description of natural mortality given in this section is highly speculative and difficult to establish. As fishes grow through a succession of predatory fields, the formulation used expresses their competitive success throughout the food chain as the cohort exploits the available food in growth in the face of mortality, whatever its cause. The cohort may replace itself on average by the critical age, but the high variability of recruitment may express also variations in the age of replacement.

The link between growth and death

The numbers in a population depend upon the food available to it. If fecundity is determined by natural selection (Lack, 1954), then the growth of individuals represents the food converted in the struggle for existence. However, in terms of the regulation of numbers, models developed by Ricker & Foerster (1948) and by Beverton & Holt (1957) require that fish larvae grow through a critical period of predation and that mortality depends upon the time taken to do so. Recently such models have been developed further in an attempt to match the mortality required to satisfy the data in plots of recruitment upon parent stock.

Jones (1973) assumed that haddock larvae died if they fail to capture a specified number of *Calanus* nauplii each day and that their searching capacity increases exponentially with age. The larvae grow at 12% per day and die at 10% per day. The numbers alive after forty-five days were calculated by numbers of food organisms and initial numbers of larvae. The mortality was shown to be density-dependent and the numbers of organisms were not much reduced. Jones writes that there would be 'considerable biological advantage in the existence of a single mechanism able to influence both growth and mortality simultaneously and hence control this balance in the long term'.

Cushing & Harris (1973) developed an analogous model: as larval plaice feed on rich food, they grow, swim more quickly and evade capture from a particular predator. As they feed, food density is reduced and so the chance of capture increases in time. If the numbers of larvae are increased the chance of capture is still greater. Hence a density dependent-mortality is generated by the availability of food. An estimate of cruising speed of the larvae is available (Ryland, 1963). The daily ration of a plaice larva is known (Riley, 1966) and Blaxter (1965) has made some estimate of the weight loss of herring larvae under starvation. The model is run in days with new estimates of length and cruising speed each day. An essential part of the model is the formulation of predation,

$$P' = 3\pi R_p^2 C_p T_l t/(3+\pi T_l R_p^3)((3C_p/R_p)-2), \qquad (54)$$

where R_p is the attack or escape speed of the predator; C_p is the cruising speed of the predators; T_l is the density of larvae; P' is the predatory mortality generated by a single predator per unit volume. The model was run for a period of thirty days, first with a constant quantity of food each day (i.e. with density-dependent mortality only) and secondly with food decreased by larval predation, so both the processes of density-dependent growth and mortality were operating. With the use of density-dependent mortality alone, the values obtained in the stock recruitment curve for the southern North Sea plaice were roughly approximated (Cushing & Harris, 1973). Because density-dependent mortality is less than that of the density-independent, the model overestimates the true value from which we may conclude that a more realistic estimate can be developed within the values of the parameters given. Figure 36 shows the model results; Figure 36(*a*) gives the dependence of larval mortality on density and the observed mortality rate of 80% per month corresponds reasonably with the observed density of plaice larvae in the sea of 3 to $5/m^3$. Figure 36(*b*) illustrates the dependence of weight on food after thirty days; the spread of final weights in the model is probably greater than those observed in the sea so the modelled decline in food density in time may be too rapid. The dependence of larval mortality on food is illustrated in Figure 36(*c*) at a given initial larval density. The change in mortality rate with food is low over most of the range but increases sharply at low levels as might be

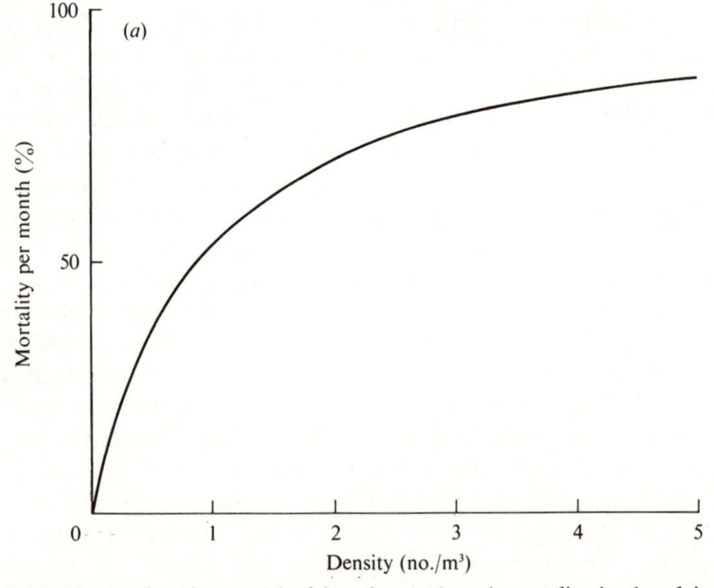

Figure 36. The results of a model of larval growth and mortality in the plaice. (*a*) Dependence of larval mortality on density; (*b*) dependence of weight on food after 30 days; (*c*) dependence of larval mortality on food.

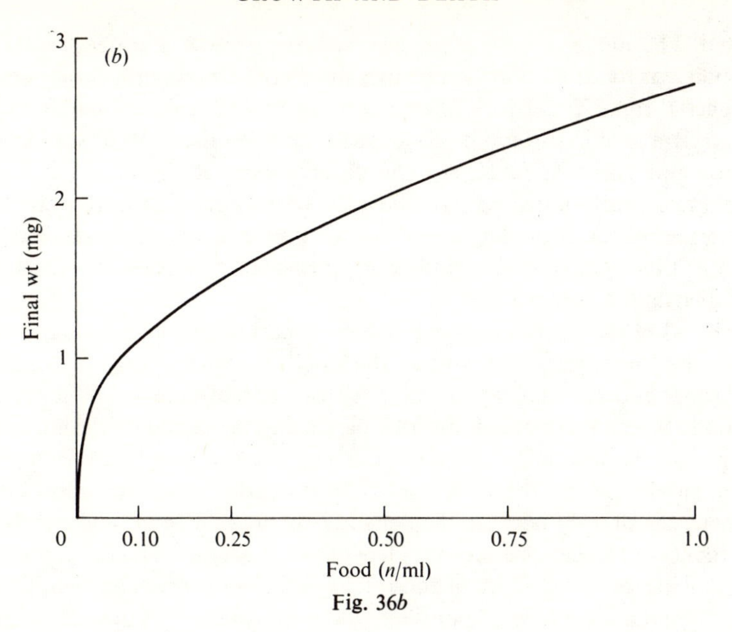

Fig. 36*b*

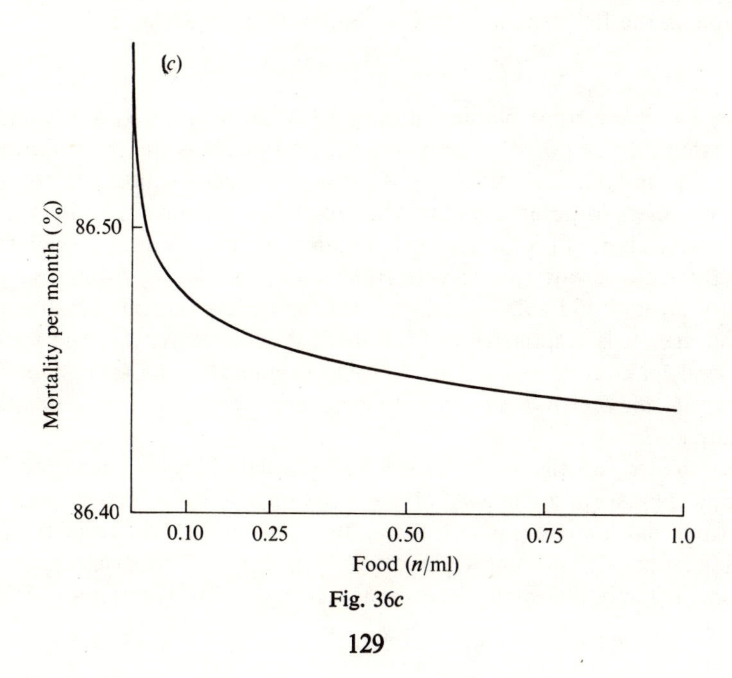

Fig. 36*c*

expected. The model establishes a link between growth and death in the availability of food. Fish grow through the trophic levels of a food chain and hence through a network of predatory fields and it is reasonable that their survival should depend on food, specifically on the ratio of speeds of predator and prey. At first sight the disadvantage of the model is that predators of nicely adjusted sizes have to be stationed at points in the life cycle. However, if predation depends on prey density, the absence of a predator at its right station would be compensated by its larger and more thinly distributed successor.

The trend of mortality with age in the early part of the life cycle depends on the initial numbers, that is upon the stock. At a later stage in the life cycle processes occur that depend on the density at that time. Let us suppose that food increases at some point of time during the larval drift and then for a period the mortality rate decreases because it depends on food. At the end of the period the food reverts to its earlier level and mortality rate increases sharply because the density is high (as if the initial number were higher). Hence, the density-dependent mortality is continuously variable with the availability of food. It is an active process and within a cohort the final recruitment represents the maximization of growth in the face of the loss of numbers. It is very interesting that patchiness of food in time provides the method by which the carrying capacity of the environment is explored and exploited.

The response of the population to the quantities of food available resembles that of the adult stock in terms of growth and mortality, as for example in the Beverton and Holt formulation or in Ricker's:

$$B = [B_o \exp(G' - M) - 1]/(G' - M), \tag{55}$$

where B is the average biomass during a period of time; B_0 is the initial biomass; G' is the growth rate within the period; M is the mortality rate within the period. The ratio (G'/M) is a convenient expression for the same processes in immature life. When food is abundant the ratio is high and conversely it is low when food is scarce. To start with, on the larval drift the ratio is only just greater than unity, but it is greatest on the nursery ground and subsequently it declines to less than unity after the critical age. It is responsive to food density within each short period of time and its course during the life cycle summarizes the success of the cohort in maximizing the density-dependent processes in generating recruitment.

The cohort and the stock have different roles. The first explores and exploits the carrying capacity of the environment in an active manner. The stock has a conservative function in which the variability of recruitment is averaged. This high variation that has dismayed fisheries biologists for so long modulates the environmental variability continuously. The climate changes in a periodic, if variable, manner through decades and centuries as will be shown below and it is clarified or even rectified, in the variability of the stock.

It is no accident that the density-dependent processes die away with age to such a degree that the density-dependent growth can rarely be detected in the adult age groups of fishes. Between recruitment and the critical age the fecund fishes may grow by a factor of twenty or so. If recruitment represents the end of a process of the maximization of biomass then the subsequent increment to the critical age at low dependence on density represents a contribution to the stock fecundity, which summarizes the processes for a subsequent generation. Adult fishes live somewhat isolated lives in an environment of fully adequate food and they put on as much weight as they can to execute the conservative functions of the stock.

Adult life in fishes is probably not a competitive phase in the life history, but that during the larval life and on the nursery ground is highly competitive. The plaice larvae in the southern North Sea live in the plankton with dab, cod, herring, whiting and sandeels. Both growth rates and death rates are maximal during this period of greatest density. The food chain is simple and comprises two or three items, copepod nauplii, copepodites and *Oikopleura*. Competition is high and in an evolutionary sense it must be maintained in each cohort. A simple way of expressing competition is in the ratio (G'/M); if high the subsequent stock is high and vice versa. Competition is then seen to be strong under conditions of rich food and weak in scarce food. A species might be eliminated more readily when food is abundant than when it is scarce, at least. The number of species in the sea is less where food is rich than where it is thin.

In an earlier chapter, a distinction was drawn between the discontinuous production cycles of temperate and high latitudes and the continuous ones of low latitudes. In temperate waters, production is high and the transfer from one trophic level to another is probably inefficient, whereas in tropical waters production is low but the transfer is more efficient. Because there is less food available in the tropical open ocean competition may be less intense than in higher latitudes. In an evolutionary time period more species would survive or become differentiated when the food was scarcer. The inverse relationship between diversity and abundance is then a function of the dependence of competition upon the availability of food. Competition is often expressed as an effect of adult life but in fishes at least such an expression would be the approximate inverse of that in juvenile life because of the compensatory dependence of recruitment on parent stock. The link between growth and death appears to be at the root of the three components of a population, the magnitude of the stock, the regulation of numbers and the maintenance of competition.

Conclusion

The problem of growth overfishing (see Chapter 7) was solved in maximizing the gain in weight and loss in numbers for the adult (or near adult) age groups. In teleological form, the same problem faces the cohort in its exploitation of the food available throughout its life cycle. The ratio (G'/M) does not only express its success or failure in generating high or low recruitment, but it also assesses the degree of competition between species. On average, $(G'/M) = 1$ at the critical age, by which the population should have replaced itself. Usually growth and mortality has been studied as almost independent processes and in one sense at least they can be so described. However, in terms of the natural regulation of numbers in the population, the link between the two is of great importance.

There is a distinction between stock and density dependence (Harris, 1974) and it acquires biological meaning in terms of the link between growth and mortality. The variability of food in time and space allows density-dependent mortality to be modulated continuously and it is an active and exploratory process which exploits the food chain to the maximum degree. Stock dependence expresses a summary of the performance of a number of successive cohorts. The one process is responsive and the other is conservative. The potential differences in (G'/M) are greatest in larval and juvenile life when there is an adaptation to differences in the availability of food. During adult life, the differences in the ratio are reduced although its magnitude has been determined earlier in the life cycle; the result of simple adult growth is to build up maximal fecundity from the sum of cohorts, or stock.

7

The population dynamics of fishes

Introduction

A medium or large population of animals comprises a number of spawning groups between each of which there is some degree of interchange. A population geneticist would specify such conditions to maximize variation (Li, 1955) and such groups or stocklets may occur in fish populations; they have been identified in the Pacific salmon and in the British Columbian herring and the degree of interchange between them has been specified. Stocklets may exist in the Downs and Norwegian herring stocks, but no degree of exchange has been established. Between stocks there is little or no interchange and the long spawning period, together with the diffusion of larvae, ensures the exclusion of stocks of the same species from the same spawning and nursery grounds. As shown in an earlier chapter the major stocks of cod in the north Atlantic are, with one exception, genetically distinct in seven transferrin and three haemoglobin alleles. Within each of the spawning groups, suffering interchange, the chances of mating should be randomly distributed. The exclusiveness of stocks may be physically determined in the nature of the migration circuit (as exemplified by the stock of plaice in the Southern Bight of the North Sea) and this may be characteristic of offshore stocks of fishes. In general, unit populations in the commercial stocks are known if not established. There are exceptions, for example the cod stocks at Iceland and SW. Greenland which appear to spawn on the same ground, as shown by the extensive Danish tagging experiments (Tåning, 1937). There are diverse methods for suggesting unity but only those with a genetic basis are likely to establish it.

The need for the control of fisheries appeared in European waters towards the end of the nineteenth century. Sailing smacks explored the North Sea throughout the century and, with the advent of steam trawling after 1881 (Cushing, 1966*a*), fishermen saw that the stocks of trawled fish were declining. Various commissions in the United Kingdom recorded evidence that the average catch of a steam trawler fell by a third in a decade and, as it is a reliable index of stock, the populations must have been well exploited; indeed, Garstang (1900–3) showed that the average catch declined as a function of increased fishing. The west coast of North

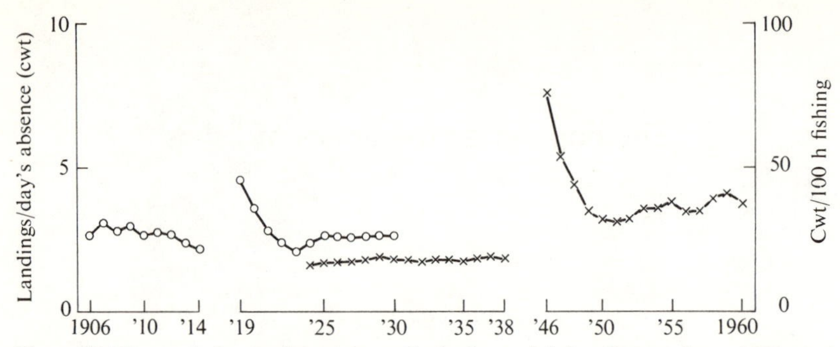

Figure 37. The stock density (as catch per day's absence) (○) and as catch per 100 hours' fishing, (×) of the southern North Sea plaice between 1906 and 1960 (Wimpenny, 1953; Bannister, personal communication).

America was opened in the first two decades of the twentieth century to eastern markets by the trans-American railways and as a consequence the stock of Pacific halibut became heavily fished. In both regions fishermen interpreted the reduction of stock as an interference with breeding, but they probably observed a deprivation of growth by capture at too early an age. In the later stages of the history of a fishery, however, the fishermen might well have been right, as will be shown below.

The effect of fishing on the plaice stock in the southern North Sea since he early years of the century is shown dramatically in average catch,

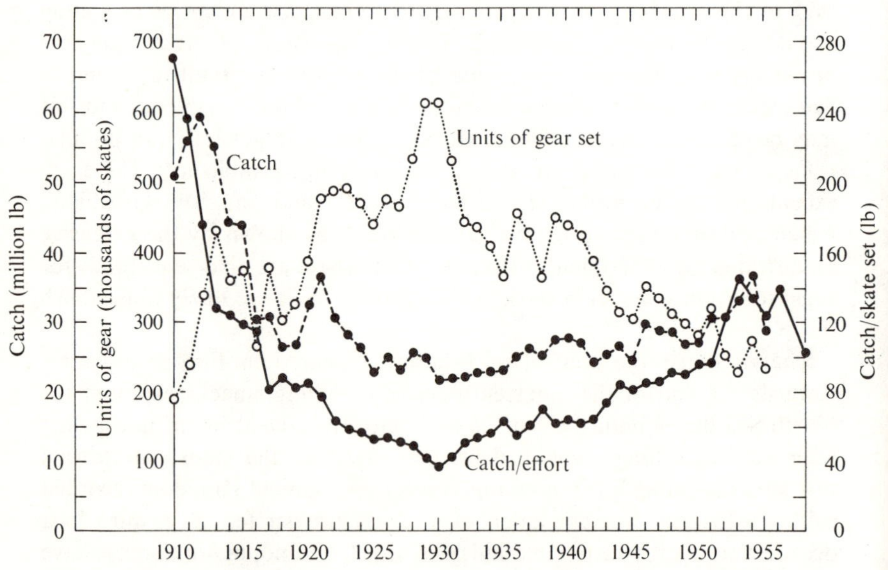

Figure 38. Catches, fishing effort and stock density (or catch per skate, which is a group of hooks on a long line) for the Pacific halibut from 1910 to 1955 (Fukuda, 1962).

catch per unit effort or stock densities (Figure 37). Fishing was relaxed in wartime and the stock densities of plaice rose during the course of each war and fell again sharply as fishing was resumed. Fishing effort is time spent fishing and stock density is expressed as catch per day's absence or catch per 100 hours' fishing. Such indices are in fact the same as the average catch of a fishing vessel which is used by the fishermen at sale as an index of profit. Two indices of stock density are shown in Figure 37, cwt/day's absence and cwt/100 h fishing and they are presented in parallel for a period of seven years. The ratio of stock densities at the end of a period of

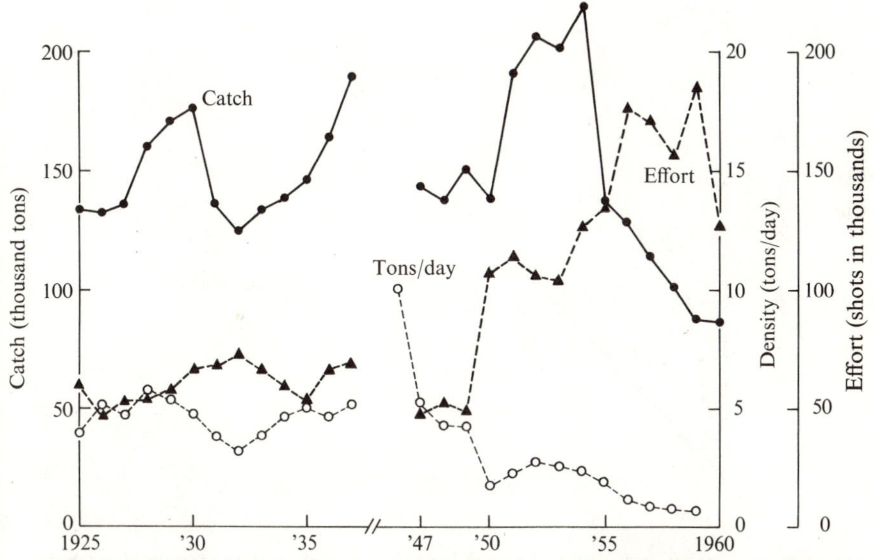

Figure 39. Stock density in tons/day (○) of the Downs herring from 1925 to 1959, together with international catches (●) and international effort (▲) exerted on the stock (Cushing & Bridger, 1966).

war to that before it started was about 2.0 in the First World War and about 3.3 in the Second. The stock density during the fifties was nearly twice that in the thirties. There are two conclusions to be drawn from Figure 37: (1) fishing reduced the stock by a factor of two to three, and (2) the stock recovered very rapidly with the relaxation of fishing, which implies that recruitment was not affected.

Figure 38 illustrates the history of the Pacific halibut fishery; the average catch, or stock density, fell by nearly seven times between 1910 and 1930. In the same period catches decreased by a factor of three and effort rose by the same factor. In 1930 fishing was restricted by a closed season and as a consequence stock density increased and, eventually, catches did so as well (Fukuda, 1962). Density-dependent changes may have occurred during the period (Southward, 1967), but this does not materially alter the issue. The real point is that the halibut stock suffered from a deprivation of growth like that of the plaice, with possibly consequential density-dependent changes.

Figure 39 shows the catches, stock densities (as catches per drifter shot) and fishing effort (as number of shots) exerted on the Downs stock of the North Sea herring during the period of its decline; in this fishery, herring were caught by drift nets about a mile and a half long and hung seven

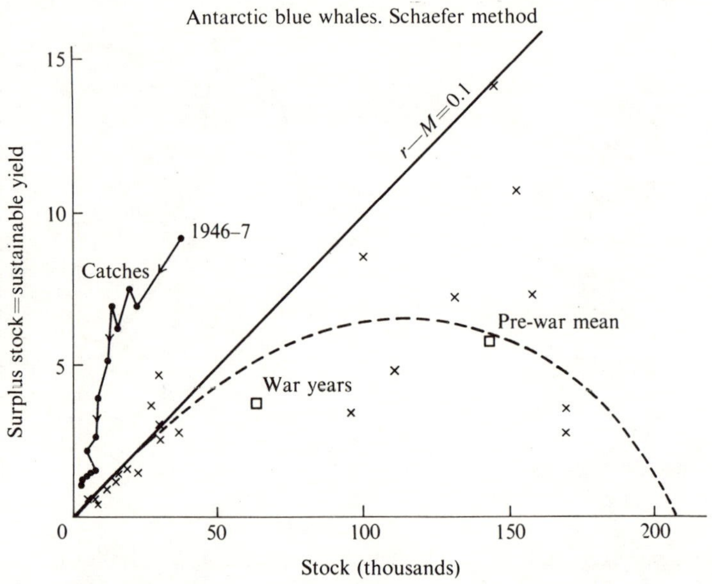

Figure 40. Catches and surplus stock as function of stock in the antarctic blue whale population, together with an estimate of recruitment less mortality (Anon. 1964).

fathoms from the surface. They were shot 'at the close' (or at dusk) and hauled some six hours later or at dawn, so the drifters' nets were shot once or more times a night. Stock density dropped by about ten times in twelve years, catches to less than half, but fishing effort increased by about three times during the period (Cushing & Bridger, 1966). After 1960 stock density continued to decline and catches fell very sharply. Herring do not grow much during their lives in the fishery and it was shown that recruitment declined by a factor of at least three during a decade (Cushing, 1968*b*). During the sixties fishing effort was reduced and by 1970 the stock had started to recover in larval numbers. Because recruitment declined as fishing effort increased and built up as it relaxed, the Downs stock probably suffered recruitment overfishing. Another example of recruitment overfishing is shown in Figure 40; catches, estimated stock and estimated recruitment of the antarctic blue whale fishery are illustrated. Stock was estimated by the Graham–Schaefer method (see below) by which the logistic curve of population growth is fitted to catches or stock density. The function of recruitment less natural mortality, fitted as a tangent to the curve from the origin, was estimated from stock densities in numbers at each age. For a long period, catches were greater than the surplus stock that can be taken without upsetting the equilibrium between stock and fishing. As catches persisted at too high a level, recruitment was reduced. Like the herring, whales do not grow much during their adult lives and they also suffered from recruitment overfishing.

Fish stocks are reduced by fishing in two ways, by growth overfishing and by recruitment overfishing. Plaice and halibut suffered from growth overfishing, that is, recruitment did not change much as fishing mortality increased and death by fishing was great enough to prevent the small fish from putting on any significant weight. Herring and whales suffered from recruitment overfishing, that is, death by fishing was great enough to reduce recruitment. It is no accident that the plaice and halibut are fishes that can grow in weight by an order of magnitude or more during their adult lives. In contrast the herring and the whale grow very little during their adult lives. Not only are the latter vulnerable to recruitment overfishing, but the former, sensitive to growth overfishing, are by their growth protected from recruitment overfishing, that is, by their greater capacity for stabilization.

The estimation of stock density

The fishing vessel kills fish during the time spent fishing (or fishing effort) and the average catch of a fishing vessel, trawler, drifter or purse seiner is a proper index of stock, shown formally as follows:

$$\text{Let } N = R'[\exp(-Z)], \tag{56}$$

where R' is the number of recruits (which may be immature) to a year class at the start of a year's fishing; N is the number of fish remaining in the stock at the end of the first year; Z $(= F+M)$ is the instantaneous coefficient of total mortality; F is the instantaneous coefficient of fishing mortality; M is the instantaneous coefficient of natural mortality.

The number dying during the year is

$$R'-N = R'[1-\exp(-Z)]. \tag{57}$$

The catch in numbers, Y_N, the proportion of deaths due to fishing is

$$Y_N = (F/Z)R'[1-\exp(-Z)].$$

$$\text{For } \lambda \text{ years, } Y_N = (F/Z)R'[1-\exp(-Z\lambda)], \tag{58}$$

where λ is the number of age groups in the year class, which is also the number of age groups in the stock in any given year.

It is named the catch equation because the ratio of catch to stock is expressed as the exploitation rate $(F/Z)[1-\exp(-Z)]$; if Z is high the rate is sometimes given as the exploitation ratio (F/Z). We will return to the catch equation in a subsequent section on cohort analysis.

The catch equation may be written in another form, that of catch per unit of effort.

Then $(Y_N/F) = (R'/Z)[1-\exp(-Z\lambda)]$. (59)

The fraction on the left of the equation is the catch per unit of effort, or stock density, and the expression on the right is the stock in numbers of a year class throughout its existence, which is the same as an age spectrum in one year. The catch per unit of effort is an index of stock.

The relationships between fishing mortality, fishing intensity, f, (or fishing effort, g', per unit area, A') stock density, d (or catch per unit of effort) and true density, D, are given by: $F = qf$, where q is the catchability coefficient; $f = g'/A'$, or intensity is effort per unit area and $FN = Y$.

Then $q = F/f = (Y/N)/(g'/A') = (Y/g').(A'/N) = d/D$ (60)

or $D = d/q = Y/fq = Y/F$. (61)

The proof that stock density is an index of stock has occupied an important place in the history of fisheries research because it is also the average catch of a fishing vessel and is a statistic that can be readily collected on ports and markets.

Variations in fishing mortality, at constant fishing effort, are variations in the catchability coefficient q. The term *availability* expresses such differences in a general way; stock may be described as *accessible* or *inaccessible* in geographical terms, vertically or horizontally and *vulnerable* or *invulnerable* to a particular gear (Gulland, 1955). Much of the technical work carried out by fisheries scientists has been the detection of variation in catchability. Because effort is merely time spent fishing, variation in catchability is really that in catch, either as a variance or as a bias. The coefficient of variation of a catch may range from 30% to more than 100%; fish may be caught efficiently or inefficiently and whatever the source of error, variability in stock density or catchability is essentially variability in catch.

Many populations of wild animals are exploited but the commercial fish stocks are numerous. A familiar figure for the number in a fish stock is 10^{10}. Large parts of such stocks are landed on the quays and markets, where the animals are counted and weighed. Such information has been collected for a number of decades, at least in parts of western Europe and

North America. No other wild populations have been so well documented for such long periods of time.

In many countries data are collected on individual catches and their positions of capture. But in others, including Britain, the information is recorded as a total catch for the voyage from a particular fishing ground. Half a century ago when a fishing vessel went to a single ground during a voyage, such a procedure yielded a proper average catch for a true position of capture. Today, fishing boats shift grounds during a voyage, even from one side of an ocean to the other and each catch should be recorded by its position of capture in log books. The quality of recent Japanese work on tuna may owe something to their detailed collection of information by 5° squares across the world ocean.

In Britain fish are measured in length on the markets in large numbers; for example, on English markets over 750000 fish are measured in length each year. From a small proportion, the otoliths (or earstones) are taken and the age of each fish is determined by the number of annual rings on them. With the use of an age/length key, the length measurements are converted to age distributions. Because the whole system is expressed in units of stock density (Gulland, 1955), the age distributions are also put in the same units. Most adult fishes in the commercial populations live for many years, if not caught. Death rates may be estimated from the ratio of numbers in successive ages in a cohort expressed in stock density. Growth rates may be estimated from mean lengths or mean weights at age from the same samples.

Before such estimates can be used in the catch equation or in any other model, total mortality must be separated into its fishing and natural components. Fish can be tagged with little plastic flags and the tagged fish may grow and survive for many years, and the proportion recaptured by fishing vessels represents the fishing mortality of the stock. Common to all tagging experiments is the axiom that the stock of tagged animals represents that stock to which they have been added. The results of such an experiment are well expressed in stock density particularly if the tagged fish survive long enough to mix well with the untagged stock.

The work on tagging developed in two ways, in lakes to estimate stock in a single season of sport fishing and in the sea to estimate fishing mortality during a number of years. There are special problems concerned with the immediate deaths due directly to tagging, with the degree of recovery from ships and markets, with deaths due to tagging in time and with the mixture of the tagged fish throughout the range of the untagged stock. There is a particular value in tagging in the early stages of exploitation, before the system of market sampling and of age determination yields an estimate of fishing mortality from the increase in fishing which always takes place as a fishery develops. If fish cannot be aged, e.g. tuna, tagging may provide the only estimate of fishing mortality. In any case, in other stocks that can be aged, growth rates and death rates may be estimated independently of those derived from market samples.

The development of models

The logistic curve

As in other branches of population dynamics, models in fisheries biology may be described as descriptive or analytic. In the descriptive ones, the constants are calculated from changes in the numbers of the population itself, and in the analytic ones, they are derived independently. The logistic equation (Verhulst, 1838; Pearl, 1930) was developed to describe the changes in numbers of the human population:

$$\mathrm{d}N/\mathrm{d}t = r_i N(1-N/k') \qquad (62)$$

where N is the number in the population; r_i is the instantaneous rate of increase (or the increment from generation to generation, averaged in year classes); k' is the saturation level, or the carrying capacity of the environment. The exponential growth in numbers is controlled by the relative saturation of the environment, N/k'. The two exponents, the increase rate and its environmental control, are usually estimated by changes in the numbers of the population. The weight of stock increases with time to an asymptote as environmental pressure increases. For example, if a small stock colonized a new area it would increase exponentially till it reached a steady-state, the rate of increase being equalled by the saturation of the environment; conversely, the rate of increase is maximal at the lowest densities. In 1879–81, 435 striped bass were released in San Francisco Bay and twenty years later, the catch amounted to 500 tons per year (Merriman, 1941). The logistic curve describes events such as these, but no insight is obtained into the nature of environmental pressure.

Hjort, Jahn & Ottestad (1933) drew an analogy between the growth in numbers in a fish stock and growth in numbers of yeast. The curve of numbers in time is sigmoid and the maximum catch would be taken at the point of inflection. Graham (1935) applied such a method based on the logistic curve to the plaice stock of the southern North Sea; Figure 41 shows the increment in weight of stock in years as for example throughout a year class or cohort. The first derivative of the curve of biomass, or weight of catch, is maximal at the point of inflection of the curve of stock growth. Later, Beverton & Holt (1957) showed that if the growth of fish were well described, the curve of stock weight would not be symmetrically sigmoid. Graham's use of the logistic curve was the first application to stocks of fish and it played a considerable part in ensuring the success of the Overfishing Convention of 1946.

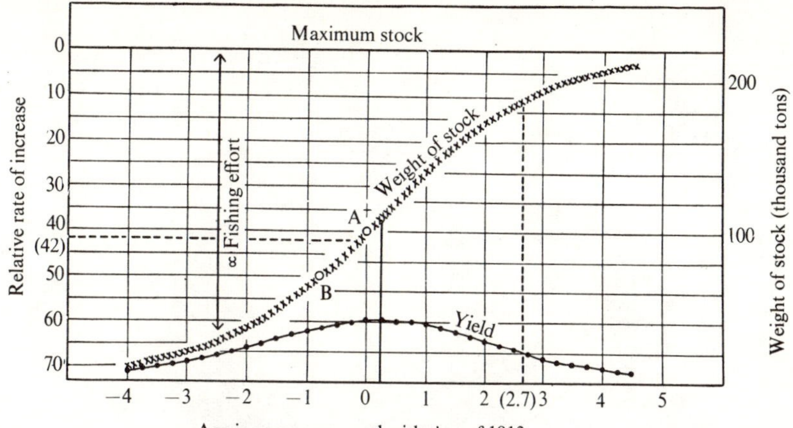

Figure 41. The increase in stock in a year class as fishing effort increases; the relative rate of increase decreases and maximum yield is found at the point of inflection of the curve of increasing weight of stock (Graham, 1935). In October–December 1919, stock density was 2.08 times greater than that in 1913. The relative rate of increase was set as equivalent to the rate of fishing, 40% per year. The curve was constructed by changing stock by an appropriate rate of increase in tenths of a year, with a fishing effort that decreased from left to right. Then the weight of stock is estimated before and after 1913.

A development of the logistic equation is the Schaefer (1954) model, which is useful for stocks of fish which cannot be readily aged:

$$dP/dt = aP(P_m - P) - FP, \tag{63}$$

where P is the stock in weight; P_m is the maximum stock in weight; a is a constant. The net rate of increase is given by $aP(P_m - P)$, and it varies with P (it is greatest at a middle stock value, zero at P_m and at $P = 0$); in practice, stock is estimated by the sum of catch and the stock increment in the year. Figure 42 shows the dependence of stock density on fishing intensity for the yellowfin tuna in the eastern tropical Pacific:

$$[P = (aP_m - F)/a = P_m - F/a].$$

Stock, $P = C/fq$, and the relation in Figure 42 may be expressed in terms of stock density and an expression for the catchability coefficient is obtained independently. Figure 42 shows the dependence of catch upon fishing intensity $(FP = Y = FP_m - F^2/a]$. The figure shows that the maximum sustainable yield was determined with relatively low confidence limits. However, in recent years, the catches of yellowfin tuna have exceeded the maximum yield considerably beyond the upper confidence limit. It is not known whether the increase was due to more recruitment, more growth or to an extension of the area inhabited by the stock. If a stock described by the Schaefer model collapses the source of failure remains unknown.

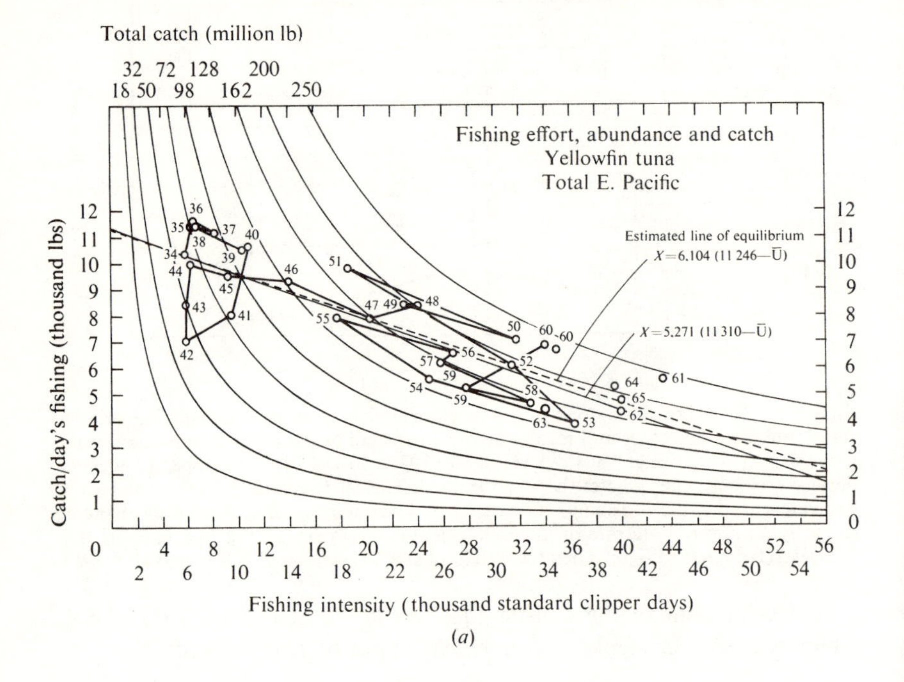

(*a*)

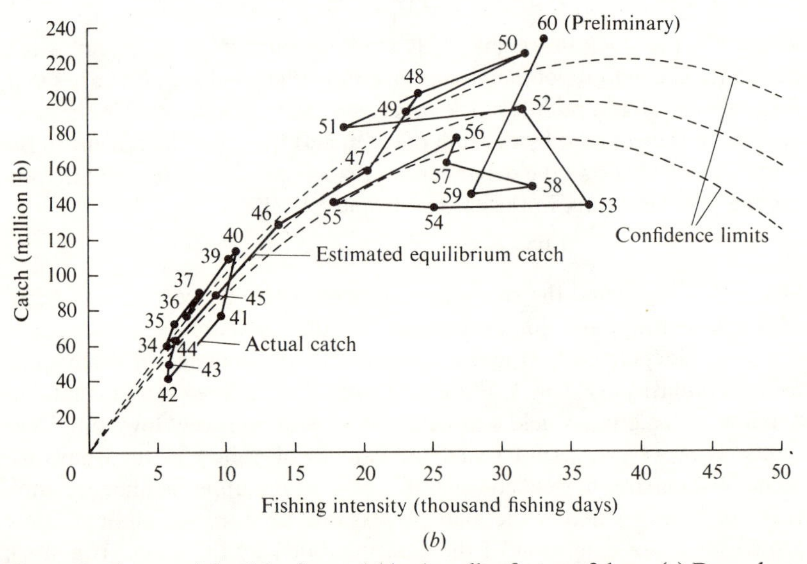

(*b*)

Figure 42. The use of the Schaefer model in the yellowfin tuna fishery. (*a*) Dependence of stock density on fishing effort; (*b*) dependence of catch upon fishing effort (Schaefer, 1957).

Pella & Tomlinson (1969) have modified the equation using a power less than 2 in the catch equation. The age distributions of most fish populations tend not to be normal, but to be extended in age. As a consequence, the maximum sustainable yield is expected to be less than half the virgin stock, whereas in the parabolic yield curve of the Schaefer model the maximum yield is half the stock. Figure 40 shows an application of the Schaefer model to the blue whale stock with an independent estimate of recruitment (less mortality). It was this independent estimate that provided the solution to the blue whale problem in the Antarctic. Gulland (1961) simplified the Schaefer model by plotting the stock density of Icelandic cod in cwt/100 ton-hours on the fishing effort averaged over the previous three years (because there are six abundant age groups in the stock).

In fisheries biology the logistic curve has been associated with the names of W. F. Thompson (who worked on the Pacific halibut and the Pacific salmon), Michael Graham (who worked on the North Sea plaice) and M. B. Schaefer (who worked on the yellowfin tuna in the eastern tropical Pacific). It is a descriptive model and is built on annual increments of catch data; it has the disadvantage that causes of failure must remain unknown but the advantage that the chance of failure by growth overfishing or recruitment overfishing is minimized.

The dependence of stock density (or catch per unit of effort) on effort is traditionally associated with the development of descriptive models in the names of Garstang (1900–3); W. F. Thompson (in Thompson & Bell, 1934) and Schaefer (1954). A recent development by Clayden (1972) has brought to this relationship the quality of an analytic model, with independent estimates of recruitment, natural mortality and so on. He took into account differences in regions between the distributions of effort, selectivity of the gear, weight at age, availability and even a form of economic weighting. A dynamic model was developed which accounted for all these effects year by year. Figure 43 shows the relation between the simulated values and the observed ones in a number of areas. In some of them, where the basic information is poor, the fit between observed and simulated values is not very good; in others, the Barents Sea and the Faroe Islands, it is very good indeed. This work suggests that fisheries biologists are slowly mastering the variability that dismayed their early predecessors.

The distinction between descriptive and analytic models is a didactic one to point the contrast between a model of the Schaefer form and that of the yield per recruit in the consequences for management; management can only come slowly into being with the first, but quickly with the second. The Clayden model shows the weakness of the distinction in that an analytic simulation based on the descriptive model can be developed to some vigorous purpose.

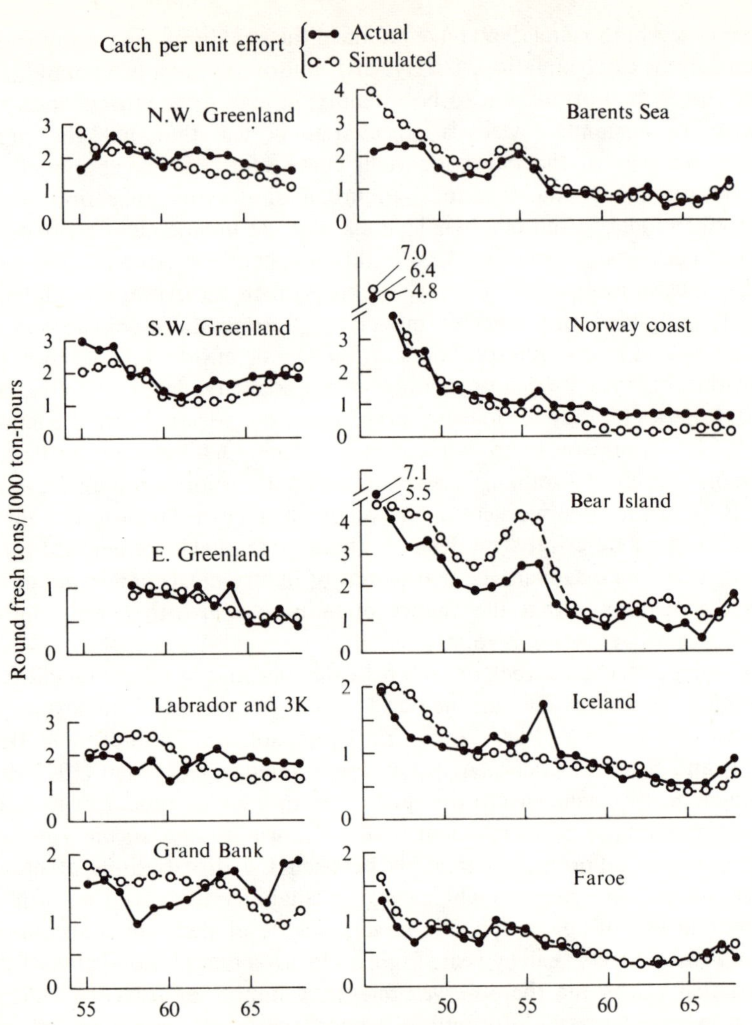

Figure 43. Trends of observed and calculated stock density of cod in a number of areas in the North Atlantic (Clayden, 1972).

The analytic model

Russell (1931) formulated an analytic model:

$$P_2 = P_1 + (R' + G') - (F + M), \tag{64}$$

where P_1 is the stock in the first year and P_2 that in the second; R' is the annual increment in recruitment; G' is the annual increment in growth; F is the annual sum of deaths due to fishing; M is the annual sum of deaths due to natural causes.

In Ross's (1910) malaria equations, the parameters were specified independently of the numbers of the population. The parasite–host relation was described with precise parameters, for example the number of mosquito bites per man. Russell's equation models the fish population in an analogous way with the independently determined parameters. An important axiom of this equation is that the population is maintained in a steady-state only in the increments and decrements of the independently determined parameters.

The Russell equation was developed more fully by Beverton & Holt (1957):

$$P_N = \int_{t_{\rho'}}^{t_\lambda} N_t \mathrm{d}t \tag{65}$$

$$P_w = \int_{t_{\rho'}}^{t_\lambda} N_t W_t \mathrm{d}t, \tag{66}$$

where P_N is stock in numbers and P_w, stock in weight; $t_{\rho'}$ is the age of recruitment (in the year class) to the adult stock or to the fished stock (which may include immature fish); t_λ is the age of extinction of a year class in the stock; N_t is numbers at age t and W_t is weight at age t.

$$\text{Then } Y_w = \int_{t_{\rho'}}^{t_\lambda} FN_t W_t \mathrm{d}t.$$

The parameters are estimated from market measurements, catch statistics and tagging experiments, which are the original bones of fisheries biology. Growth is estimated from the average lengths (or weights) by age from the length measurements and age/length keys. The length measurements are graduated by the von Bertalannfy growth equation (in eqn 34).

From the dependence of Δl on l, by ages, the slope is $[\exp(-k)]$ and the intercept on the abscissa is L_∞; for the equal intervals of age year by year, the constants may be estimated with ordinary statistical procedures (Gulland, 1969). The von Bertalannfy equation describes growth in fishes quite well, particularly for the abundant age groups, but the constants have no biological meaning beyond those cited above.

Total mortality, Z, is estimated from the age distributions in stock density from year to year within a year class:

$$Z = (1/t)\ln(N_0/N_1), \tag{67}$$

where N_0 is the stock density in numbers in the first year; N_1 is the stock density in numbers in the second year. The same result is obtained from estimates of average abundance, $\bar{N}_0$ and $\bar{N}_1$. Estimates of fishing mortality can be obtained from tagging experiments or from the dependence of differences in total mortality on changes in fishing effort, the time spent fishing; such a regression of total mortality on fishing effort separates fishing from natural mortality. The latter may also be obtained by $(Z-F)$.

The Beverton and Holt yield equation is obtained by integrating the rate of change of catch in weight in time, that is, in the year class from recruitment to extinction:

$$Y_w/R' = FW_\infty \sum_{n=0}^{n=3} \frac{\Omega_n \exp[-nk(t_{\rho'}-t_0)]}{Z+nk}\{1-[\exp(-Z+nk)\lambda]\}, \quad (68)$$

where Y_w is yield, or catch, in weight; t'_ρ is the age of entry into the fishery; t_0 is the age at which growth starts as calculated from the estimated constants (it is rarely zero); Ω is the summation constant in the cubic expansion and n is the number of summations [i.e. $\Omega_0 = 1, \Omega_1 = -3, \Omega_c = 3, \Omega_3 = -1$]. The equation may be expressed in Y_w or Y_w/R'. The latter is the yield/recruit formulation, which expresses the dependence of catch upon the stock parameters. Conveniently, the relationship is established by evading the high variability in recruitment. In the next section, the dependence of recruitment on parent stock is discussed. However, it must be remembered that the yield/recruit relationship remains valid within these limits of constant recruitment.

Figure 44 shows the dependence of catch as yield per recruit (Y_w/R') upon the amount of fishing, expressed as fishing mortality (because $F = qf$); changes in recruitment are ignored. In most fish stocks, recruitment is very variable, from a factor of three to ten in the Downs herring (Cushing & Bridger, 1966) to one of ten to a hundred in the North Sea haddock (Parrish, 1957). However, if there were no trend of recruitment with time, the variance of catch would be that of recruitment divided by the number of abundant age groups. The variance of growth is a small proportion of that in recruitment. That due to fishing is expressed in the changes in concentration of fishing vessels on the stock. The high variance in recruitment led to the supposition that it does not decline with falling stock within its fishable range (Thompson & Bell, 1934; Graham, 1935). Then the simplification which expresses yield as per unit recruit has a great operational advantage. A management decision can then be reached when the parameters of the catch equation are established; Thus, in Figure 44, an increase in catch of 20% could be obtained by a reduction in fishing to one third. The Schaefer model illustrated in Figures 40 and 42 needs observations from perhaps a decade or so with changing stock values before the curve can be established. But the yield per recruit curve can be established once the vital parameters, K, W_∞, Z and F, have been estimated.

There is a further advantage with the analytic model. In model terms Beverton and Holt examined the effects of different levels of natural mortality, together with the effects of density-dependent growth (in the recruited stock) and density-dependent recruitment. Their conclusions for the North Sea plaice and the North Sea haddock were that, within the variability of growth and recruitment and under a moderate fishing pressure, density-dependence could be ignored, for management purposes. However, because of the variability in recruitment, failure in recruitment may not be detected until it is too late. There is a sense in which the solution of the growth overfishing problem has generated that of recruitment overfishing. For clupeids and gadoids density-dependent recruitment is of great importance at high levels of exploitation.

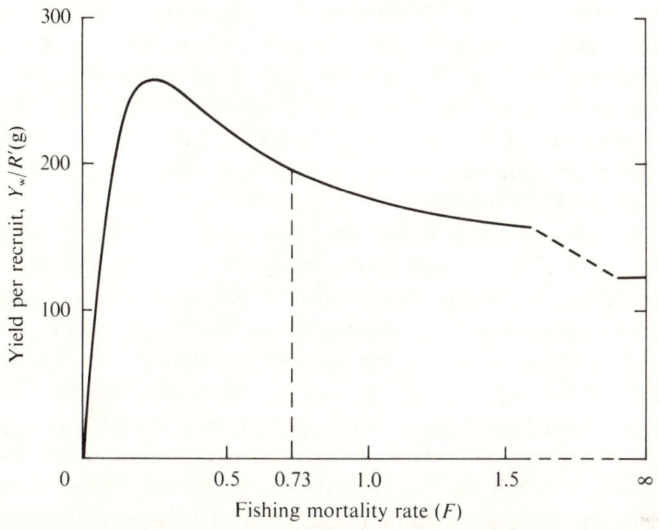

Figure 44. Dependence of catch as yield per recruit in weight on fishing mortality (Beverton & Holt, 1957); the dotted ordinate at $F = 0.73$ represents that exerted on the plaice stock during the thirties. The maximum sustainable yield is shown at about $F = 0.2$.

The demersal stocks in the North Atlantic comprise fishes which grow twenty to thirty times during their adult lives. The Beverton–Holt model describes, at different levels of fishing mortality, the gain in weight due to growth and the loss in numbers in the stock by death. In catch, as shown in Figure 44, there is a middle level of fishing mortality at which the greatest catch is taken, the maximum sustainable yield. In management, the problem was to reduce fishing mortality until the greatest catch was taken. The solution was to put larger meshes in the cod-ends of the trawls, allowing the little fish to escape and grow, in order to be caught later in their lives. Mesh size governs the age of entry to the fishery and there is a maximum catch, for a given fishing mortality, at any age of entry. Similarly, catch increases with fishing mortality to an asymptote so long as the age of entry increases to a middle age, which is not too high. The study of the selectivity of trawl meshes is one of the technical problems facing fisheries biologists.

The Beverton–Holt solution to the problem of growth over-fishing has been the scientific basis for the conservation of demersal fish in the North Atlantic. It is not yet the best, because large fishes like turbot and halibut, which are not very abundant, are not fished at the best age of entry. Further, as will be explained below, stocks can become vulnerable to recruitment failure under heavy fishing pressure. Yet despite these limitations North Atlantic conservation is successful, particularly since international enforcement has started in that ocean. Recently, Gulland (1968*a*) has introduced the concept of marginal yield, that is, an increment of yield becomes more costly as the maximum yield is approached. Hence, the best economic yield is taken at some level of fishing mortality less than that at which the greatest catch is taken. The problem is to define such a level within the population parameters, i.e. independently of economic factors, which vary between countries.

The scientific advantage of the Beverton–Holt model lies in its analysis of growth and death by age from the age/length keys and the market statistics of stock density. Instead of a single observation of catch and stock density in each year, as demanded in the Schaefer model, there are observations of stock density by age; so the annual information is increased by the number of age groups in the stock. But this numerical advantage is trivial as compared with the independent estimates of growth, recruitment and mortality. Part of the argument during the thirties in the International Council for the Exploration of the Sea, rested on density-dependent growth; Bückmann (1932) had suggested that the loss in numbers by fishing could be compensated by a very sharp increase in growth as stock declined. Although the point was answered at the time by Russell (1932) in model terms it was not decisively answered until Southward (1967) published the results of his investigations on density-dependent growth in the Pacific halibut. However, the most important result of the study of the vital parameters has been the discovery that recruitment may decline under heavy fishing (Cushing & Bridger, 1966; Garrod, 1967) and that it may rise or fall with climatic trends.

Cohort analysis

A development of the catch equation has led to an analysis that resembles the estimation of a virtual population, or sum of catches throughout the life of a year class. The study of virtual population has a long history. Derzhavin (1922) published a study of the Kura sturgeon (*Acipenser fulvescens* Rafinesque) based on year class data for more than half a century. The sum of catches within a year class from recruitment to extinction represented a least estimate of stock or virtual population; the sturgeon lives for a very long time and Derzhavin exploited this fact to establish the method. Fry (1949) applied it to the stock of trout in Lake Opeongo, in eastern Canada, and in 1957 he wrote that the virtual population represented 'the sum of the fish, belonging to a given year class, present in the water at any given time that are destined to be captured in the fishery in that year and all subsequent years'.

$$\text{Let } E_t = C_t/V_t,$$

where E_t is the exploitation rate $\{= (F/Z)[1-\exp(-Z)]\}$ during the year, t; C_t is the catch in the year t; V_t is virtual population in the year t. Because V_t is a least estimate of stock, E_t is the greatest estimate of exploitation rate, provided that F and M are assumed constant with age. Then $E_{t\max}/E_t = Z/F$ and $E_{t\max} = 1-\exp(-Z)$, which is the annual estimate of mortality and it is biased upward. Ricker (1958) and Bishop (1959) consider a number of biases more rigorously in the estimates of mortality.

The study of the stock of Californian sardine introduced the concept of availability (Marr, 1951; Widrig, 1954). Murphy (1965) analysed the failure of the fishery with a development of the catch equation in which fishing mortality is estimated as the fraction of catch to stock, rather than the product of fishing effort and the catchability coefficient. In the ith year

$$C_i = N_i \frac{F_i}{F_i+M}[1-\exp(-F_i-M)], \tag{69}$$

where N_i is the number in stock at the beginning of the ith year.

$$C_{i+1} = N_{i+1} \frac{F_{i+1}}{F_{i+1}+M}[1-\exp(-F_{i+1}-M)]$$

$$= N_i[\exp(-F_i-M)]\frac{F_{i+1}}{F_{i+1}+M}[1-\exp(-F_{i+1}-M)].$$

The ratio of catches in the two years is given by

$$\frac{C_{i+1}}{C_i} = \frac{[\exp(-F_i-M)]F_{i+1}(F_i+M)\{[1-\exp(-F_{i+1}-M)]\}}{F_i(F_{i+1}+M)\{[1-\exp(-F_i-M)]\}} \tag{70}$$

There are three unknowns in this equation, F_i, F_{i+1}, M and for each year of catch added an extra unknown is added but an extra observation on catch is also added. If F and M in one year are given independently, the full set of equations can be solved for a number of years as for example in a cohort or year class.

Gulland (in an appendix to Garrod, 1967) introduced a modification to the method. Catches in numbers are tabulated from the last age group, λ, back in time to the first recruiting age group in the year class. The natural mortality M, is estimated independently and with it, E is calculated because $F = Z - M$.

Then $N_\lambda = C_\lambda/E_\lambda$ and $V_\lambda = (C_\lambda/E_\lambda)F_\lambda/Z_\lambda = C_\lambda/\{1-[\exp(-Z_\lambda)]\}$.

In the subsequent analysis there is an important separation in time between $C_{\lambda-1}$ and N_λ. The catch, $C_{\lambda-1}$, is caught until the end of the year $t_{\lambda-1}$, whereas the stock N_λ is exploited from the beginning of the year t_λ. The ratio $N_\lambda/C_{\lambda-1}$ is expressed in quantities at the same point in time. Then

$$\frac{N_\lambda}{C_{\lambda-1}} = \frac{N_{\lambda-1}[\exp(-Z_{\lambda-1})]Z_{\lambda-1}}{N_{\lambda-1}F_{\lambda-1}[1-\exp(-Z_{\lambda-1})]} = \frac{\exp(-Z_{\lambda-1})Z_{\lambda-1}}{F_{\lambda-1}[1-\exp(-Z_{\lambda-1})]}\cdot \quad (71)$$

From tables of $[\exp(-Z)]Z/F\{1-[\exp(-Z)]\}$, $F_{\lambda-1}$ is determined.

Then $E_{\lambda-1} = (F_{\lambda-1}/Z_{\lambda-1})\{1-[\exp(-Z_{\lambda-1})]\}+\exp(-Z_{\lambda-1})E_\lambda$,

and $N_{\lambda-1} = V_{\lambda-1}/E_{\lambda-1}$; thus estimates of N, E and F are obtained by ages within the year class.

Simple models have shown that as calculation proceeds from the last age group to the first in the year class, the estimates of F tends towards a true value, within the limits imposed by the estimate of M. The adjustment improves as more catches are added. The abundant age groups at the start of the recruited year class are those in which F is well estimated. The method finds its application in stocks with fairly long age distributions. Where measures of stock density are absent and when fish live for a long time in the fishery, the method is of great value. Because the estimate of F improves during the calculation, recruitment is well estimated. Although the variability in stock density is equivalent to that in catch, recruitment is less well estimated in stock density than in cohort analysis because the mortality is more variable. This is because $F = C/N$ in cohort analysis and not $F = qf$, as in stock density studies. Further, the numbers in the stock are estimated absolutely, which has considerable advantages if catch quotas are used in regulation.

Recently, Pope (1972) has developed an approximation:

$$N_i = C_i\exp(M/2)+N_{i+1}\exp(M), \quad (72)$$

$$\text{and } N_i = \{C_i[\exp(M/2)]\}+[C_{i+1}\exp(3M/2)]+[C_{i+2}\exp(5M/2)]+ +\{N_t\exp[(t-i)M]\}, \quad (73)$$

where N_i is the population at the ith birthday; C_i is catch at age i; t is the last age of the year class for which catch data are available, $N_t = C_tZ_t/F_t$ the last age of the year class for which catch data are available,

$$N_t = C_tZ_t/F_t[1-\exp(-Zt)]$$

in the year t and this expression may be substituted in the last term of equation 73. Then $F_i = \ln(N_i/N_{i-1}) - M$ for the year t or for the cohort. Essentially, stock on one birthday represents that on the next added to the fish that have died naturally and the catch (part of which would have died naturally had they not been caught). Pope showed that bias in F is reduced sharply with the number of age groups in the year class and more quickly at higher values of F. The advantage of this method is that the population and its parameters can be disentangled perhaps more readily than with Gulland's method.

The methods described in this section were developed in recent years because in many areas of the north Atlantic (particularly the north-west) catches are sampled but stock densities are not. In long age distributions the estimates of fishing mortality become truer as more catches are added in the year class. An estimate of fishing mortality made from a ratio of stock density is more variable merely because it is a simple ratio and not a self-correcting one.

The dependence of recruitment on parent stock

In the descriptive models the contributions of growth and recruitment were not distinguished and the model improved year by year as additional data were added. The great advantage of the analytical model is that management advice could be given quickly once a fairly small quantity of information became available. The problem of growth overfishing was solved in the yield-per-recruit solution; it is an unhappy accident of history that this solution really generated the problem of recruitment overfishing and a solution to this problem is needed.

Ricker (1954) introduced the problem of stock and recruitment and formulated it:

$$R'/R'_r = w\exp[a(1-w)], \tag{74}$$

where R' is recruitment; R'_r is recruitment at replacement stock, P_r (where $R' = P$); $w = P/P_r$; $a = P_r/P_m$ (where P_m is the stock at which maximum recruitment occurs). The shape of the curve is described in the constant a; if a is small the curve is lightly convex and if a is large it is a dome (Figure 45). It has been used with considerable success in studies of the Pacific salmon, where catches are expressed as $R' - P$ and exploitation rates as $(R' - P)/P$. The equation was developed from

$$R' = \alpha' P[\exp(-\beta' P)] \tag{75}$$

where α' is a coefficient of density-independent mortality; β' is a coefficient of density-dependent mortality. The coefficients cannot be estimated independently, but the equation expresses most simply our intuitive knowledge of the dependence of recruitment on parent stock. The density-

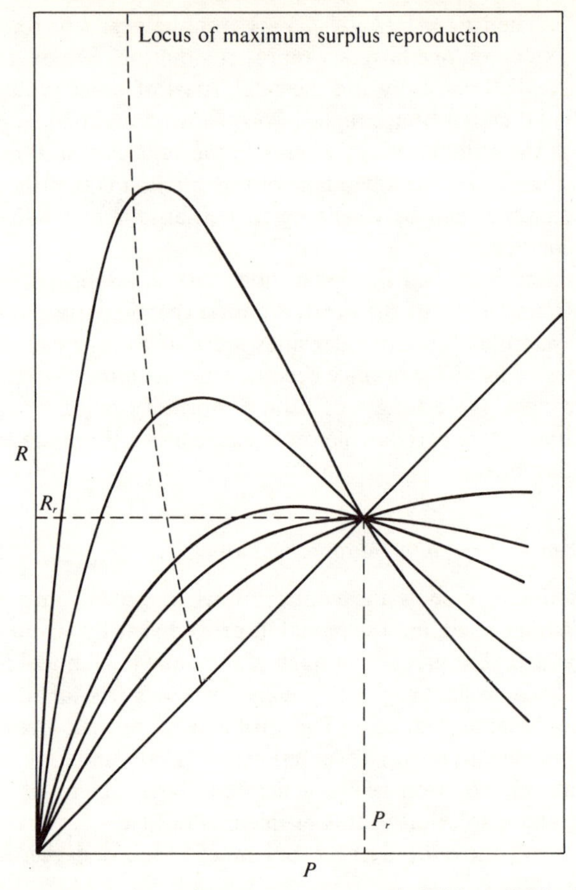

Figure 45. Dependence of recruitment on parent stock (Ricker, 1958); the family of curves represents different capacities for stabilization, ($a = P_r/P_m$) increases with density dependence.

dependent mortality is supposed to be generated by the aggregation of cannibals (or more generally predators) on the eggs or very young larvae and hence a control is established from generation to generation. The interesting point is that the density-dependent function is in mortality and not in fecundity, as in some other animals. However, if recruitment were determined by fecundity (which is a function of stock weight and is thus potentially density-dependent) the natural death rate of fishes would have to be much higher and much more variable than it is presumed to be to generate the observed variation in recruitment by varying the amount of biomass. Fish stocks are most numerous in the very young stages and density-dependent mortality would then be highest.

Ricker developed the equation by maximizing recruitment in terms of stock and by expressing it as a proportion of its replacement, i.e. at which recruitment is equivalent to stock. At stock magnitudes greater than replacement, the ratio is less than stock. At magnitudes less than replacement, it increases to a maximum and at lower magnitudes still, it declines to zero at zero stock.

When $P_r > P_m$, density-dependent processes are important and when $P_r < P_m$, they are less important relative to environmental factors. As density-dependent mortality increases with stock, the curve is generated by the ratio of density-dependent mortality to density-independent mortality. The equation describes how an unfished stock stabilizes itself about its replacement point. Below P_r in stock, recruitment is greater than stock and, after a natural recruitment failure, stock reduces and generates higher recruitment and then stock is returned to P_r. Conversely, after a strong year class, $P > P_r$ and recruitment is reduced until $P = P_r$. These processes constitute the stabilizing mechanism. Cushing (1971*b*) has suggested that at low stock the variance of recruitment may be enough for recruitment to fall below stock (see Figure 69). If two or three such year classes occurred in sequence, stock would trend irrevocably downwards perhaps towards a lower stable level. At high stock ($P > P_r$) a sequence of high recruitments could generate a higher stable stock level, if the environments could sustain it. Such a mechanism explains the expansion of colonizing stocks (like the large-mouthed bass in San Francisco Bay, as noted above) and the restriction of stock numbers under severe environmental conditions (such as the atlantoscandian herring stock). The Californian sardine stock was reduced in the late forties and it assumed a lower level of stock when the anchovy stock replaced it (Ahlstrom, 1966).

Because recruitment is very variable, large quantities of data are needed to fit Ricker's curve and, with one observation of recruitment taken each year, many many years are needed to collect enough material. Any decision on how much fishing should be allowed is delayed for statistical reasons and such delay may inhibit conservation. If recruitment were assumed to be independent of parent stock, the decision could be made more quickly and a start to conservation could be made. Such a procedure can be dangerous because any decline in recruitment is attributed to natural causes rather than to fishing.

The nature of the larval drift was discussed in the last chapter. Three points were noted: (1) it provides a geographical base to the stock in temperate waters where spawning grounds may be relatively fixed in space or in time; (2) numbers may be regulated during the period when the baby fish live in the plankton and (3) during the period of larval drift, differences between stocks of the same species are maintained and competition between species may be at its most intense. In Ricker's first equation, the generation of recruitment is expressed in the simplest possible way, that is, in density-dependent mortality (generated by aggregating predators). There are additional reasons for preferring the initial equation to the developed one. First, Ricker defined the replacement stock, P_r, as that at which stock in the absence of fishing is equivalent to the whole recruited year class between recruitment and extinction. The trend of natural mortality with age in a virgin stock is not yet established, although it is expected to partly follow the Gompertz Law (Gompertz, 1825; Beverton, 1963); that is, natural mortality in the older adult age groups must increase exponentially with age at some stage in the history of a year class. The precise estimate of the replacement stock in a multi-age one is inaccessible at the present time. A second reason for preferring Ricker's initial equation is that in a multi-age stock, the annual recruitment is only a small proportion of the stock; in an extreme example, twenty recruitments are needed in successive years to replace the stock. The stock is in a steady state with its annual increment of recruits equivalent to the annual loss by death and it may be considered to replace itself at any stable level of population.

Ricker (1973) has tabulated statistics for some stock/recruitment curves; for his initial curve (eqn 75), some of the important ones are:

Equilibrium catch, $C_E = P\{[\alpha' \exp(-\beta' P)] - 1\}$.

Exploitation rate, $E = C_E/R'_E = 1 - \{1/[\alpha' \exp(-\beta' P)]\}$.

Maximum equilibrium catch, $C_s = P_s[\alpha' \exp(-\beta' P_s) - 1]$.

Exploitation rate, $E_s = \{1 - [\exp(-\beta' P_s)]/\alpha'\}$.

The stock at which the maximum equilibrium catch is taken, P_s, is estimated from

$$(1 - \alpha' P_s)\alpha'[\exp(-\beta' P_s)] = 1.$$

Limiting exploitation rate, $E_l = 1 - 1/\alpha'$.

The maximum catch and the exploitation rate needed to obtain it are calculable and the limiting exploitation rate is readily determined. If the stock is to survive, the limiting exploitation rate should never be reached and to make a stock secure the variance of E_l should be established in stock units.

Beverton & Holt (1957) developed two stock/recruitment equations. In the first, recruitment is an asymptotic function of stock and the coefficient of density-dependent mortality was not restricted to a very early period in the life cycle. They were aware that the observations on the Georges Bank haddock (Herrington, 1948) were not fitted by their first curve. Their second curve depended on a critical period; if the larval fish grew through this critical period of predation quickly, density-dependent mortality was

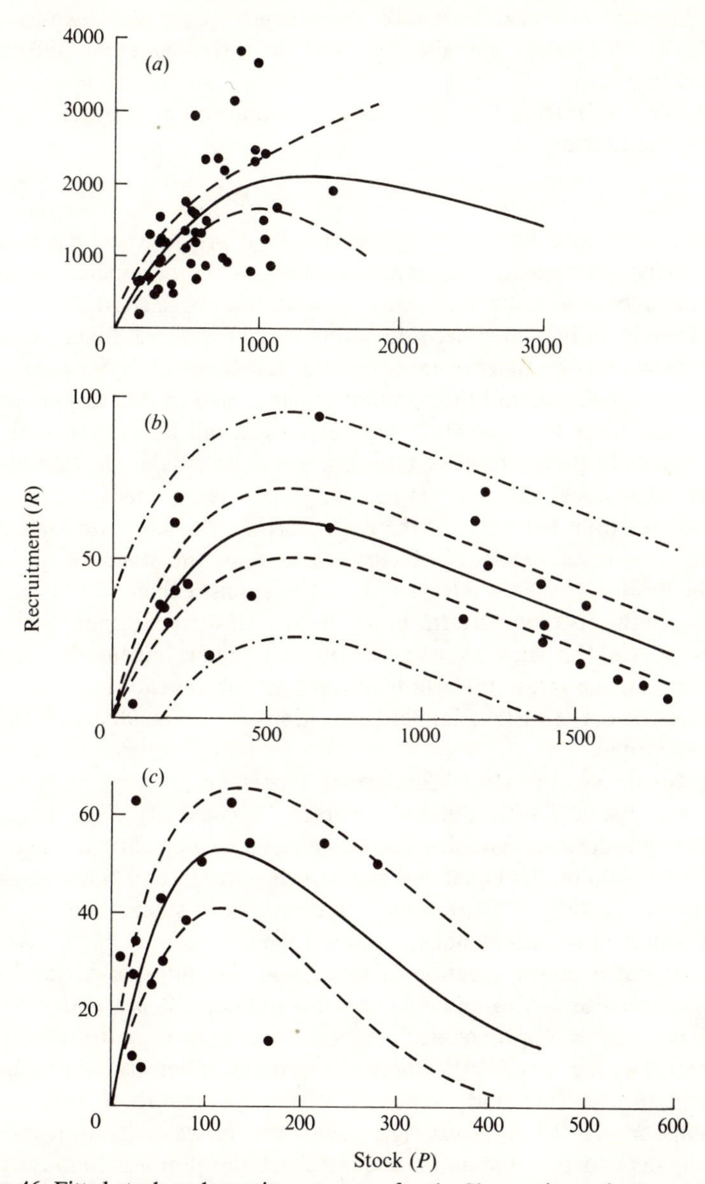

Figure 46. Fitted stock and recruitment curves for the Skeena river salmon, arctonorwegian cod and North Sea haddock (Cushing & Harris, 1973); the dotted lines show the 95% confidence limits to the curve. The broken line shows the standard deviation of the residuals (for the arctic cod only).

low and vice versa because the critical period depended inversely on food and directly on stock. Formally, the second equation is identical with Ricker's initial equation although the derivation is quite different in biological terms.

Cushing & Harris (1973) fitted Ricker's initial equation by least squares, i.e. by minimizing

$$\Sigma[R' - \alpha' P \exp(-\beta' P)].^2$$

Having calculated $b^2 = \Sigma[R' - \alpha' P \exp(-\beta' P)]^2/n-2$ (where b is an index of density dependence) the standard errors of the parameters α' and β' and the calculated value of R at any value of P can be derived. A confidence band can be drawn on either side of the calculated curve. Figure 47 shows such curves with confidence limits fitted to data for the Fraser river salmon, North Sea haddock and the arctonorwegian cod. The dotted lines are the 95% confidence limits to the fitted curve (in a full line); that is, there is one chance in twenty that the fitted curve will lie outside the dotted lines. Hence the stock which generates maximum recruitment, P_m, may be defined if within the range of the data. Further, the left hand limit of the curve is also well defined. The curves also show the standard deviations of the residuals. Where this curve cuts the abscissa there is less than one chance in twenty that recruitment will be zero. If there are enough observations to establish such a curve, the object of management should be to remain near the greatest recruitment; then any observations of recruitment outside the curve of the standard deviations of the residuals must be of natural origin.

In Figure 46, the stock/recruitment curves for salmon, haddock and cod are distinct, with different degrees of convexity. The increasing convexity indicates increasing density-dependent mortality and it rises in the order, salmon, haddock and cod, which also represents an increase in fecundity. Cushing (1971*b*) examined a number of stocks and fitted data of recruitment to parent stock, and used the equation $R = aP^b$. Recruitment is not a power function of stock, but the index, $b<1$ (and tends towards zero or -1.0) and trends in density-dependence with fecundity may be roughly approximated; the equation may be more realistically expressed as $R/P = aP^{(b-1)}$, where b is an index of density dependence. In general, for herring and salmon, $b \simeq 0.5$; for flatfish, $b \simeq 0$ and for gadoids, $b = -0.5$ to -2.0. The coefficient, b, was called an index of density dependence, and an inverse linear relationship was found between it and the cube root of the fecundity. The simplest interpretation of this result is that the magnitude of density-dependent mortality is a function of the distance apart of larvae in the sea. In Cushing & Harris (1973) the same point was made by comparing $(-\beta\bar{P})$ (in $R' = \alpha P \exp(-\beta' P)$) with b for those stocks in which s.d.$_\beta/\beta<40\%$, as shown in Table 11.

Table 11 *Correlation between the coefficient of density-dependent mortality at average stock* $(-\beta\bar{P})$ *and the index of density dependence*

	$\beta\bar{P}$	b
Pink salmon	0.61	0.45
Red salmon	0.77	0.53
Chum salmon	1.13	0.77
Atlantic herring	0.45	0.56
Pacific herring	1.06	0.84
Californian sardine	1.14	0.55
Flatfish	1.02	−0.18
Gadoids	1.75	−2.17
Yellowfish tuna	1.41	−1.34

$r = 0.85$; d.f. $= 7$; $P = 0.01$.

At the mean level of exploited stocks, density-dependent mortality increases with fecundity; as exploitation rates in fecund fish like the gadoids tend to be higher than in the less fecund clupeids, the trend shown in Table 11 is underestimated. Salmon eggs lie for five months or so in the redds and density-dependent mortality has been detected at the early stage in life history (Hunter, 1959; McNeil, 1963). The trend in density-dependent mortality in the purely marine species rises by a factor of three or more.

If the relationship between density dependence and cube root of the fecundity is established, the dome shaped curve of gadoid stock and recruitment is not a special case but is part of a more general one. Then the first equation of Beverton and Holt, the asymptotic one, should be rejected. Although the initial equation of Ricker and the second one of Beverton and Holt are derived from quite different premisses, they are formally identical and can be used directly.

In the previous chapter it was suggested that the natural mortality of some fishes might be expressed as a density-dependent function of age. The best data on the dependence of recruitment on parent stock in a commercially important marine species are those of the arctonorwegian cod stocks (Garrod, 1967). Figure 35(*c*) shows the trend of natural mortality, *M*, with age in the Baltic cod. Figure 46 shows the stock/recruitment curve in weight, the error to the curve (at 95% confidence limits) and curve of the error of the residuals; where the latter cuts the abscissa there is one chance in twenty that recruitment will be zero. At a given value of stock, recruitment can be calculated in numbers and with estimates of natural mortality, fishing mortality and weight at age, the stock can be shown reconstituted. However, $M = Z - F$ but $F \neq Z - M$ where M is derived from the trend shown in Figure 35(c). Therefore the stock, and that reconstituted, are not identical and have to be iterated with different values of F until they are. With this method a yield curve can be constructed directly from the stock/recruitment curve, as shown in Figure 47. The values of fishing mortality obtained in this way are true and unbiased. With a flat-topped stock/recruitment curve, like that of the plaice, the

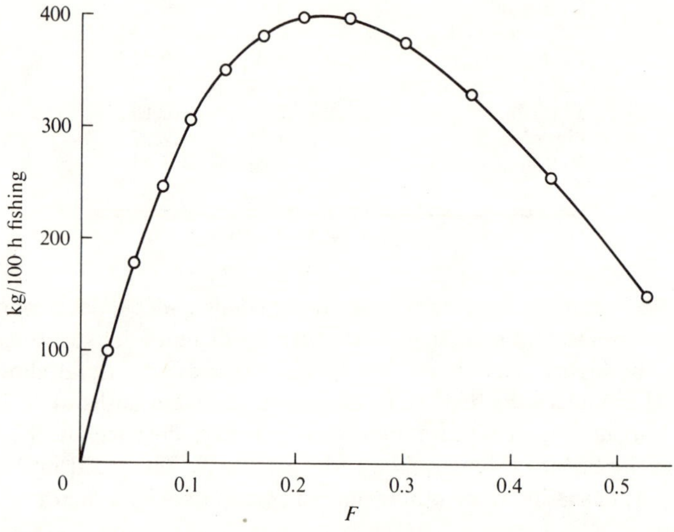

Figure 47. The yield curve derived from the stock and recruitment curve of the arcto-norwegian cod.

maximum of the yield curve and that of yield per recruit are not very different. However, for the herring-like and cod-like fishes the maxima of the yield curves are less than those of the yield-per-recruit ones, because the stock/recruitment curves of these fishes are not flat-topped. Management based on such yield curves demands less fishing activity than those based on yield per recruit curves.

Recently, Harris (1974) has distinguished stock-dependent processes from density-dependent ones; a stock-dependent mortality is generated by the initial number of eggs or larvae, as in Ricker's equation, whereas density-dependent mortality can occur at a later stage in the life cycle. Although the distinction is useful and necessary the term 'density-dependent' has a wider connotation in more general usage. Harris showed that stock-dependent mortality can generate a dome-shaped stock/recruitment curve and that density-dependent mortality can only generate an asymptotic curve. Harris examined the growth during the critical period, t_c, in differentiating t_c with respect to L_∞; then integrating (for a small t_c),

$t_c = L_t/KL_\infty$ therefore $t_c \alpha 1/L_\infty \alpha 1/W_\infty{}^{\frac{1}{3}} \propto$, $1/(\text{food eaten})^{\frac{1}{3}} \propto (\text{density})^{\frac{1}{3}} \propto N_0{}^{\frac{1}{3}}$. Therefore $t_c \alpha N_0{}^{\frac{1}{3}}$. Then

$$R' = \alpha' P[\exp(-\beta' P)^{\frac{1}{3}}]. \tag{76}$$

Again there is some indication that the processes are related to the distance between larvae. Harris then enquired whether mortality could be a function of both stock and density, by differentiating recruitment with respect to the initial numbers and showed that the density-dependent mortality was an essential component in the development. He went on to show that a dome-shaped curve can be generated by a combination of the two forms of mortality. Ricker's curve and the second one of Beverton and Holt are both stock-dependent. The aggregation of predators seems unlikely as a control mechanism because a stabilization mechanism might be independent of other populations. Perhaps the critical period of Beverton and Holt should be extended to account for later changes in growth that might be density-dependent in immature life.

From the shapes of gadoid and clupeid curves, a reduction in stock yields a greater increase in recruitment in the fecund fishes. Under fluctuations in recruitment, the gadoid stocks have greater capacity for stabilizing themselves than have the clupeids. Consequently, fecund fishes recover in stock more quickly. The gadoid stocks tolerate greater rates of exploitation than the clupeids, without endangering recruitment. Rollefsen (1955) has stated that the stock of arctic cod has been exploited in the Vestfjord continuously since the twelfth century. The stock of Norwegian herring, which lives in the Norwegian Sea has suffered periodic eclipses, associated with climatic variations (Beverton & Lee, 1964). The contrast between the two stocks of fish living in the same region confirms the thesis that the fecund animals have greater capacity for stabilization. An extension of the argument is that the elasmobranchs with a few large eggs cannot stand very much exploitation (Holden, 1973).

There are two consequences of the relationship between fecundity and density-dependent mortality. First, if two stocks of the same numbers are compared, the fish with the bigger gonads must generate more larvae in the sea during the period of larval drift. It was suggested in an earlier chapter on quite other grounds, that numbers are regulated at this phase of the life cycle. The second consequence is that growth may be regarded as an agent of fecundity, i.e. fish grow big in order to be fecund in order to stabilize their stocks in adverse environments.

The somewhat artificial distinction between growth and recruitment overfishing is starting to disappear. It originated historically in the development of the small plaice problem in the North Sea and in the related dogma that recruitment was independent of stock within its fishable range. The yield per recruit model developed from the recognition that growth overfishing was the first form to be tackled by international management. Its failure to account for the decline of recruitment under the heavy pressure of fishing was foreseen in the self-regenerating yield model developed by Beverton and Holt. Any contemporary development of an analogous model will dissolve the distinction between growth and recruitment overfishing.

Independent estimates of stock

Stock density is, in principle, a true index of stock but it may be biased. The stock may be improperly sampled by the catch because a proportion is absent or not accessible, or present but not caught, that is, is not vulnerable to the gear. A more general source of variance is the poor recording of catches. Towards the end of the last century, fishing vessels in Britain tended to work a single ground on each voyage and the total catch for the trip by the number of hauls represented a good average catch from the position of the single ground. Today, fishing vessels within the North Sea change grounds frequently within trips and in distant waters in one voyage they may travel from one side of the ocean to the other. There is a distinct need for all individual catches to be recorded properly by positions of capture. Only then can the information be fully exploited.

Very few branches of science depend exclusively upon one form of information; any developed science derives its principles from disparate sources. At certain times in its history, fisheries biology has suffered from anxieties due to its dependence upon catches and stock densities. There were arguments in the past on the need for management, whether stock decline was due to fishing or to natural causes. Today discussion centres upon the nature of the dependence of recruitment upon parent stock. Such debates might have been resolved more readily had there been reliable methods for estimating stock independently of catches.

Such an independent method was devised by Hensen (1887) by counting the number of eggs in the sea. The eggs of most fish species are planktonic and, if sampled properly, their numbers just after spawning divided by the fecundity indicate the numbers of female fish in the stock. Plaice stocks in the southern North Sea (Simpson, 1959), halibut stocks in the North Pacific (Thompson & van Cleve, 1936), mackerel stocks off the eastern seaboard of the United States (Sette, 1943) and pilchard stocks in the English Channel (Cushing, 1957) have all been surveyed by Hensen's method. It has, however, been used infrequently. The spawning grounds and spawning seasons are often not very well known, apart from those noted above and one or two others like the fish stocks sampled off California during the fifties and early sixties (Ahlstrom, 1966) or the arcto-norwegian cod stock in the Vestfjord (Hjort, 1914). Cod spawning grounds are known elsewhere in the North Atlantic, but are not well described, because the eggs of cod and haddock are indistinguishable in the very early stages of development and haddock spawn close to cod. Thus, the most important demersal stocks in the North Atlantic are inaccessible to egg survey.

The second reason why the method of egg survey has fallen into disuse is a statistical one. The error on an estimate of stock made in this way is rather high. Beverton & Holt (1957) estimated that the stock of plaice in the southern North Sea amounted to 3×10^8 to 4×10^8 fish calculated both from stock densities and from egg surveys. The numbers of pilchards in the English Channel amounted to $10^{10} \times 2.2 \times 0.45$ (Cushing, 1957); Saville (1964) has shown that reasonable estimates of fishing mortality for the Faroe haddock can be made by substituting the numbers from egg survey in the catch equation. However, English (1964) has pointed out that the time-curve of egg production is often unknown. The differences in time are the greatest source of variation. However, the true curve of production could be constructed from a study of stock density and a study of maturity stages. Differences in time can then be expressed as deviations from the true curve of production and perhaps the error in time would be reduced.

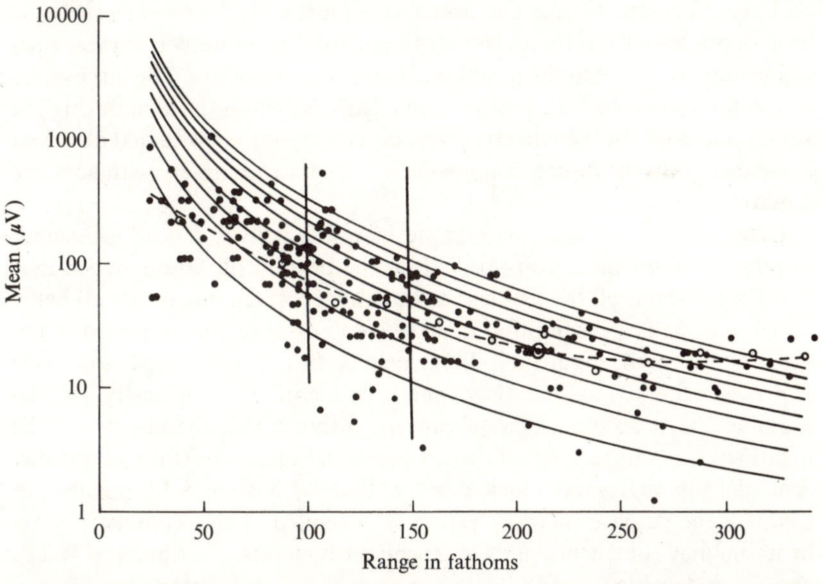

Figure 48. Distribution of signals received from single fishes within one fathom of the sea bed on the continental shelf between Cape Town and Walvis Bay off South Africa. The full lines represent calculated values in range for different sizes of fish between 20 cm and 90 cm. The points represent observed averaged signals. The dotted line represents the trend of average signal with range. The large circle in 210 fathoms shows the average size of fish as sampled during the period of survey (Cushing, 1968*c*).

The method of egg survey provides an independent estimate of stock on the spawning ground. At other seasons on other grounds, no such independent estimate can be made. However, a primitive one can be made with an echo sounder. The method depends on the sonar equation:

$$E_L = S + T_a - 2H', \tag{77}$$

where S is the source level of the transducer in dB ref. 1 μV; T_a is the target strength of a single fish in dB; H' is the one way propagation loss due to spreading or attenuation, in range, in dB; E_L is the echo level observed (Cushing, 1968*c*). The source level is the output of the transducer in units of pressure; the transducer converts electrical energy into acoustic energy in transmission and in reception it converts acoustic into electrical energy. The transducer should be calibrated, but the calibration is fairly stable for quite long periods. A decibel is a unit expressed as a logarithmic ratio. Target strength is the signal received from a single fish at a range of 1 m from it and it is proportional to a power of length of between two and three. Signals received from fish are variable, but with adequate averaging procedures fish of different sizes may be discriminated fairly readily. Figure 48 shows the signal received from single fishes of 20–90 cm with a Kelvin Hughes 'Humber' gear; the points are average observations made on hake off South Africa. The dotted line represents the average signal at each depth and, from this evidence, bigger fish live in deeper water, which is a general rule within the illuminated layers of the ocean. The large circle at 210 fathoms shows the average size of fish caught in the area during the survey and it will be seen that it corresponds fairly closely with that observed acoustically. In the future, fish may be sized quite accurately with acoustic methods.

Given the size of fish, the volume in which it lives can be estimated. Figure 49 shows the density of fish on the bottom off South Africa in a band one fathom off the sea bed; as the trawlers catch practically all hake, it is a reasonable presumption that the stock of hake is represented in the indices shown in Figure 49. In principle, fish can be sized and their densities estimated at sea. They cannot be identified acoustically and the estimates must be on a zone of capture within which the targets can be identified, e.g. one fathom off the sea bed as in Figure 49. At the same time, echo density represents stock density, that is, it should be possible to estimate the catch in a trawl before it is hauled. The technique is only in its infancy, but it provides independent methods of estimating stocks and stock densities.

If it is believed that independent estimates of stock and stock density are needed, the methods of egg survey and of acoustic survey may be used. Neither is fully developed but provides a chance in the future of estimating the catch before it is made and the stock before it is exploited.

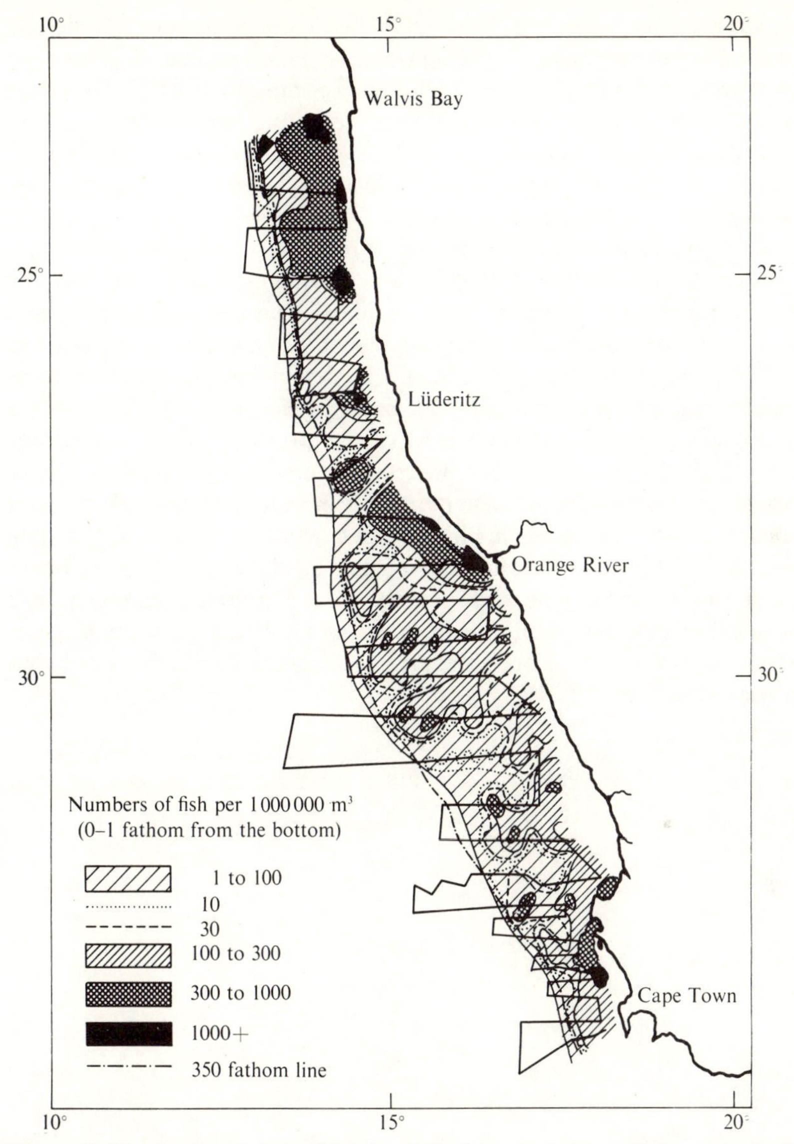

Figure 49. The density of single fishes in $n/10^6 m^3$ in one fathom off the sea bed off South Africa (Cushing, 1968*c*).

Conclusion

The present world catch of fish is about 60 million tons. Gulland (1970*a*) has made an extensively based projection of future world catches and suggests that they will not exceed 100–140 million tons. During the last decade or so, world catches have been expanding at a rate of about 6% per year, so the maximum to be taken with conventional gear will be reached in a decade or so. Even under the best conditions, it may take a number of years after the start of exploitation before reliable assessments can be made. The prime needs are to establish the best ways of collecting the primary data (i.e. records of individual catches by positions of capture) to develop models in which all the vital parameters in the population are adequately described; at the present time, growth is quite well accounted, but the mechanisms of natural mortality and of recruitment are poorly known.

Since the Second World War, the descriptive model of fish populations has been largely replaced by the analytic one developed by Beverton and Holt or its descendants. The logistic model, or Schaefer model, remains useful when the ages of fishes cannot be determined. A recent development is cohort analysis (and its successors), which eliminates some of the earlier biases in that method and which can provide estimates of recruitment and of fishing mortality. In recent years, fisheries biologists have realized that the axiom of constant recruitment within the fishable range of stock can no longer be maintained in all circumstances. Indeed, Cushing & Harris (1973) have shown that, in some stocks, the stock/recruitment curve can be well described. Hence the great recruitment failures in fish stocks should become a thing of the past. As our data improve, more complex models will be developed and greater understanding of the population processes will emerge.

8

Fish stocks and the production cycles

Introduction

In a general sense the numbers in an animal population are regulated by the availability of food. As indicated in an earlier chapter the larval drift in fishes may be the critical period during which numbers are regulated. The production cycle is the source of all food in the sea and the food is most abundant at the lowest trophic level. In an earlier chapter a distinction was drawn between adult and juvenile life: the cohort before recruitment exploits the carrying capacity of the environment and food may then be limiting, whereas in adult life food is conserved for the subsequent generation and does not limit adult growth. The life cycles of some fishes are adapted to the form of the production cycle because the small animals that are the most abundant provide the food for the larval fish. Fish appear to spawn to gain most from the food available in the production cycle.

The different forms of production cycle in temperate and high latitudes, in upwelling areas and in the open tropical ocean were described earlier. Algae can subsist heterotrophically (Fogg, 1966) in the photic layer and far below it in the deep ocean (Bernard, 1957) and at the bottom of arctic lakes (Rodhe, 1963). Plankton animals including fish larvae can probably survive on detritus in mid-water or aggregated by the surface bubble layers (Riley, van Hemert & Wangersky, 1965); indeed Lasker (1970) believes that anchovy and sardine larvae must eat detritus in order to survive. However, production in the sea depends primarily upon the photosynthesis of the planktonic algae and if fishes depend upon the cycle of production their life histories should be linked to the main outburst. Not only are numbers probably regulated during the larval drift but recruitment is mainly determined then. The magnitudes of the recruiting year classes can be forecast from the numbers of metamorphosed fish, so the processes that determine recruitment must have taken place before metamorphosis. In temperate waters the larval drift may be in a fixed position from year to year at the same season; in upwelling areas both position and season may vary, as the plumes drift offshore, and in tropical seas the phenomenon is diffuse and inconstant, but detectable if the scale is broad enough. For example, the tuna that spawn off the Philippines father immature fish in the Kuroshio extension, a larval drift of many thousands of miles. If there is a link between fish populations and production cycles it should be found in temperate seas, and conversely it might be rather difficult to establish in tropical seas.

The link between spawning and production cycles

The plankton recorder survey (now carried out by the Institute of Marine Environmental Research at Plymouth) is based on monthly samples from fixed merchant ship lines across the North Sea, the western approaches to the British Isles, and part of the North Atlantic. The recorder is towed at a depth of 10 m and the plankton is retained on silk mesh of 350–400 μm aperture. Colebrook & Robinson (1965) have studied the seasonal cycles of phytoplankton estimated arbitrarily on the silk as 'greenness' and also those of the older stages of copepods. The spool of nylon moves rather slowly and algal cells are probably retained by a mild clogging towards the end of the passage of the nylon past the filtering aperture. Neither algae nor zooplankton are fully sampled but the average timing of the onset of the production cycles can probably be well described from the trends of the fractions retained. Colebrook and Robinson distinguished three forms of production cycle in the waters around the British Isles. The first is the central North Sea type that is characteristic of the region between the Fladen ground north-east of Aberdeen and the Dogger Bank between northern England and Denmark; it is also found south of Iceland and in the Bay of Biscay. There are spring and autumn peaks of about the same amplitude and the spring one occurs in April or May. The second type of cycle is the coastal Atlantic one found in the English Channel, the Celtic Sea, the northern North Sea and within the Hebrides. The spring peak is early, in March or April, and it is of much greater magnitude than the one in autumn. The third cycle is the oceanic Atlantic type that is found anywhere in the north-east Atlantic off the continental shelf. There is either a late spring peak of high amplitude and an early low autumn peak or a single outburst in mid-summer. As will be shown below the three categories of production cycle are linked to three groups of herring in the north-east Atlantic, autumn, winter and spring spawners. They are distinct races and each comprises a number of stocks; although their genetic characters have not yet been studied to any degree, some of the differences between races are profound, as will be shown below.

In the North Sea, at least, the herring larvae eat some algae as the yolk-sac is being resorbed and as they grow they eat nauplii and young copepodite stages, often of *Pseudocalanus minutus* Boeck (Hardy, 1924). Baby plaice, just after yolk sac resorption, often have large diatoms like *Coscinodiscus* in their guts, that have probably been caught during training attacks (observed in culture; J. D. Riley, personal communication). They take the large algal cells for a few days but the first food on which the little fish (of most species) depend are the copepod nauplii. This conclusion emerges from an analysis of the very extensive work on the food of larval fish by Lebour (1918,1919,1921). If herring populations did depend upon the production cycles their larvae, like those of other fishes, would eat nauplii and young copepodite stages. Larval cod feed almost exclusively upon the nauplii of *Calanus finmarchicus* in the region between Iceland and Greenland (Bainbridge, 1968). The nauplii are hatched at an early stage in the production cycle and indeed their appearance is dynamically linked to the course of the cycle (Cushing, 1959). The adult copepods spawn when the algal food reaches a fixed level (Marshall & Orr, 1952), so the appearance of nauplii indicates not only that level in food, but also a specific point in this production cycle. In the Southern Bight of the North Sea baby plaice and sandeels do not usually feed on copepod nauplii, but on *Oikopleura* (Shelbourne, 1957; Wyatt, 1973). The index of greenness in the plankton recorder samples is probably a sufficient index of the appearance of the nauplii. If the production of fish larvae were timed to that of their food the baby fish would grow with their private food supply, as Jones (1973) has suggested.

The plankton recorder survey grid has been established in the North Sea since 1948 and the monthly samples are expressed in statistical squares in many parts of the north-east Atlantic. In Cushing (1967*b*), the data were grouped in certain smaller squares which were selected along the tracks of the larval drifts of various stocks of autumn, winter and spring spawning herring. The spawning dates of these stocks were known as were the times for larval survival, the periods between the time of fertilization and the point of no return. Blaxter & Hempel (1963) defined this point in development as that by which half the larvae have died of starvation after the yolk has been resorbed. If the survivors are given food few of them can make use of it. The time between the peak date of spawning (if precisely known) and the point of no return is a critical period of survival and there is a distribution of such periods depending upon the spread of spawning time and the temperature at which development takes place.

The herring stocks considered were as follows (Cushing, 1967*a*):

Autumn spawners (including the summer spawners off Iceland)

1. The Dogger herring spawns on patches of rough ground around the north, west and southern edges of the Dogger Bank (they are named the Middle Rough, the North-West Rough, the South-West Patch, the Hospital and the Outer Silver Pit). The stock also spawns on the English coast near Whitby and Scarborough, on the north-east Bank off North Shields

and perhaps on various coastal banks off Lincolnshire. The spawning season lasts from September to October and the larvae drift away eastward across the North Sea towards the Danish coasts.

2. The Buchan stock spawns on gravel banks off the north-east coast of Scotland, on the Aberdeen Bank, at various points in the Moray Firth and on the eastern coasts of the Orkney Islands in August and September. The larvae drift away eastwards across the North Sea probably well to the north of the Dogger Bank and they may reach the Skagerrak, north of Denmark.

3. The Icelandic summer spawners spawn on rough ground near the Vestmann Islands off the southern coasts of Iceland in July and August and the larvae probably drift away in the Irminger current to their nursery grounds.

Winter spawners

1. The Downs herring spawn on patches of gravel in November and December in the Southern Bight of the North Sea and in the eastern English Channel. The names of the grounds are Galloper, Sandettié, Vergoyer, Creux St Nicholas and the Varne. The larvae drift north-eastwards towards the Flemish, Dutch and German coasts.

2. The Plymouth herring used to spawn (and may do so again) between Bolt Head and Bolt Tail off the coast of Devon in December and January. The direction of the larval drift and the position of the nursery ground are both unknown. From the clockwise circulation in the western Channel in spring and summer (Dietrich, 1950), the nursery might be on the north coast of Brittany.

3. The Dunmore herring spawn close to the coast of south-east Ireland (off the town of Dunmore East, near Waterford) and the larvae drift away south-westerly along the Irish coast and presumably the nursery ground lies on the southern coast of Ireland.

Spring spawners

1. The Hebrides herring spawn off the islands for which they are named, in February and March and the larvae drift away north-easterly; with the exception of one ground near Barra Head, the spawning grounds are not exactly known.

2. The Norwegian herring, during the thirties and early fifties, spawned in gravel gullies near the island of Utsira in February and March and larvae drifted north-easterly along the coast; later a spawning further north off Stadt became prominent, followed by one off the Lofoten Islands in the few years before the fishery was extinguished.

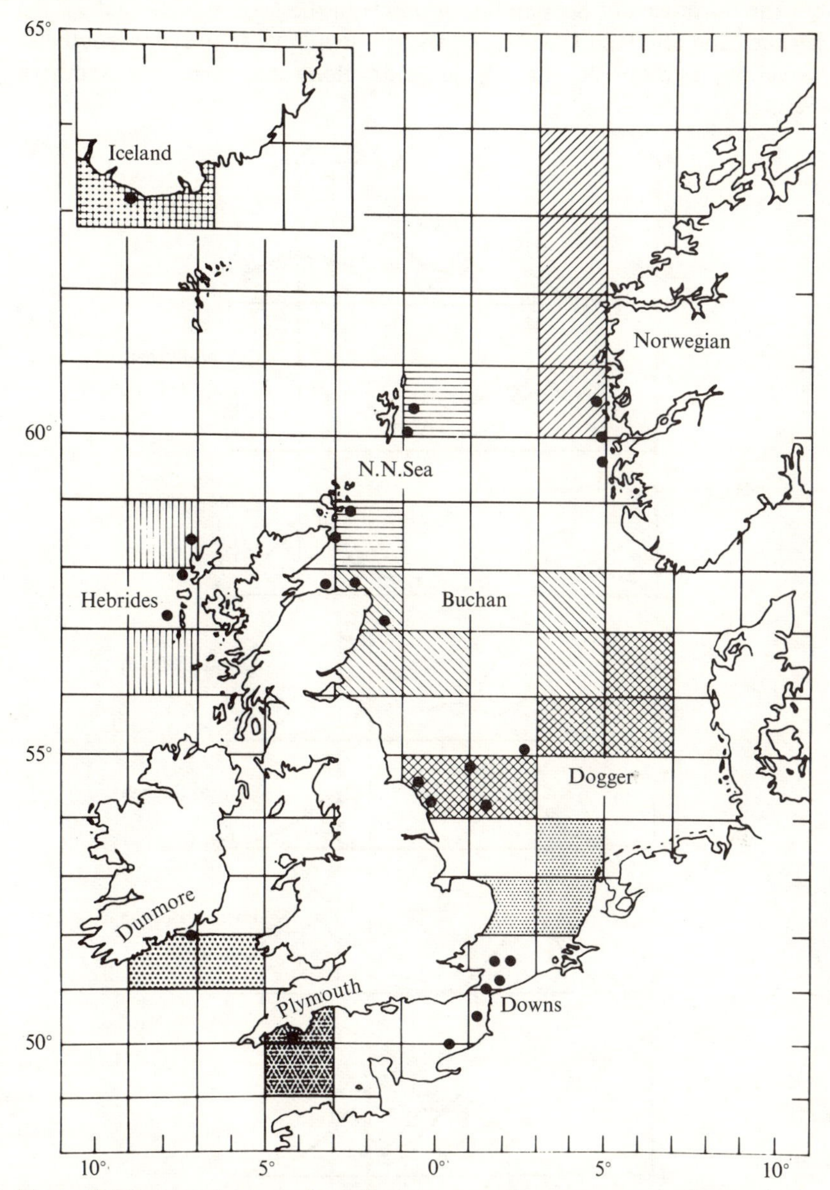

Figure 50. The spawning grounds of all nine groups of autumn spawning herring in the north-east Atlantic; the statistical squares used to sample the nine larval drifts are also shown in appropriate symbols (Cushing, 1967*a*).

3. The northern North Sea spring spawners spawn in March and April off the north-east of Scotland on grounds in the Moray Firth and off the Orkney and Shetland Islands, and the larvae may drift away in a southerly direction, perhaps to nursery grounds along the coasts of southern Scotland.

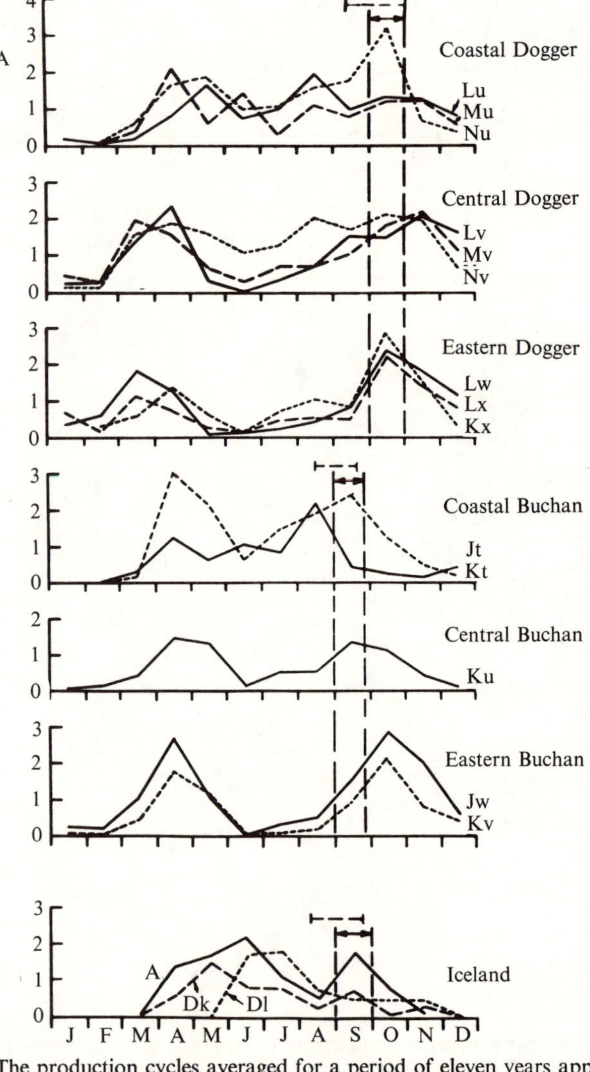

Figure 51. The production cycles averaged for a period of eleven years appropriate to each of the nine herring stocks (Cushing, 1967*a*). The spawning period is shown (|- - -|) and the point of no return (⟷). The letters indicate the statistical squares used (for example, Nw, Ow, Ov for the Downs stock); they are marked but not named in Figure 51.

The spawning grounds of all nine groups are illustrated in Figure 50 together with the statistical squares that were used to sample the larval drifts for the eleven years for which data were available. Figure 51 shows the production cycles averaged for each herring stock. The horizontal broken lines indicate the spawning periods, the median spawning date is shown as full vertical lines and the periods of larval survival are shown as arrows. If the larvae have the opportunity to feed before they reach the point of no return, they will survive. If the production of larvae is matched

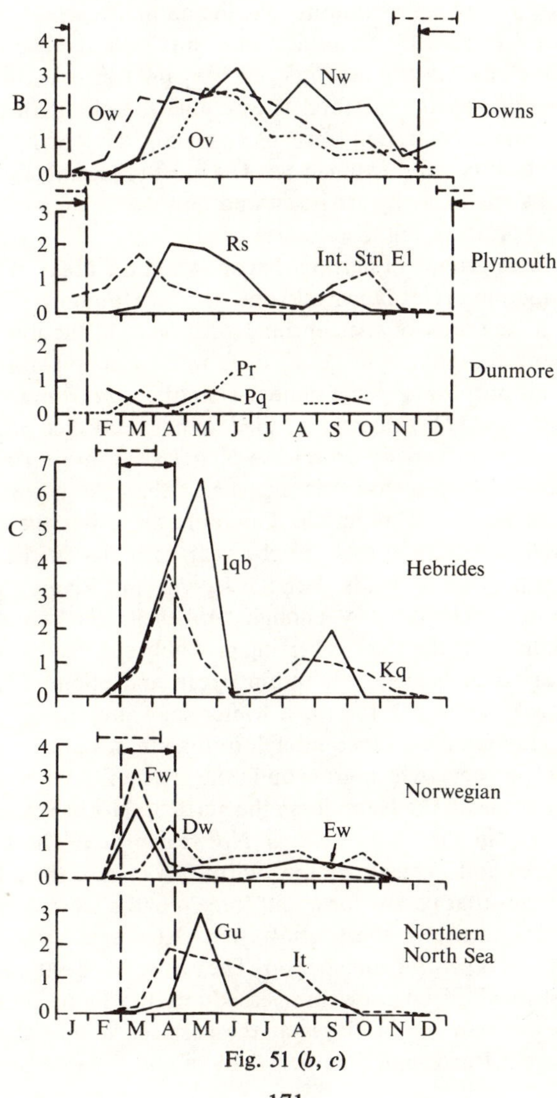

Fig. 51 (*b*, *c*)

to that of their food, it might be estimated in the distribution in time of the periods of larval survival.

The autumn spawners are linked to the central North Sea production cycle. In the western, central and eastern Dogger the average period of larval survival ends at or near the peak of the autumn outburst. As the larvae drift across the North Sea towards the Danish coast, food is available for three months at least. In the central and eastern Buchan area, food is also always available and the peak amplitude of the production cycle shifts later by about a month as the larvae drift eastwards. Food is available for about three months, but the autumn cycle is earlier than that in the Dogger area by about a month, just as is the spawning time. Off Iceland, the autumn production cycle is again a little earlier, as is the date of spawning; again, the food is available for about three months. There is a detailed correspondence of spawning date with the average timing of onset of the production cycle. The most remarkable point is that the autumn-spawned herring are not found outside the area of the central North Sea except off Iceland.

There are three groups of herring that spawn in the areas of the coastal Atlantic production cycle: Downs, Plymouth and Dunmore. The data are poorer than in the area of the central North Sea production cycle. The observations of the Dunmore larval drift were probably taken too far offshore for this purpose. For the same reason, the data from International Station E1 (Harvey, 1934) have been added to those collected off Plymouth. The winter spawning herring do release their larvae into an early spring production cycle, three to four months earlier than the oceanic one. No conclusion can be drawn about the Dunmore area, but the production cycle may well occur quite early in the year as it lies in the lee of the prevailing south-westerly winds. For the Downs and Plymouth area, the coastal Atlantic cycle is early enough to benefit the winter-spawned larvae, particularly as the Downs herring probably spawned a month later before the period of heavy exploitation began and eliminated the older fish in the stock. Thus, for the three winter spawning groups of herring there is a production cycle close enough in time to sustain it.

The spring spawners, which grow up in the areas of the oceanic production cycle, are those off the Hebrides in the northern North Sea and atlanto-scandian herring in the south-western Norwegian Sea. Those spawning off the Hebrides and in the northern North Sea release their larvae into production cycles that persist for about three months from April or May. The cycle starts earlier in the south-western Norwegian Sea, in March, and it persists for a shorter time period. The larvae of the atlantoscandian stock that spawned off Utsira were released into the Baltic outflow which is a cool, freshish and stable layer at the surface. Within this layer the production cycle starts much earlier than in the oceanic areas further offshore (Steele, 1961*a*). Then the larvae drift north and north-easterly out of the Baltic outflow into the Norwegian Sea, a region where production must start later because the water is deeper. The spring spawning groups of herring are associated with the oceanic production cycles.

The nine groups of herring listed above include all the important stocks in the north-east Atlantic. They have long been grouped as autumn and spring spawners (Johansen, 1924) and in recent years the third group of winter spawners has been added (Parrish & Saville, 1965; Cushing, 1967*b*). The three groups are associated with the three forms of production cycle, autumn spawners with the central North Sea cycle, winter spawners with the coastal Atlantic one and spring spawners with the oceanic type. Temperate fishes spawn at fixed seasons in the same positions from year to year and the larval drift is of great biological importance as a geographical base for the stock and as the period during which much of the natural regulation takes place. The evidence presented here secures the argument because the link between spawning and production is established. If spawning is fixed in space and time, the variability of the production cycle is contained to some degree and the little fish take the best advantage of it. If the stocks depend upon the production cycles in an evolutionary sense, the greatest part of the numerical control must occur during the period of larval drift. To turn the argument upside down, it follows that fish in temperate and high latitude seas are bound to return to their native grounds to spawn.

The same arguments cannot be sustained in the same detail for demersal fishes for two reasons. They do not spawn on the bottom on grounds that can be specified precisely from year to year. With one or two exceptions the spawning grounds of demersal fishes are generally not very well described; for example, the best information on cod spawning grounds in the North Sea is that of Graham (1934) based partly on maturity stages and partly on the distribution of gadoid eggs in the international cruises in the first decade of the century; however, work in progress at present will chart the cod spawning grounds. Among demersal fish, as with the north-east Atlantic herring, there are spring and summer spawners. In the Southern Bight of the North Sea, sandeels, whiting, plaice, dab and cod (Simpson, 1949*b*) all spawn in December, January and February and the implication is that the larvae drift away towards the Dutch and German coasts and enjoy the same production cycle as the Downs herring that spawn in the same region.

The mean peak spawning date of the plaice in the Southern Bight of the North Sea is 19 January (Cushing, 1970) and the eggs take about seventeen days to hatch, depending on the temperature. On average, the larvae need food after the first week in February and Figure 51 shows that the production cycle (i.e. that quoted for the Downs herring) has started by then, on average. This plaice stocklet is linked to the same production cycle as is the local stock of herring.

Cushing (1972*a*) estimated the critical depth in the Southern Bight using Steele's (1965) formula (i.e. eqn 6) and measurements of transparency (Lee & Folkard, 1969). It was found that production should start from the Dutch coast outwards and, if the weather were calm, it should start in late January in a depth of 20 m and in late February in 40 m. Joseph (1957) has published a vertical section of particle content in the water off Texel Island. Particles of sand from the bottom were swept more than half way up the water column by tidal mixing, so algal cells must be distributed throughout the column. Then the depth of mixing is the depth of water. Because turbidity increases with wind strength, production is limited to the shallower water in rough weather, and when the weather is very rough there is probably no production at all. With the use of Strickland's (1960) estimate of energy at the compensation depth, D_c, the production ratio (D_c/D_m) is effectively (D_c/Z) where Z is the depth of water. In calm weather the ratio can only increase slowly with increased radiation, and in rough weather its development is delayed.

If there is a link between recruitment and production a negative correlation would be expected between wind strength and the magnitude of the year classes. For the years 1929–50 recruitment of plaice was inversely correlated with the wind strength in March, irrespective of direction ($p = 0.05$; $n = 17$); a limited quantity of material has been used because, before 1950, the Lowestoft fleet operated in the Southern Bight only, but since then the fleet has exploited plaice elsewhere.

There is a little evidence that the recruitment of the plaice can be described roughly with a model in which the production cycle is modified by variability in wind strength; there is a link between the stock of fish and the production cycle. Similar links might well exist between the same cycle and the other species of fish that grow up in the same region. After the cold winters in north-west Europe there are strong year classes of plaice, cod and whiting. The fish larvae develop more slowly and, because the cold weather is anticyclonic, the sea is calm. Hence, the production of larvae is delayed (particularly as larval development is a power function of temperature) and the production of their food is accelerated; perhaps an optimal match of larval production to that of their food is obtained.

From weekly ring trawl catches at International Station E1, Russell (1930) published a diagram of the seasonal occurrence of post-larvae of a large number of fish species. In 1933 and 1934, the algal production cycle peaked in early April (Harvey, 1934) and the nauplii and copepodites appeared about three weeks later, that is, at the end of April or the beginning of May. Figure 52 shows the average distribution in time of fish larvae, all of which peak at this season. Those that depend upon the spring production cycle include *Gadus minutus* L., *G. callarias*, *Platichthys flesus* (L.), *Solea solea* (Risso), *Solea variegata* (Fleming), *Onos* sp., *Gadus merlangus* L., *Limanda limanda* (L.), *Zeugopterus punctatus* (Bloch), *Microstomus kitt* (Walbaum), *Phrynorhombus norvegicus* (Gunther), *Ammodytes lanceolatus* Lesauvage and *Trigla* sp. Note; the names of fishes are those given in Russell's original figure, e.g. *G. callarias*, the cod, is now *G. morhua*.

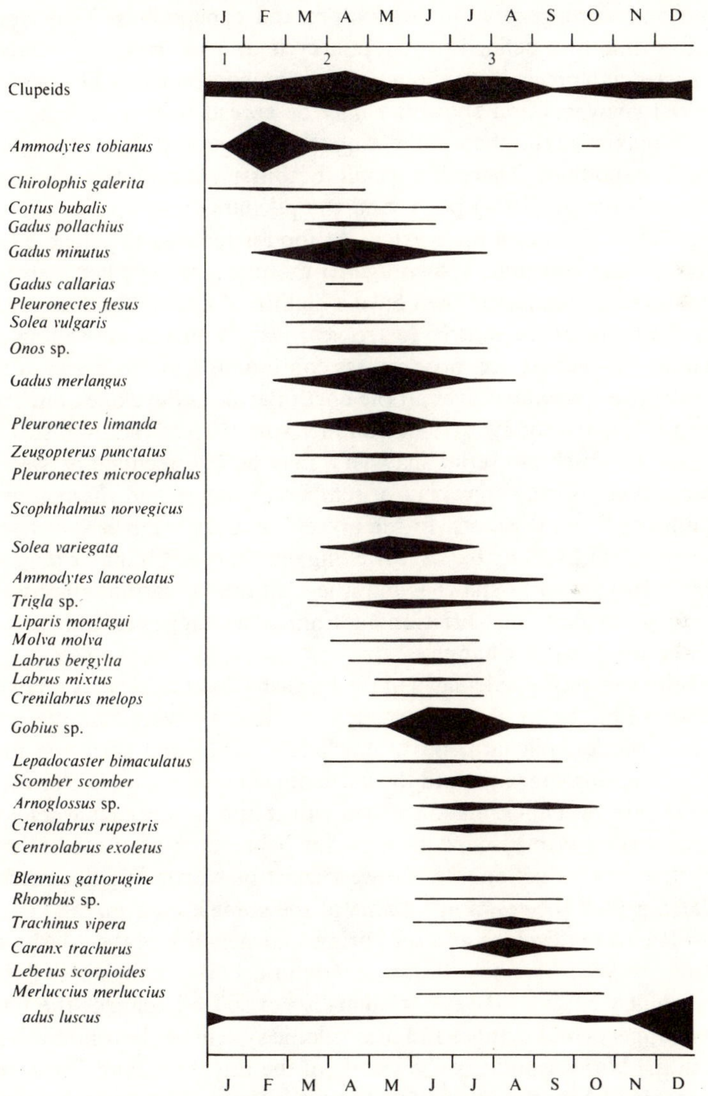

Figure 52. The distribution in time of the larvae of a number of fish species in the western English Channel (Russell, 1930).

Because the gadoids and the flatfish are well represented in this list, most spring spawners in temperate and high latitude waters depend upon a spring production cycle in much the same way as do the spring spawning herring.

The summer spawners include *Sardina pilchardus* (Walbaum), *Callionymus* sp., *Scomber scombrus* L., *Ctenolabrus rupestris* (L.), *Ctenolabrus exoletus* (L.) and *Trachurus trachurus* (L.). The algal production cycle has passed in April, but during the period of May to August there are the second and third generations of nauplii and copepodites. They are not linked in time to a defined production cycle and so spawning would not be expected at a precise time. Some of the summer spawners, like pilchards, are serial spawners and spawning may be spread over a long period in order to maximize the chance of eating the later generations of nauplii and young copepodites. There are small outbursts during the summer as shown in Harvey's (1934) paper and the pilchards may spawn after they have aggregated on to a food patch. As food is reduced the algae increase briefly and the copepods spawn again; the pilchard eggs hatch in forty-eight hours and the larvae then have a chance of finding food before they reach the point of no return. In tropical waters outside upwelling areas spawning and feeding are more or less continuous and the larvae of tuna-like fishes are distributed over all the north Pacific anticyclone at nearly all seasons (Matsumoto, 1966). The summer spawners of the English Channel, most of which are serial spawners, may be intermediate between the precisely timed spring spawners of temperate waters and the continuous spawning of tropical waters. In the upwelling areas some fish may spawn at the point of upwelling but as upwelling may vary in timing or may occur intermittently, a serial spawner like the Californian sardine may well be adapted to exploit the variation in timing, as suggested above for its cousin in the English Channel.

In British waters, particularly in the English Channel, there is an overlap in distribution between northern species, herring, cod and plaice and southern species, pilchard, horse mackerel, turbot and sole; the former tend to be spring spawners and the latter are summer spawners. The spring spawners are probably linked to the distinct spring outburst whereas the summer spawners may depend upon small predator–prey oscillations in the system of production. On the west coast of North America there is a similar array of species, and in general the summer spawners live in the North Pacific anticyclone and the spring spawners live in the Alaska gyral (Ahlstrom, 1966; Alverson, Pruter & Ronholt, 1964). In the north Atlantic the boundary between the two systems is not so well defined; the Canary current appears off Portugal and the cyclonic system lies beyond the British Isles in the Norwegian Sea. The waters of the English Channel may represent a region where the two biological systems meet. The southern summer spawners, serial spawners, may be characteristic of steady-state production cycles, whereas the northern spring (and autumn) spawners may be characteristic of the discontinuous high amplitude production cycles of high latitudes.

The adaptations of the fishes

Colebrook (1965) has suggested that because the variability of algal production increases with depth that of copepod production must do so as well. If the magnitude of recruitment depends upon the food available, then the year classes should become more variable with increasing depth of water in the region of the larval drift. In those stocks that survive in deeper water, the individuals tend to live longer and the stock supports itself from one good year class to the next; for example, the 1904 year class was the mainstay of the atlantoscandian herring for nearly twenty years. When fish live long they attain great size but their growth rates tend to be slow. The growth characteristics of winter, autumn and spring spawning herring in the north-east Atlantic are shown in Table 12. The greatest recorded age, T_{max} is less in an exploited population than in a virgin one; all nine stocks were exploited, but the true differences in T_{max} are probably greater than those due to exploitation.

The variation in onset of the production cycle depends on the depth of water over which the larvae drift. The coastal Atlantic type, in which the winter spawners grow, is found in the shallowest water; the central North Sea cycle, in which the autumn spawners develop, is found in moderate depths and the oceanic cycle, that supports the spring spawners, occurs in deep water. As variance in timing and amplitude of the production cycle increases with depth so does T_{max}, L_{∞} (and incidentally the age of recruitment, not shown in Table 12), whereas for the same reason K declines. The spring spawners live twice as long as the winter spawners, but their maximum length is only one-fifth greater. However, because herring do not grow much in their adult lives such a difference in weight is considerable. The spring spawners grow at about half the rate of the winter spawners and they recruit and mature at a later age. When fish recruit partially at four, five or six years of age the variability of year classes is damped in the age distribution to some extent. If the variance of recruitment increases with depth, it is met partly by the slower growth rate and partly by the partial recruitment over three or four age groups.

There are also reproductive differences between the stocks in fecundity, egg size, gonad weight and in the length of maturity stage V, which is the immediately pre-spawning stage in the cycle of maturation. Fecundity and gonad weight are functions of body weight but there are considerable differences in egg size between the autumn spawners and the others; indeed the spring and winter eggs are twice the volume of the autumn eggs

Table 12 *Biological differences between the autumn, winter and spring spawning herring in the north-east Atlantic*

	Winter	Autumn	Spring
Growth			
T_{max} (yr)	12	16	23
L_∞	28.5–29.5	30.5–31.5	35.0–37.0
K	0.35–0.50	0.35–0.43	0.17–0.25
Growth rate	Fast	Moderate	Slow
Reproduction			
Gonad weight at T_{max} (g)	44	55	88
Fecundity at mean length ($n \times 10^3$)	38	80	51
Egg diameter (mm)	1.4–1.5	1.1–1.2	1.5–1.7
Duration of maturity stages IV and V	Long	Short	Long
Production cycle			
Depth of water on larval drift	Shallow	Moderate	Deep
Variability in timing of production cycle	Low	Moderate	High

(Hempel, 1967). The autumn egg is small because there is little yolk and this is presumably linked to the high availability of food for the autumn spawning herring in the central North Sea production cycle. Gonad weight increases with L_∞, but the autumn spawners are much more fecund than the other two groups because the eggs are smaller. Then, if density dependence is a function of fecundity as suggested in an earlier chapter, the autumn spawners should have a greater capacity for stabilization and be more efficiently insulated from environmental changes. The maturity stages IV and V are short amongst the autumn spawners and long in the winter and spring spawners. All herring feed most during the spring outburst and, during the early maturation stages, the development of the gonads proceeds at about the same rate in all three groups of spawners (Iles, 1964). The autumn spawners pass a month or so in stage V, but the winter and spring spawners pass many months in the same stage waiting for the time to spawn. The pre-spawning maturity stages and the egg sizes are reproductive characteristics that are independent of the growth indices. Such reproductive differences, like the differences in growth, are profound ones that indicate a considerable degree of isolation.

There are differences between cod stocks, in the Barents Sea, off Iceland, Newfoundland, Labrador, Greenland, Faroe Islands, in the North Sea, the Baltic Sea and the Celtic Sea. The biological information is not so detailed as that for the herring populations, but there are differences in growth rate and age of maturation which vary with latitude. The cod in the Celtic Sea grow to 70 cm in three years and mature then, whereas in the Barents Sea a 70-cm cod starts to mature at seven years of age. In the benign and southern environment the age distribution is short, up to eight years of age and the year classes are not very variable. In the harsher northern seas the age distribution is long, more than twenty years, the growth rate is slow and the recruitment varies greatly. There are stocks intermediate between the two extremes and they are all probably linked to production cycles in the same way as the herring. This brief account is taken from Letaconnoux (1954) and Rollefsen (1954).

Within any species of fish there are often considerable differences in the characteristics between stocks, particularly in growth and fecundity. Herring stocks in harsh environments grow slowly to considerable ages and so their variable recruitments are spread amongst a number of age groups. The same effect can be detected in an array of cod stocks. The reproductive capacity of fish stocks seems to be linked to their capacity to stabilize their numbers and so the fecund cod is more stable than the less fecund herring. Even within the groups of herring stocks the autumn spawners appear to be more fecund and more stable than the winter ones.

In the north-east Atlantic, the production cycles are discontinuous and differences in adaptation of growth and fecundity to them bear little relation to adaptations in tropical or subtropical waters. The tuna are very fecund and their spawning seasons are not well-defined, but they may not live for very long, as compared with cod, size for size. Because of the short life span, the variability of recruitment is probably less than that of the cod; if fecundity is high, as it is, the index of density dependence is high and so their capacity for stabilization must be considerable. In the unexploited state the variations in tuna populations may be low, as might be expected in the continuous production cycle of tropical and subtropical waters.

Three groups of fish have been examined. All three survive in deep water, but their capacity for stabilization increases with fecundity, that is from herring to cod and from cod to tuna. The winter spawning herring that live in shallow seas grow quickly but are not very fecund. Autumn and spring spawning herring are both more fecund; the autumn spawners sacrifice yolk to obtain high fecundity, whereas the spring spawners obtain the same effect by growing more slowly. Similarly, cod survive in the Barents Sea by growing much more slowly than they do in the Celtic Sea. With the probably greatest capacity for stabilization tuna live in the subtropical ocean and do not penetrate high latitudes. Perhaps the ratio of growth to mortality in the younger stages is greatest in the fecund fishes and in the course of evolution tuna have excluded the cod and herring from the subtropical ocean.

The match of spawning to the production cycle

The variability in timing of the production cycle is governed by two factors, the date at which the critical depth exceeds the depth of mixing and the rate at which the production ratio develops. In the temperate spring, the radiation received by the water increases during the season but wind tends to decline. The two climatic factors, wind strength and radiation, govern the starting date of production and the rate at which it continues. As a rough guide, the algal reproductive rate, as modified by these factors, increases by about an order of magnitude during the first half of the year.

Figure 53 shows how the amplitude, spread and timing of the production cycle might vary with the development of the production ratio. The algal

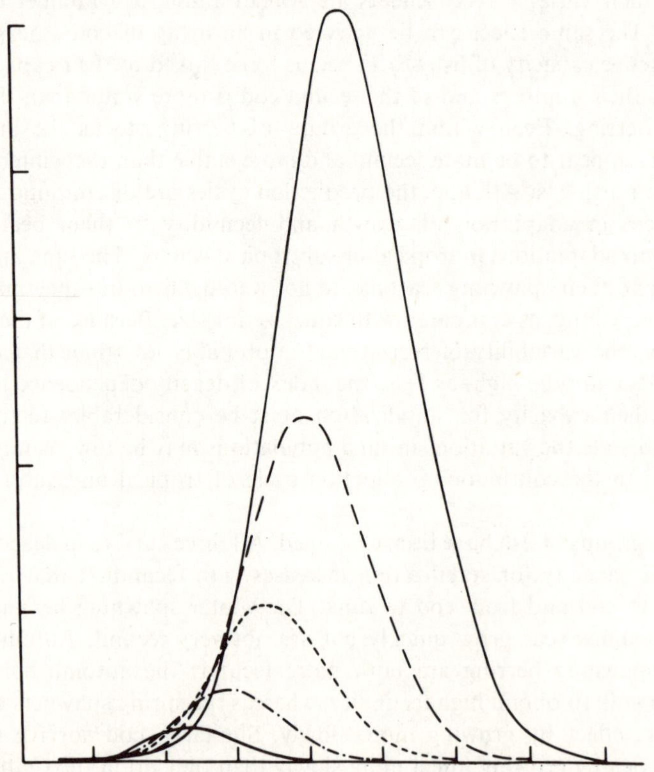

Figure 53. The variation of amplitude, spread and timing of onset of the production cycle with the rate of development of the production ratio, D_c/D_m (D_c is the compensation depth and D_m is the depth of mixing; Cushing, 1973*a*).

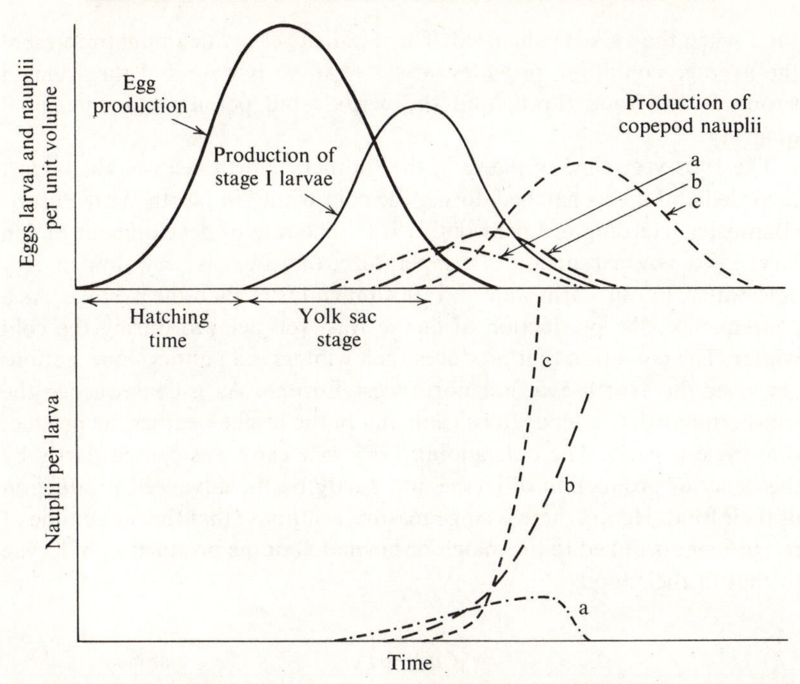

Figure 54. The match or mismatch of larval production to that of their larval food. The numbers of nauplii/larva represent the degree of feeding success. The three curves represent three conditions of production and hence three conditions of feeding success (Cushing, 1973*a*).

reproductive rate was increased from 0.1 to 1.0 in steps of a variable number of days (2.0, 2.5, 3.0, 3.5 and 4.0). Grazing mortality was started at $R = 0.35$ and it was increased also in steps to equal R at $R = 1.0$ and subsequently to exceed it by 25%. The required bell-shaped curve was simulated and the cycle was controlled when $R < G$. In the figure the timing of production is delayed as the production ratio develops more slowly (with longer steps in the development of the algal reproductive rate) and both amplitude and spread are modified, probably in an exaggerated manner.

A picture of the effect of match or mismatch is given in Figure 54. The larval production is shown as a distribution in time. Nauplii are represented as the main food of the fish larvae and the number of nauplii/larva represents the degree of feeding success under each condition. In the early production cycle, food is available early and it increases in time at a low level. In the late cycle food is available later but increases quickly to a high level. In the early cycle, all survival may be low because although the larvae can feed they do not grow very fast and they may be subject to predation. In the late cycle, survival is initially very low, but is subsequently high; it is possibly dangerous because the larvae might not find food when the yolk is exhausted. The middle cycle, which might represent the average condition, provides larvae at an early date and they remain supplied with food throughout the period – but perhaps not a maximal quantity.

The 1963 year class of plaice in the southern North Sea was the largest recorded and it was hatched during the cold winter in North West Europe (Bannister, Harding & Lockwood, 1974). The rate of development of fish larvae is a power function of temperature, that is, it is very slow in very cold water, but in warm water it is not much faster than the average. As a consequence, the production of larvae was well delayed during the cold winter. The condition that produces such winters is an anticyclone stationary over the North Sea and north-west Europe. As a consequence the southern North Sea tends to be calm and in the bright weather the production cycle is early. The outstanding 1963 year class was caused partly by the delay of production of larvae and partly by the advanced production of their food. Hence, there is some reason to suppose that the magnitude of recruitment is linked to the match or mismatch of the production of larvae to that of their food.

Conclusion

The herring stocks in the north-east Atlantic are linked to their local production cycles. Autumn, winter and spring spawning herring relate to the three distinct forms of cycle that are themselves found in three different depth regions of the sea. A whole array of fishes other than herring may be linked to the same production cycle as the herring in the southern North Sea and this correspondence may well occur in other areas. Similarly, the spring spawning fishes, both demersal and pelagic, in the English Channel, appear to be linked to the spring outburst, like that off Plymouth. The herring were shown to be adapted quite closely in growth and reproduction to the variability of the cycles and perhaps the cod were similarly adapted.

The fish examined were the cod and herring of the north-east Atlantic, the biology of which is well known, as is the distribution of production cycles. In tropical waters, both within and outside the upwelling areas, the link with cycles of production is probably quite different. In the deep ocean of the subtropical anticyclones no timing link is needed because food is continuously available. The phenomenon of serial spawning may be an adaptation to this condition. In this view the summer spawners of the English Channel may be strays from the subtropical anticyclone; the alternation of pilchard and herring (Cushing, 1961; see below), southerly and northerly species, anticyclonic and cyclonic species, suggest that both are living on the edges of their range.

The production cycle in an upwelling area is in structure a temperate one, but it does not necessarily occur at exactly the same season. If the fish are serial spawners they can exploit such an extended season very well. In fact, the Californian sardine spawns in any of the months of spring, summer and early autumn. This suggests that serial spawning occurs and is accelerated when food becomes superabundant. It is then a tropical adaptation to a temperate production cycle, of high amplitude, but is highly variable in timing.

The match/mismatch thesis, combined with that of density-dependent mortality modified by the availability of food, provides some basis for a theory of the generation of recruitment. It is reasonable to suppose that the magnitude of recruitment depends upon food availability. The importance of the match/mismatch theory lies in showing how a fish stock can respond to climatic variation and to climatic trends. In distinguishing stock dependence from density dependence, the responsive function of the latter is contrasted with the conservative function of the former. Climatic variation is perhaps exploited by the exploratory density-dependent processes, but the conservative stock dependence provides some insulation from climatic trends. Indeed, the combination of the two processes might be considered to rectify the long periodicity of climatic change.

9

Food webs in the sea

Introduction

The study of food webs in the sea probably started when Petersen (1918) established that fish production depended on the benthos. Any food comprises many species at each food level or trophic level, a category of animals with common food choices in the Eltonian pyramid of numbers. This concept presupposes that animals can be satisfactorily classed as herbivores, carnivores and so on. Some animals can be so classified quite easily but others, for example, copepods, may be herbivorous or carnivorous by opportunity. Hence, the distinctions between trophic levels can become blurred. Historically the study of food networks has developed in two ways, first in the recognition of a general yet simple chain-like structure and secondly in the establishment of the detailed and complex web-like structure.

Herring eat copepods and in their turn they eat algae. Stated in this way it is a simple chain, but Hardy's (1924) classical diagram (Figure 55) shows how complex it is in detail. To start with, the larval herring eat *Coscinodiscus* (a diatom), *Peridinium* and *Prorocentrum* (dinoflagellates) and later, larval molluscs, tintinnids and the nauplii and copepodites of *Pseudocalanus* and of some harpacticoids. The post larvae feed on *Pseudocalanus* and *Temora* and the juveniles take *Pseudocalanus*, *Acartia*, *Temora*, decapod larvae and mysids. The adult herring eat young sandeels, arrow worms, the copepods *Calanus* and *Temora*, the cladocerans *Evadne* and *Podon*, the euphausid *Nyctiphanes*, the amphipod *Apherusa* and the larvacean *Oikopleura*. As the fish grow, their prey become larger and during their life cycle each animal is supported by the food web and similar temporal arrays must support each other animal in the sea.

In more general terms than Hardy's detailed account, Blegvad (1916) showed how the food fishes, flounder, dab, plaice, cod and eel all depended upon crabs, the little fish of the eelgrass community, and upon a benthos composed of *Mytilus*, *Mya*, *Macoma*, *Pectinaria*, *Cardium*, *Arenicola*, *Nereis* and *Nephthys*. The animals of the eelgrass community eat copepods and other small crustacea, *Rissoa* and *Hydrobia*, and the benthos community subsists on detritus (and perhaps phytoplankton). As in the herring food web, the copepods eat the planktonic algae. It is a more extensive

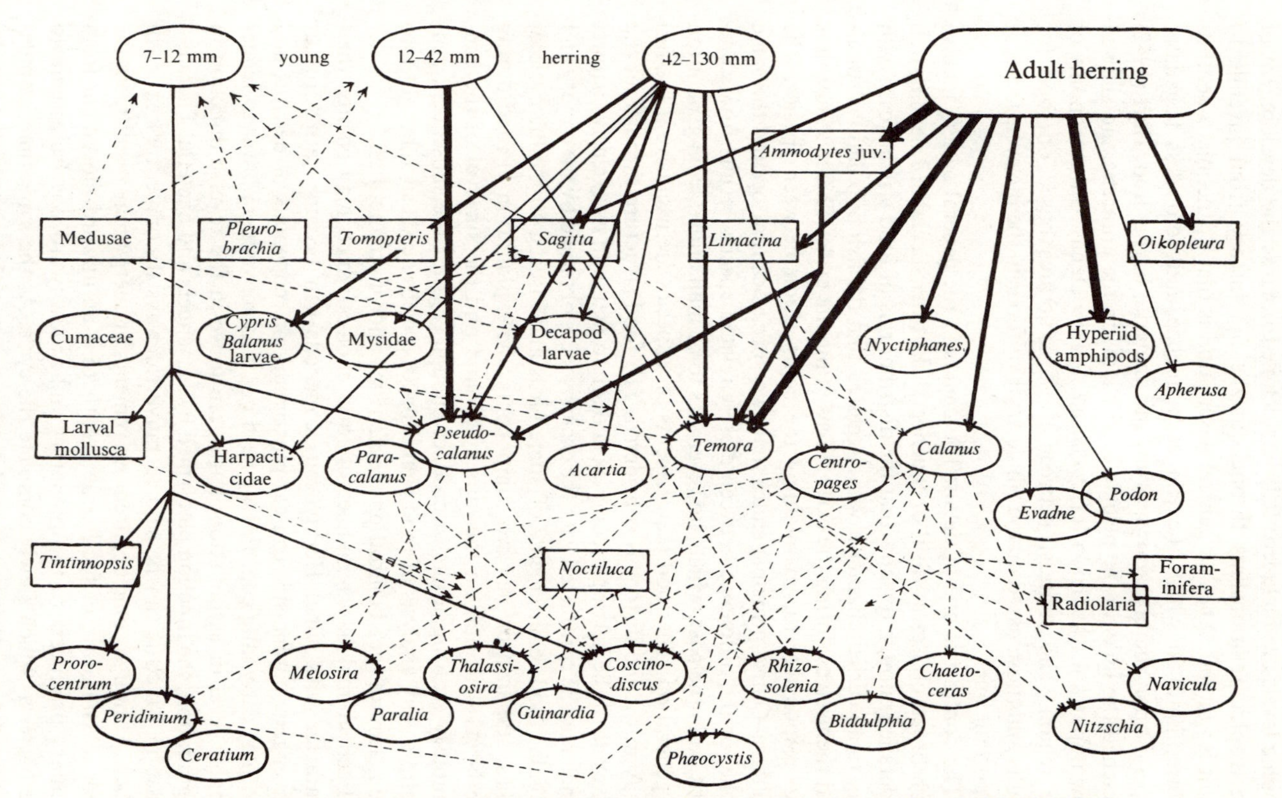

Figure 55. The food web upon which the North Sea herring depends during its life cycle (Hardy, 1924).

description than Hardy's because a number of animals at the highest trophic levels are included. It is less detailed because the development of the trophic relationships throughout the life history from juvenile to adult is ignored. Another general description is that of the micro-benthos (Mare, 1942) in which the ciliates, amoebae and colourless flagellates depend upon the breakdown of organic material in the mud. Hart (1942) sketched the links in the pelagic food chains in the antarctic and showed how important was the krill (*Euphausia superba* Dana) to the economy of the region. The contrast between those structures based on food for adults and those based on the whole array of foods in the life cycle is a very important one.

The contrast is one of detail. The chains in the general system involve such broad classes as 'fish' or 'nutrients'. But the specialized herring web is detailed enough to show that the distributions of food size increase with age. In an earlier chapter it was suggested that the fish populations stabilized themselves during the early stages in the life cycle. This mechanism is also responsible for maintaining the competition between species. The Volterra–Gause principle states that two species cannot occupy the same niche, or the difference between closely related species results from competition in the past, which, however, has to be maintained from generation to generation. The food requirements of a species are exclusive if the full array in the life history is considered. The food structure throughout the life cycle of the herring is complex enough to maintain its exclusive nature, which may be contrasted with the exclusiveness obtained by the choosy and singular food requirements of some birds and insects. In the life cycle of fishes, stabilization has two components, the exploitation of food during the larval drift and on the nursery ground and the survival of the continuously growing adults to produce enough eggs for the subsequent generation. Any analysis of food chains or food webs needs such insight into the nature of the stabilization mechanisms and of competition.

Each animal has a number of food choices which may change their character with time. Herbivorous copepods can become carnivores or scavengers, euphausids can eat phytoplankton, copepods or detritus and cod may eat young cod. Energy need not only move from lower to higher trophic level in assimilation, but it can move in the reverse direction in degradation, the function performed by scavengers and bacteria. There is a sense in which it can be said to move sideways if competitors eat each other. Although some apparent movements of energy may result from the poor classification of food structure, there is an upward movement from primary producers to top predators and a downward one from them to the bacteria, which slowly complete the cycle. At the same time, populations maintain themselves within this system of energy exchange by sacrificing most of the gonad weight to the maintenance of the system. In reproducing, a population of second-order carnivores may generate a very numerous population of juvenile herbivores most of which are eaten in the system, but the survivors grow and the biomass of the original population is maintained at the expense of the system. Thus, there is an upward movement of biomass, accompanied by the concomitant death of adults, a downward movement of biomass. The two systems of energy transfer and of biomass transfer are not identical because they may be out of phase with each other. An interesting point is that the primary producers are all transferred upwards as energy and the top predators conserve their biomass by eliminating most of the mortality. The ecosystem thus depends upon the interlock of carnivorous and herbivorous populations.

The quantitative study of food chains starts with rough estimates of the transfer coefficients from one trophic level to the one above it. Not only should they be based on more detailed information, but on a longer time period, perhaps more than one generation. The growth efficiency of a young animal is greater than that of a mature one with a reproductive load so the coefficients should be spread across the whole generation. The short food chains with few choices at each trophic level transfer a lot of energy in an inefficient way for a short time, say three months, as in the temperate production cycles. The somewhat longer chains with many food choices transfer less energy continuously and efficiently as in the tropical cycle.

The transfer coefficients

Food webs have a structure based on levels of food, or trophic levels, expressed as herbivores, carnivores, producers, consumers and so on. To understand the structure in such terms the transfer of energy or material from one level to another must be described and measured.

Consider three trophic levels, A, B and C, A higher and C lower; a high trophic level is one of carnivorous predators and the lowest comprises the algae. Let g_1 and g_2 be the proportion transferred from the levels, B and C, respectively; the coefficient expresses the movement of energy into maintenance, growth and reproduction in the higher level. Each quantity (in calories) in a trophic level represents production, the sum of respiration and yield. Growth efficiency and protein utilization decrease with age in fish (Gerking, 1959) and in crustacea (Reeve, 1970) and if the quantities were expressed in numbers or in weight, such size-specific changes in energy conversion would not be taken into account. Further, the energy content of fat is greater than that of protein and a transfer coefficient based on weight or numbers would underestimate one based on energy. Usually the transfer is estimated in calories, if possible.

Lindeman (1942), probably much influenced by Juday's (1940) energy budget in Lake Mendota, estimated the production at each trophic level in Cedar Bog Lake, Minnesota and in Lake Mendota, Wisconsin, USA. From the work of Juday and Ivlev (1939) he estimated that the proportion of respiration to growth was 33% at the primary level, 62% at the herbivore level and >100% among carnivores; analogous estimates were made of losses due to predation and decomposition. Table 13 summarizes his results in g cal/cm² per year.

Lindeman defined biological efficiency as the ratios of net productions or ratios of stocks, which is not the same as ecological efficiency unless $g_1 = g_2$. He suggested that efficiency increased up the food chain because the activity of predators increases their chance of finding their prey.

Table 13 *Production in g-cal/cm² per year at three trophic levels in two Wisconsin lakes*

	Net Production	Respiration	Predation	Decomposition
Cedar Bog Lake				
Primary	70.4	23.4	14.8	2.8
Secondary	7.0	4.4	3.1	0.3
Tertiary	1.3	1.8	–	–
Lake Mendota				
Primary	321.0	107.0	42.0	10.0
Secondary	24.0	15.0	2.3	0.3
Tertiary	1.0	1.0	1.0	0.0
Quarternary	0.12	0.2	0.2	0.0

Note: 'net production' is really standing stock. Lindeman called production the sum of respiration and yield, together with (stock × turnover time), which is superfluous. The energy budget for a trophic level is simply the sum of respiration and yield in cal/cm² (Slobodkin, 1962).

Earlier Lindeman (1941) had published details of 'production' of different components in the food chain in Cedar Bog Lake for four years. The importance of these observations is that a little evidence is given of the variability in the structure. The coefficients of variation of total producers and of secondary production are both 14% whereas that of the carnivores is higher at 30%. In more detail, the coefficients for predatory plankton, browsers, predatory benthos and swimming predators lie between 35% and 50%. The variability of primary production appears to be less than that of components higher in the ecosystem. Intuitively one might have expected the reverse because the higher elements live longer and might possess mechanisms for reducing variability. The secondary 'production' is about one-tenth of the primary 'production', but the tertiary production is nearer one-fifth of the secondary 'production'.

Table 14 *Annual 'production' in four years in Cedar Bog Lake in cal/cm²*

	1937	1938	1939	1940	Mean
Nanoplankton	19.6	20.4	16.7	9.9	16.7 ± 2.2
Net phytoplankton	1.5	9.8	18.9	6.3	9.1 ± 3.7
Pondweeds	63.0	10.5	35.0	70.0	44.6 ± 9.7
Zooplankton	8.5	5.3	7.4	3.2	6.1 ± 1.2
Predatory plankton	0.1	0.3	1.9	0.9	0.8 ± 0.4
Browsers	1.0	0.3	1.0	1.0	0.8 ± 0.2
Predatory benthos	0.1	0.2	0.1	0.4	0.2 ± 0.07
Swimming predators	0.5	0.1	0.4	0.4	0.3 ± 0.13
Total producers	84.1	40.7	70.6	85.9	70.3 ± 10.14
Secondary production	9.5	5.6	8.4	4.2	7.0 ± 1.07
Carnivores	0.7	0.6	2.4	1.7	1.3 ± 0.43
Total consumers	10.2	6.2	10.8	5.9	8.3 ± 1.22

Three transfer coefficients were defined by Slobodkin (1960):

$$E_1 = g_1B/g_2C \quad \text{(ecological efficiency);}$$

$$E_2 = g_1B/C \quad \text{(food-chain efficiency);}$$

$$E_3 = g_1B_i/I'(1-P'/P) \quad \text{(population efficiency with constant food intake);}$$

where I' is the energy consumed by the population in unit time; i is a subscript indicating age; P' is the caloric content of the population in a steady-state under predation; P is the caloric content of the population with no predators.

If there were only one population at each trophic level, a number of predator generations would be needed to form a good average and the same is true for an array of populations. Ecological efficiency is the ratio of yield from B to that from C; if $g_1 \simeq g_2$, it is equivalent to the ratio of production in B to that in C. Food chain efficiency is less than ecological efficiency (because $C > g_2C$) and is a ratio of yield to production. If the ratio is high the food available is well exploited and vice versa, but of course, the competitive mechanisms remain unrevealed.

Ricker (1969) defined transfer efficiency as the ratio of production of predator to that of prey:

$$P_n = I'_nK_n = Y_{n-1}K_n = (F_{n-1}/Z_{n-1})P_{n-1}K_n.$$

Therefore $(P_n/P_{n-1}) = (F_{n-1}/Z_{n-1})K_n$.

Therefore where n is the predator trophic level and $n-1$ that of the prey,

P is production; I is intake of food; K is food conversion efficiency; F is mortality rate generated by the predator in prey numbers; Z is the total mortality rate suffered by the prey; and Y is the yield of the prey to the predator.

The ratio (F/Z), analogous to the exploitation ratio in fisheries work, was named the ecotrophic coefficient. Dickie (1972) extended the argument:

$$Y_n = (F_n/Z_n)P_n = (F_n/Z_n)K_n I'_n.$$

Therefore $(Y_n/I'_n) = (F_n/Z_n)K_n$.

$$Z_n = P_n/B_n \text{ and } (Y_n/I'_n) = F_n . (B_n/P_n)K_n,$$

which is Slobodkin's ecological efficiency defined in terms of a production–biomass ratio. Further, $Y_n = F_n(K_n I'_n) . (B_n/P_n)$ which provides a series of measurable parameters. The most important point, however, is that the processes of growth and reproduction are separated.

Thus, the earlier coefficients, g_1 and g_2, may mask understanding. Let $g = (KI')(P/B)$ where (KI') is a coefficient of energy conversion from food to growth and where (P/B) in years is one of energy conversion to growth in numbers, i.e. turnover time or reproduction. Variations in (KI') are affected by differences in the fat–protein ratio, in maintenance and in searching efficiency. Variations in (P/B) express the possible differences in year class strength, which reflect the capacity of the population to respond to or isolate itself from environmental changes as suggested earlier. Variations in numbers, the regulatory capacity of the population, are perhaps controlled by the ratios of growth rates to death rates. The processes govern the magnitude of recruitment, the regulation of numbers from generation to generation and the competitive advantages (or disadvantage) in any population. The important question is not so much the quantity of energy transferred but its partition amongst the three processes within the population. In general terms, the predator should maximize its food consumption so (P/B) must be neither too high nor too low. However, (P/B) could vary widely in response to the environment (particularly the effects of climatic change) and initiate the trends in numbers which are the results of competition. To the student of energetics in ecology, the separation of the energy transfer coefficients into two components may appear unnecessary and pedantic. The variability of the system as shown in Lindeman's study in Cedar Bog Lake must be due to the interlocking responses of the different populations.

The difference between ecological efficiency and food chain efficiency in the sea might be made manifest in the time period for examination. The first might be determined in a single generation, but the second needs more than one generation because variation in available food should include variation due to reproductive differences. The simple food chain in high latitude cycles is transitory, whereas the more complex one of the low latitudes endures for many generations. The quantities which would describe the differences between the two types of food chain are averaged and perhaps lost in the transfer coefficients. The amplitude of the high latitude production cycle is high, which means that the net rate of increase must vary considerably. In low latitudes, the net rates may vary very little because the amplitude is low. The differences in net rate are the results of competitive effects. It is possible that the transfer coefficients conceal some desired information that is basically numerical.

In the end, the whole system must be limited by the growth efficiency of the animals. That of an individual (or a small group in a tank) is expressed in the ratio of growth to intake and in the metabolic cost of growth (which is the ratio of the proportion metabolized to the proportion retained in growth). High conversion efficiencies of up to 45% have been obtained for chickens in batteries, because they do not run about and they are killed before they reproduce. The average growth conversion coefficient of a searching and reproducing animal cannot be very high. Let us suppose that the greatest possible growth efficiency is 40%; the gonad weight of fish is about one-fifth of body weight and for about one-quarter of the year the fishes starve, which reduces effective efficiency to 26%. If the least growth is 5% or less, the proportion devoted to search could be as much as 20%. Hence, ecological efficiency on such a basis could range from 5% to 25%. If the proportion of gonad weight is higher, as in smaller animals, the growth in adults is also less, which means that the same range is probably valid.

Slobodkin (1960) and Richman (1958) made the fullest analysis of the first two transfer coefficients (ecological efficiency and food chain efficiency) for parthenogenetic populations of *Daphnia*. They were fed on the alga *Chlamydomonas* and a proportion of the larger animals was removed periodically to represent predation. The quantities of *Daphnia* stock, algal stock, food eaten and *Daphnia* removed were expressed in calories. The number of animals removed was a fixed percentage of new-born young per four-day period. Richman had shown that the number of young produced was linearly related to food supply, so predation was a direct function of food. As expected, yield increased with increased predation, but the yield per unit food was greatest at the lowest quantity of food. Ecological efficiency was a linear function of predation, i.e. g_1 increased, but g_2 did not. Food chain efficiency increased with predation to a maximum beyond which it declined, because if g_1 increases B must eventually decrease and C increase. Conversely, at the lowest food level, food chain efficiency was greatest. Slobodkin (1962) showed how growth efficiency depended upon algal density and on the size of *Daphnia*. The highest growth efficiency (0.2) was found at a relatively low density of 20000 cells/ml, in young animals, about a quarter grown. Densities of algae in the sea do not much exceed 20 000 cells/ml (converted to mg/l wet weight). The growth coefficients ranged from 0.07 to 0.19 and they may have been low due to searching, death independently of predation and possibly to clogged food filters. The food chain efficiency observed was 0.085; if the average growth efficiency was about 0.12, then $g_2 = 0.70$. Ecological efficiency was 0.12 to 0.13, probably close to the average growth efficiency, so $g_1 \simeq g_2$. Slobodkin (1962) quotes a number of field determinations of ecological efficiency between 0.055 and 0.133, which correspond roughly to the experimental measurements. Lindeman's original estimates implied stock ratios of about 10:1 between lower and upper trophic levels and the ratios of production were a little less. Slobodkin's experiments confirmed these ratios in general terms.

In a study of the Indian Ocean, Cushing (1973*b*) expressed secondary production as a fraction of primary production and if $g_1 \simeq g_2$, this ratio estimates ecological efficiency. It will be recalled that radiocarbon measurements were averaged in statistical squares across the ocean separately for the south-west monsoon and the north-east monsoon. Secondary production was expressed as the stock of zooplankton raised by the number of generations in each monsoon period and it represents an underestimate of the true production. This is because no general measure of mortality is yet available. The generation time was estimated roughly from the dependence of the duration of copepodid developmental stages upon temperature; there is not really enough information yet to establish the relationship definitively, but sufficient to form a preliminary estimate. Figure 13 shows the transfer coefficients as a function of primary production. In the deep tropical ocean, in the quasi-steady state production cycle, the transfer coefficients are three times higher than those in the richer upwelling areas. The median value of 10% is close to that suggested by Slobodkin as an average. If secondary production is underestimated, the true value of the coefficient would be higher. If the maximum food conversion efficiency is 40%, the greatest underestimate is by a factor of 2.66. If ecological efficiency could be as high as 25%, secondary production is underestimated by at least a factor of two-thirds. In the rich areas of the Indian Ocean, the coefficient is about 0.05, perhaps because much of the production is spilled during feeding and thick deposits of algae accumulate on the sea bed, forming green mud. In the poor areas, the coefficient is about 0.15 because the animals are unable to graze to excess and very little is spilled – the bottom of the deep ocean is often bare red clay. The difference in production intensity (in g C/m^2 per day) between the rich and poor areas is about one order of magnitude. But in secondary production, the difference is reduced to a factor of three, so, in transfer, variation in primary production is damped.

The population efficiency expresses the loss of energy by age in predation as a proportion of the energy in the prey population. When population efficiency was calculated by age for the *Daphnia* population, the oldest were the most efficient and the youngest the least. In an estimate of population efficiency, energy devoted to growth in a given age group is a loss to that age group. The population efficiency is the inverse of the growth efficiency which is greater amongst younger animals. Slobodkin (1962) wrote that a predator should not only maximize yield from its prey, but should also maximize the prey's population efficiency. Growth efficiency declines with age and population efficiency increases with age. The maximization of both processes must result in a compromise. The predator must take the prey when growth efficiency has started to decrease with age, i.e. during its reproductive phase. In fisheries biology, the concept of maximum sustained yield is the same as the population efficiency in principle, for the largest animals must be taken for the least loss in numbers. The virgin biomass of a fish stock reaches its peak in age during the adult age groups. It is this quantity of material or energy that is transferred at any trophic level higher than the primary one. Of course, the loss in numbers of larval and juvenile animals is very high, although the transfer in weight is not the greatest. The high numerical loss is probably part of the stabilization mechanism, but the transfer of energy is in mature animals.

The structure of some food chains

A number of food chains have been studied quantitatively in the last two decades. A small spring with a muddy bottom in Massachusetts, USA was examined by Teal (1957) and he estimated the quantities at each trophic level; the ecological efficiency of the transfer of algal energy to herbivores was 0.09. Odum (1956) examined the food structure of a deep spring at Silver Spring, Florida, USA, where there was a complicated web of carnivores and secondary carnivores living amongst large water plants. The ecological efficiency of the transfer of energy to the herbivores was 0.114 and that for the transfer from herbivores to carnivores was 0.055. Two main food chains were distinguished in the system, one of grazers that transferred energy upward and the other that exploited detritus. Most benthic food chains are based on scavengers and detritus feeders and in the deep ocean itself, scavenging euphausids are most abundant at certain depths where they may fulfil a role analogous to that of the benthos. Indeed, the distinction between the two forms of food chain may be artificial in marine studies; some algae, if not all, can live both autotrophically and heterotrophically, herbivores probably eat detritus when algal food is not available and the input of detritus to the seabed itself must depend upon the grazers in the water above it. Just as herbivores can become carnivores within their life cycles, so the food web itself changes in time, particularly in high latitude waters. The categories of functions (producer, herbivore, carnivore, scavenger) do not change but they do not necessarily correspond with groups of animals, because the functions can change within a life cycle.

Petipa, Pavlova & Mironov (1970) have studied the biology of the plankton animals and plants at different depths in the Black Sea, in age and in

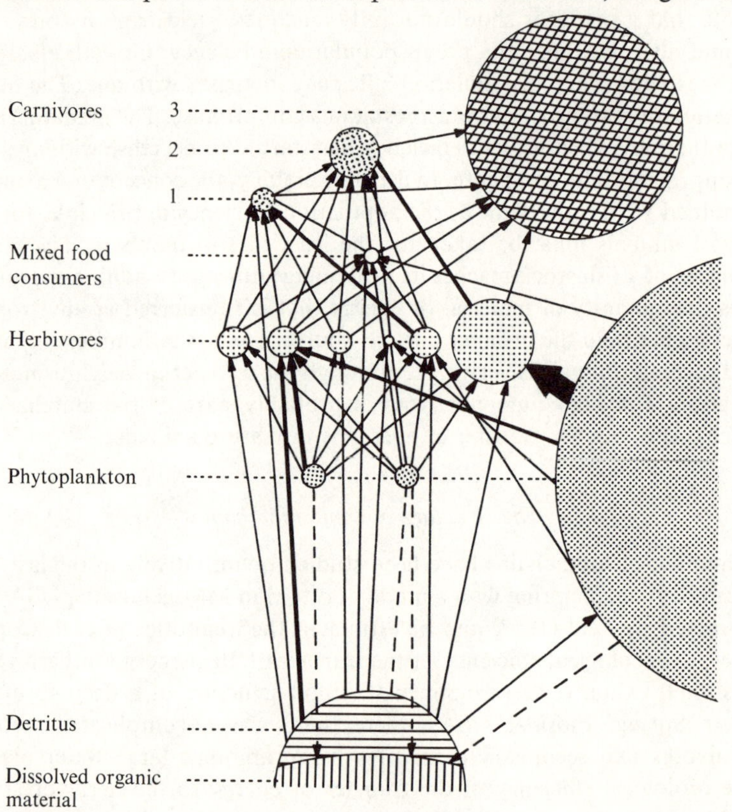

Figure 56. The food web in the epiplanktonic community in the Black Sea (Petipa *et al.*, 1970).

energy transfer for a three-week period. Figure 56 shows the food web; stock sizes are represented by the areas of the circles and the thickness of each arrow represents the daily rates of consumption. There is a category of mixed plant and animal consumers because some herbivores change their trophic status, as noted above; some copepods become carnivorous as algal food disappears and some primary carnivores grow into secondary carnivores. In the epiplanktonic community, the ecological efficiencies were greater than 0.20. Perhaps the transfer of energy occurs mostly among young and actively growing animals at high density, as Steele (1965) suggested. In the bathyplankton community, the average ecological efficiency was almost as high. The estimates are high probably because the observations were averaged over a short time period. In a complete food web, much of the energy is transferred to juvenile animals, but this early energy transfer must be set against losses in the reproductive phase of older animals. Figure 56 shows the relative importance of different parts of the food chain structure. The main transfer of energy takes place from the small and medium algae to three groups of herbivores and then to first-stage carnivores. There are six categories of herbivores, one of which subsists to a fair degree on detritus, so, together with mixed food consumers and some facultative carnivores, the higher carnivores are provided with many choices. It is possible that the detailed analysis of food webs will reveal age-specific differences in the transfer coefficients. Such an analysis is needed for a longer time period to assess the changes in more detail.

Steele's (1965) description of the food chains in the North Sea is probably the most adequate because it is based on the long-term averages of catches and upon the estimated food requirements of the stocks that support the catches. He suggested that the estimated value of 10% for ecological efficiency is possibly too low. With reasonable values of natural mortality, the annual yield to man and predators of North Sea pelagic and demersal stocks are 0.60 and 0.17 g C/m^2 per year, respectively, with an ecological efficiency of 10%. The annual fish production is 6.0 and 1.7g C/m^2 per year, respectively. The annual primary production is 100 g C/m^2 per year from which 8.0 and 2.0 g C/m^2 per year might represent the yield from the second trophic level. If one compares the yields with the estimated food production one concludes that the ecological efficiencies of 10% must be low. Steele made careful estimates of production at each trophic level which may be summarized in Figure 57.

Obviously the ecological efficiencies are higher than 10%. Steele also tried to express it in detail for individual fish species with a different form of transfer coefficient ($Y/R+B$, where Y is yield, R is the quantity lost in respiration and B is that lost in reproduction; which is very nearly yield/production). He showed that as haddock grew older (from 2.25 to 4.25 years) this coefficient dropped from 0.22 to 0.10 and that when the herring experienced the planktonic changes in the early fifties (see Chapter 10) it increased from 0.087 to 0.141.

Lindeman and his successors suggested that the transfer coefficients were about 10%. Steele showed that in the North Sea the ecological efficiency must have been greater than 10% through a number of generations. His second coefficient, yield per production, decreased with age, but increased during climatic change. This second coefficient reflects primitively the complexities of population analyses in contrast to the somewhat too simple

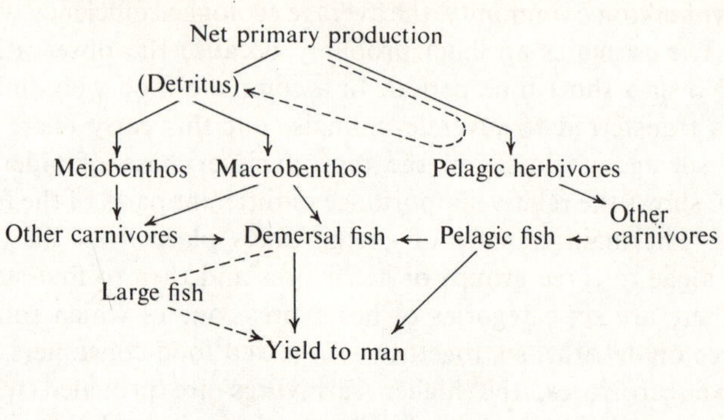

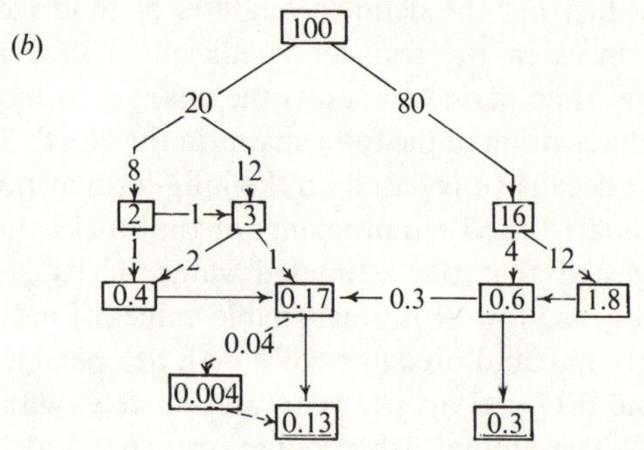

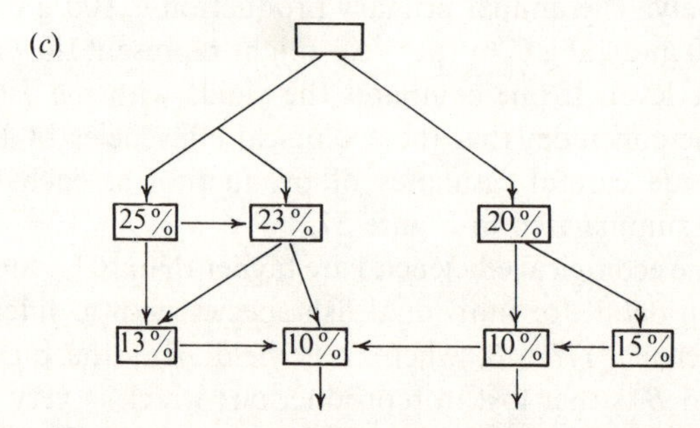

Figure 57. The food web in the North Sea (Steele, 1965). (*a*) Simplified food web; (*b*) production in each group; (*c*) ecological efficiency in each group.

concept of ecological efficiency. In an earlier section, it was concluded that the true transfer coefficients in the Indian Ocean were greater than 10%, as Steele suggested, by a quite different route in the North Sea. The question then arises why these estimates should differ from Slobodkin's experimental ones.

There were two points in common in the observations made at sea: (1) high values can be found in the exploitation of algae by herbivores; (2) the transfer coefficients were higher than in Slobodkin's experiments. The transfer coefficients shown in Figure 13 are low because secondary production is underestimated and it was suggested that the true values lay between 10% and 20%, but perhaps not much more than 10% in the very rich areas. There is no conflict between this range of estimates and Steele's, but Petipa's may be too high because the short period of study concentrated the sampling on the younger age groups. If Steele's estimates are right, a further conclusion is that the transfer coefficients are higher at the lower levels of the food chain. Such a conclusion might not be unexpected if it is realized that animals in the higher levels of a food web tend to live longer and a greater proportion of material is stored not only in fat or carbohydrate, but also in bone, scale and muscle. The proportion temporarily stored in gonads may decrease up the pyramid, but the total store remains higher in the upper levels.

In fisheries biology, which is slowly emerging from its obsession with the dynamics of single stocks, the use of such analyses as Steele's is considerable in that the limits to marine production can be stated even if the processes that govern productivity are not yet fully understood. It has been argued that stocks should be exploited when the fish are small and growing most efficiently. But, as Gulland (1970) pointed out, such efficiency in growth should be conserved and exploitation restricted to a later age when the fish are larger and when they convert energy less well; further, if fish were taken when small, the chance of zero recruitment would be very high. Fishing should tend to maximize the population efficiency. Not only is the energy obtained in growth conserved, but that in reproduction should generate the maximum recruitment.

Gulland (1970) extended the argument raised by Steele. He examined the efficiencies in the food chain that supports the Peruvian anchoveta (which may feed on algae or on small zooplankton or on both). An efficiency coefficient (fish production/food eaten) was calculated for different proportions of algal food and it was shown that even if the anchoveta takes 60–80% of the phytoplankton, the effective efficiency approached that of a three-stage food chain. Gulland also examined two particular fisheries problems: should one not fish dogfish in order to save herring that are their prey? and should one not fish sandeel in order to save the predator cod? In either case, the value of the predator catch should be greater than that of the catch of prey, i.e. the efficiency of utilization should be greater than the relative products of value, exploitation or growth. Thus, the theory of food chains has a direct use in fisheries biology.

Some of the work of fisheries biologists is of direct interest in the more academic study of food chains. A study of the possibility of transplanting plaice from the continental coast of the North Sea to the Dogger Bank led to the detailed analysis of part of a benthic food chain. Birkett (1970) studied the transfer of energy from a patch of *Mactra* to its predators.

Table 15 gives the main result in g cal/m^2 per day in the first and second years of the life of *Mactra*. The proportion of energy turned to growth is high in the *Mactra* population (55%) whereas that of the predators is less (20%). The energy in growth is 160% of the heat output in *Mactra*, but in the predators it is only about 30%. Both comparisons contrast the sedentary habit of the clam with the active one of the predators. The ratio of energy consumed at the two trophic levels dropped from 70% in the first year to 12% in the second; the ratio of growth energy dropped from 27% in the first year to 13% in the second. Presumably in the second year, as the *Mactra* became less abundant, the predators shifted partly to other foods. The whole system changes sharply from one year to the next.

Table 15 *Benthic production on the Dogger Bank*

	Mactra		Fish		Asteroids		Natica		All predators	
Years	I	II	I	II	I	II	I	II	I	II
Biomass (g/m^3)	37	50	13.8	0.1	0.8	1.5	0.4	0.7	14.9	2.3
Energy consumed	993	608	660.0	4.9	27.8	53.0	14.4	26.7	702.0	84.6
Faecal energy	50	30	53.0	0.4	1.4	2.7	0.7	1.4	55.1	4.5
Energy of metabolites	50	30	53.0	0.4	1.4	2.7	0.7	1.4	55.1	4.5
Heat output	348	213	429.0	3.2	10.0	18.5	5.0	9.3	444.0	31.0
Growth	546	335	125.0	0.9	15.0	29.1	8.0	14.6	148.0	44.6

Trevallion, Edwards & Steele (1970) have studied the dependence of the 0-group plaice stock in Loch Ewe (in northern Scotland) upon the regeneration of *Tellina* siphons. They show that the total energy requirement of the plaice population may be limited by the regeneration rate of the siphons (also in energy). Experimental work showed that *Tellina* can only reproduce when enough food can be channelled away from siphon regeneration; consequently, the reproduction of *Tellina* may depend on the slackening of predation, so the two population control mechanisms are linked. An approach to the formulation of the mechanisms of search and those of stabilization is starting to emerge from the studies initiated by the fisheries laboratories with purely practical ends in view.

The stabilization mechanisms in populations result in stable communities, that is, stable numbers, and consequently a determined position in the food chain. In the sea the disbalanced production cycle of high latitudes may be contrasted with the quasi-steady state one of low latitudes. In the former, the community is simple with considerable variability in numbers and in the latter it is diverse, yet numerically stable. The difference between the two systems is that the stabilization mechanisms in high latitudes never have a long-enough period to do more than make coarse control adjustments, let alone generate the diversity of an evolutionarily stable structure. Exceptions are the perennial fish populations that have evolved mechanisms to stabilize their numbers in high latitudes by spawning with great fecundity at fixed seasons; indeed, they may exploit the discontinuous production cycle by growing up with a private food supply, as Jones (1973) suggested. The number of species at any food level is the number of food choices and the greatest structural stability is found with the greatest number of species and then the predators should be omnivorous. This is because all the density-dependent mechanisms that control the numbers of each species are interlocked in a competitive manner. Indeed, it was suggested in an earlier chapter that competition in the sea was generated during larval and juvenile life in the ratio of specific growth rates to specific mortality rates, with the consequence that competition becomes effective in differences in recruitment or indeed in the net rate of increase. Indeed, the large number of species under conditions of low abundance may be the consequence of muted competition, and where competition is sharper under conditions of high abundance there are fewer species.

In fisheries biology, the need has arisen recently to understand and formulate the mechanisms of stabilization in the problem of the dependence of recruitment upon parent stock. From the arguments presented in an earlier chapter, it is suggested that the control of numbers takes place during the larval drift by the agency of density-dependent growth and mortality. The same processes are at the source of the generation of year classes, which if they rise and fall in sequence, express the effects of competition. There are three processes that dominate the dynamics of any population, the magnitude of recruitment, the control mechanisms that establish stable numbers and the maintenance of competition between species. The three processes are effectively linked in one mechanism, which may be completed during the early life history. Adult fishes lead exclusive and uncompetitive lives and the food they eat is not taken in competition, but is devoted to the conservative function of the gonads.

Conclusion

The study of food webs originated in the work of Lindeman (1942) and Ivlev (1945). More recently, Slobodkin (1960,1962) defined the three transfer coefficients, ecological efficiency, food-chain efficiency and population efficiency and extended their use and estimation. A recent international symposium (Steele, 1970) was established to examine the use of transfer coefficients in the estimation of the potential yield of fish in the world oceans. Whatever animals comprise the third trophic level, they may be as large as fish, for example squid, and as edible. In an examination of the world's upwelling areas (Cushing, 1971*a*), the animal production in the third trophic level amounted to about 100 million tons wet weight using transfer coefficients of 1% of the primary production and 10% of the secondary production. A much more extensive study of the production of fish throughout all the world ocean was made by Gulland (1970). Several methods were used, but estimates of primary (and sometimes secondary) production were employed to set limits to the projected values of tertiary production in the unexploited areas. The potential world catch (that is, ignoring antarctic krill or the myctophids of the deep ocean) will not be much greater than 100 to 140 million tons. The most important consequence of this practical use of food chain studies was the discovery that the fish resources in the world ocean are limited and that they would probably be fully exploited within the next decade or so.

The further study of food chains will lie in the analysis of processes at sea with models rather than in simplified experimental observations or in the rough carve-up of an ecosystem. What is needed are estimates of growth efficiency, of predation and of recruitment under different conditions of density over rather long time periods. Some ecological processes can be studied in great detail on land, but for many reasons the study of their interaction in the soil or in the variable climate is much harder than in the sea. The sea looks inaccessible and foreign, but it is simple, and the problems of sampling the animals therein are not as difficult as they appear to be. To analyse the food chains in a community properly is hard, however, because the problems of competition, stabilization and predation are all interlaced and it is possible that the work on food chains as such may become submerged in studies under other names.

10

Temporal changes in the sea

Introduction

Temporal changes of numbers in the sea are sometimes slow and sometimes rapid; some are of small scale but persistent whereas others are of considerable magnitude. Perhaps as a consequence of the numerical changes there have been alterations in the structure of communities. There are observations on plankton in time series in the Atlantic and in the Pacific, but most information for long periods is provided by catches from the commercial fish stocks. Statistics on fish catches go back sometimes for as much as a hundred years particularly in Europe, Japan and in the North American Pacific salmon fisheries. Beverton (1962) showed that the catches of many demersal stocks in the North Sea remained at a steady level for about seventy years, the period during which landings had been recorded in England and Wales until 1962. However, the catches of haddock declined by more than an order of magnitude and the catches of sole increased by about the same degree towards the later part of the period. By peculiar chance the haddock eggs that were hatched in the year in which Beverton's paper was published generated a year class that was twenty-five times larger than any of its predecessors.

However, such changes, although considerable, are small as compared with the dramatic events in the pelagic fisheries; the extinction of such fisheries implies changes in stock of many orders of magnitude. For example, catches of the distinct Norwegian and Swedish herring stocks have alternated in a periodic manner for centuries and the fishery for one disappears completely during the period of abundance of the other. Beverton & Lee (1964) correlated the periodic events with the extent of ice cover north of Iceland, where the Norwegian herring used to feed. They suggested that it was the area occupied by the stock that varied periodically but the variation in ice cover may indicate other forms of climatic change. There is a contrast between the high variability of the stocks of the herring-like fishes in the long term and that of the flatfish and cod-like fishes that tends to be rather low. It is a difference that is probably a function of the different forms of stabilization mechanisms.

The detailed analysis of time series requires specialized statistical techniques (see Kendall & Stuart, 1966) that are not employed here or in the papers cited. There are two reasons for this, first that some events are linked which are contemporaneous but disparate and secondly, that a decline in catches can occur quickly and the number of observations may be small. The changes are usually of sufficiently large scale to be obvious and they can be adequately described with rather few observations.

Studies of observations in time series are used here for two purposes. First, they reveal the variability of the numbers of populations. As will be shown in the next chapter, such long-term changes may be modulated by climatic change and they are associated with profound changes in the structure of the ecosystem. The second aim is to study the extent to which the stabilization mechanism can damp or rectify the environmental variation. There is, of course, no real distinction between the two purposes because they are different facets of the single process by which recruitment is generated and populations are stabilized.

Planktonic changes

In British waters long-term changes in the plankton are described in three time series. The first was started off Plymouth in 1925 when collections of fish eggs, fish larvae and the associated macroplankton were made each week at International Station E1 near the Eddystone Lighthouse. The catches have been taken more or less continuously since then with some gaps in the series (Russell, 1930,1969). The second series was based on a line of stations, the Flamborough Line, occupied between the Yorkshire coast and the Dogger Bank each month between 1935 and 1960, except 1940–6 (Wimpenny, 1944; Cattley, 1950,1954). The third series is the plankton recorder network established by Sir Alister Hardy in the thirties and which has been extended since the Second World War. The recorder is towed by merchant ships at a depth of 10 m on fixed lines at monthly intervals across the North Sea and the North Atlantic. Animals are counted in ten-mile samples and monthly estimates of abundance are available by species in statistical squares since 1948.

Off California, Allen (1941) made a long-term study of the net phytoplankton. More recently, after the collapse of the sardine fishery, an extensive area of the California current was charted regularly for more than a decade. As a consequence the changeover from the sardine to the anchovy stock off California was documented in some detail, as will be described below.

The changes in the western English Channel

Between 1925 and 1926 the samples taken off Plymouth with a 2-m ring trawl showed a number of remarkable changes. In the autumn of 1931 one species of arrow worm, *Sagitta elegans* Berrill, was replaced by another, *Sagitta setosa* J. Müller. The first was considered an indicator of mixed Atlantic water of north-west origin (Southward, 1963) and the second is a neritic animal that tends to live in the Channel, east of Plymouth. At about the same time, the phosphorus content of the seawater in winter fell by nearly one-third and it did not recover until 1970 (Russell & Demir, 1971). In the autumn of 1931 when one species of *Sagitta* succeeded the other, the total quantity of macroplankton declined by about four times and some species disappeared. In the autumn of 1970, the quantities of macroplankton increased again. The changes in phosphorus, in *Sagitta*, and in macroplankton were abrupt and they were noticed immediately, as was the recovery in macroplankton and in phosphorus. Subsequently, some more gradual changes became apparent (Cushing, 1961).

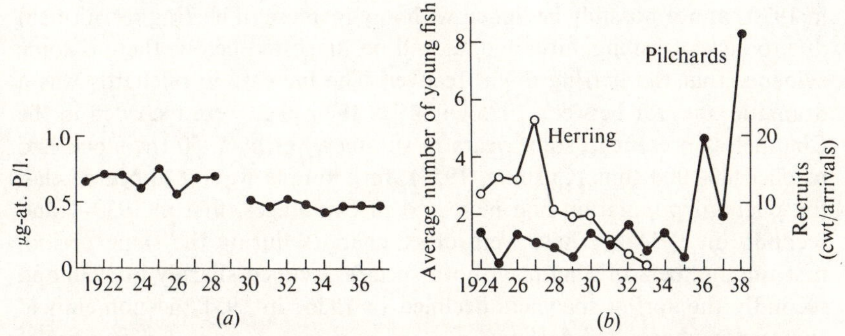

Figure 58. The changeover from herring to pilchard in the western English Channel (Cushing, 1961). (*a*) The decline in phosphorus in μg-at. P/l; (*b*) larval clupeids caught between May and July, predominantly pilchards and herring recruits as cwts/arrival.

The first change was a decline in recruitment to the Plymouth herring stock that started with the 1925 year class which entered the fishery in 1929. The Plymouth stock was probably a small one confined to the British coastal waters of the western English Channel. The second change was an increase in pilchards, recorded first as eggs in the plankton samples in the early autumn of 1926, then as significant numbers in the summers of 1930, 1931 and 1934 and lastly by very large numbers right through the summer of 1936 from April onwards. The pilchard remained the dominant pelagic fish in the area probably until the late sixties. However, it is likely that towards the end of the period the fish had withdrawn into the western English Channel with spawning delayed until the autumn.

Figure 58 shows the changeover from herring to pilchard in the western English Channel between 1926 and 1936. About the same period of time elapsed between the collapse of the Californian sardine and the rise of the anchovy stock, which will be described in more detail below. As a generation of pilchards or anchovies lasts about three years, the takeover from one species to another occurred in three or four generations.

There is no evidence that fishing caused the decline of recruitment to the stock of Plymouth herring and, indeed, the older age groups remained abundant (as if fishing mortality were low) until they were replaced by the failing year classes, so the fishery itself became extinct in 1938. The spawning ground (between Bolt Head and Bolt Tail in Devon, England) could be seen from the surface by the phosphorescence of the spawning shoal and Mr Pengelly of Looe has told me that French trawlers did exploit the ground during the twenties. If trawling affected recruitment by destroying eggs on the ground, it remains possible that the events off Plymouth were started by fishing. However, there is no evidence that such fishing generated the decline in herring recruitment. In particular, the reversal of the more general changes in phosphorus and macroplankton in 1970 cannot possibly be linked with any increase in herring recruitment due to lack of fishing, although, as will be suggested below, there is some evidence that the herring might recover. The increase in pilchards was a dramatic one, for between 1925 and 1929 their eggs were recorded in the Channel as present in some years in summer, yet by 1950 the stock had reached 800 000 tons (Cushing, 1957). In a simple way, it appeared that the pilchard population had increased in two stages, first in 1930–1 and secondly in 1934–6. There were other changes during the same period, first the numbers of *Sagitta* of both species declined sharply in 1936 and secondly the spring spawners declined in 1936; in 1931 the non-clupeid summer spawners had declined.

Cushing (1961) interpreted the events as a two-stage change, the succession of two pilchard generations in 1931 and in 1934–6; the eggs and larvae that appeared in 1934–6 were said to be the progeny of an increase in larval population in 1930–1 that recruited to adult stock in 1934 and 1935 and spawned in their turn. The Plymouth herring spawned in January and February (Ford, 1933). A possible link between herring and pilchard was put forward by relating both to the winter phosphorus values. Herring recruitment was correlated directly with winter phosphorus one year after hatching (but not with that in the same year). The pilchard spawned in the western English Channel mainly in May, June and July (Cushing, 1957). The number of pilchard eggs was inversely correlated with winter phosphorus six months after hatching (but not with that six months before hatching). With the winter phosphorus values as a link between the herring and pilchard stocks, the quantity of herring hatched in winter is inversely correlated with the numbers of pilchards hatched in the following summer. But the decline in herring recruitment by year classes was probably complete by 1930–1 when the sharpest increase in pilchard eggs might have taken place. The decrement of winter phosphorus in 1930–1 (as compared with earlier winters) supposedly represented an increment of larval pilchards. Lastly, winter phosphorus was directly correlated with the stock of pilchards, as number of eggs, three and a half years later, that is, the decrement in phosphorus was inversely proportional to the subsequently recruited stock (Figure 59). Hence, the decrement of winter phosphorus might represent a quantity of larval pilchards during very early larval life before density-dependent mortality had become effective.

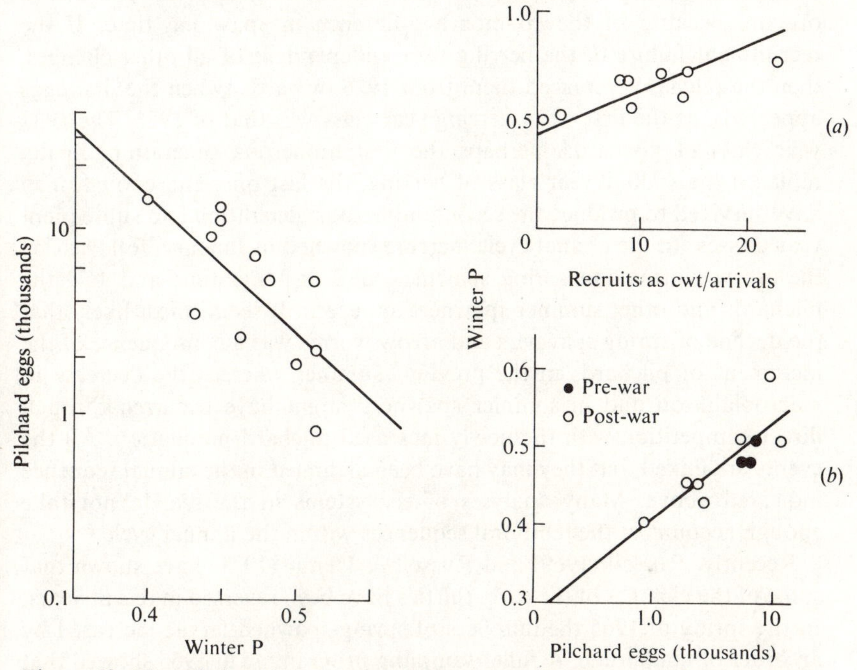

Figure 59. The relationship between herring and pilchards in the Western English Channel (Cushing, 1961). (*a*) Regression of winter phosphorus on herring recruitment as cwts/arrival. (*b*) Regression of winter phosphorus on numbers of pilchard eggs in the previous summer. (*c*) Regression of pilchard eggs on the winter phosphorus three and a half years earlier.

The interpretation is not as simple as it looks. First, the two periods of abundance of the two populations, herring and pilchards, were separated in time and the correlation could therefore suffer from serial errors. Secondly, an increment of pilchard larvae was not noticed in 1930 or 1931, but it must be recalled that the larvae were classed as clupeids and were not identified to species. It was at first thought that the pilchard larvae had to survive for six months to satisfy the decrement of winter phosphorus. The inverse correlation shown in Figure 59 must have represented larvae and not fish of six months old. During the period 1931–4, the lack of an increment in larvae might have been due to predation (or hidden in the classification as clupeids); however large the increment in numbers, nearly all of them must be eaten. If the decrement of phosphorus represented an increment of newly hatched pilchard larvae, which were eaten, then the decrement is transferred to another account, but it remains a decrement. Perhaps the observed increase in pilchard larvae in 1934–6 was really a consequence of reduced predation, perhaps by animals in the macroplankton that declined in numbers by a factor of four.

The nature of competition between the herring and the pilchard is obscure because of the six-month difference in spawning time. If the recruitment failure of the herring were independent of all other changes, then the pilchards replaced them from 1926 onwards (when the first eggs appeared), as the first weak herring year class was that of 1925. The 1931 year class of pilchards, perhaps the first numerous generation, finally replaced the 1930–1 year class of herring, the last one, and enough may have survived to produce the second numerous generation and subsequent year classes. In the annual cycle, herring spawned in January, followed by the arrow worms and spring spawners, and in May, June and July the pilchards and other summer spawners appeared. It seems most likely that the decline of spring spawners and arrow worms was a consequence of the increment of pilchards in the previous summer whereas the decrease in macroplankton and in summer spawners might have occurred through direct competition with the newly increased pilchard population. All the events are linked, but they may have been arranged in the annual sequence indicated above. Many analyses of ecosystems in the sea do not take enough account of the temporal sequences within the annual cycle.

Recently, Russell (1969) and Russell & Demir (1971) have shown that some of the changes of the early thirties have been reversed in recent years. In the spring of 1965 the numbers of spring-spawned larvae increased by an order of magnitude. A fuller sampling programme in 1966 showed that the numbers of spring-spawned larvae were as high as they were in the twenties and early thirties and the high numbers have persisted since 1965. In 1970 the quantities of macroplankton increased and the winter phosphorus increased during the winter of 1970–1, and both returned to the pre-1931 values. The most extraordinary point is the sequence of events, first the recovery in numbers of spring spawners and then five years, or a generation, later came the rise of macroplankton and the increase in winter phosphorus: it is the mirror of the earlier changes, when the decline in spring spawners happened, five years after the decline in phosphorus and macroplankton. During the second half of the sixties the numbers of pilchards in the western English Channel have declined. The sequence of events in the sixties and seventies, the reverse of that in the twenties and thirties, suggests that they are linked in a structural way that may well depend upon the annual cycle, as indicated above.

It is surprising that such a change in the fish populations had such deep and extensive effects upon the structure of the ecosystem. A community of herring, macroplankton and some demersal spring and summer spawners was replaced by a population of pilchards and perhaps this replacement is now in process of reversal. The elasmobranchs in the western Channel were probably reduced after 1930 (Cooper, 1948) and the average density of the benthos declined (Holme, 1953). If these events were also linked to the competition between herring and pilchard, all the ecosystem was involved in these changes. Both Steele (1965) and Gulland (1970) concluded that the link between the commercial catches of demersal stocks and primary production was closer than had been hitherto imagined. The suggestion is not only that fish catches depend upon the production of algae, but also that changes in the fish populations may dominate the ecosystems.

The temporal replacement of one pelagic fish species by another is not limited to the western English Channel. The periodic alternation between

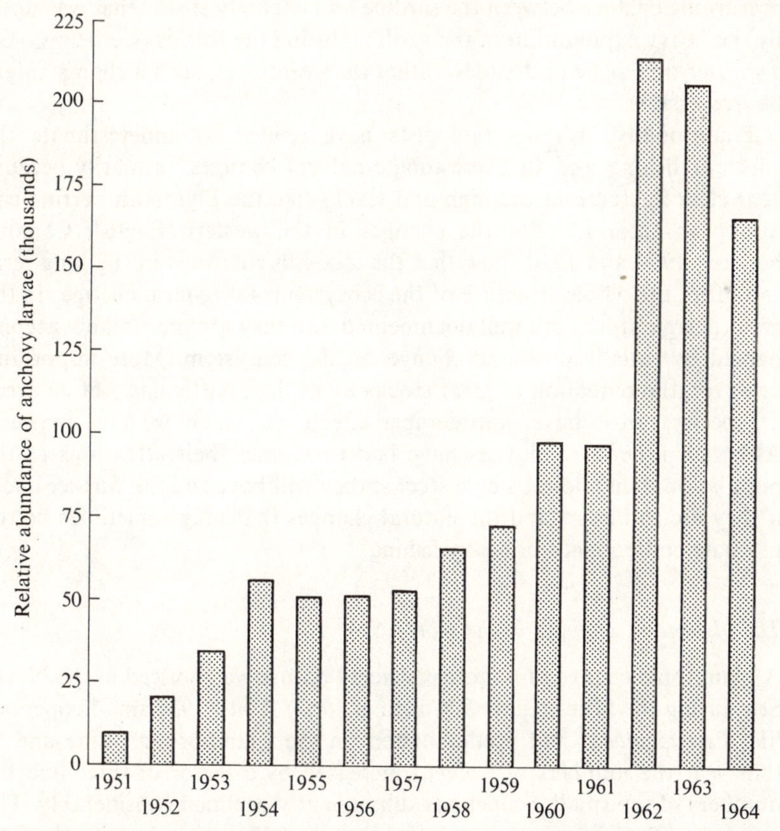

Figure 60. The changeover from sardine to anchovy off California (Ahlstrom, 1966).

the Norwegian and Swedish herring stocks has already been noted. After the failure of the Californian sardine stock with the 1950 year class, and the failure of the fishery in 1952 and 1953, it has been gradually replaced by a stock of anchovies (Ahlstrom, 1966). From the extensive egg counts, it can be seen that in 1950–3 the two stocks were both low; since then, the sardine stock has declined by a factor of five and the anchovy stock has risen by a factor of twenty (Figure 60). The rise of this stock lasted for about a decade. The area occupied by the spawning anchovy off southern California is now very extensive, whereas that of the sardine is very limited. Before 1945 the spawning area of the sardine was probably extensive. The replacement is now complete and it involved very large stocks reckoned in millions of tons in the virgin stock (in numbers, $>10^{10}$). In a study of scales trapped in anoxic sediments off the west of California, sardine and anchovy stocks have been shown to alternate for a period of more than a thousand years (Soutar & Isaacs, 1969) at least in the sedimentary pockets; for a long period the anchovy predominated. Perhaps there is a long-term precarious balance between the sardine and anchovy stocks that was upset by the heavy exploitation of the sardine during the forties (see below). On a shorter time scale of decades, rather than centuries, such a change might be irreversible.

Traditionally, fisheries biologists have tended to underestimate the effect of fishing and to overestimate natural changes, primarily because year class fluctuations are high and stocks like the Plymouth herring can disappear naturally. But the changes in the western English Channel between 1924 and 1970 show that the stock fluctuations in the long term can affect the whole structure of the ecosystem. Long-term changes in the great pelagic stocks are well documented and they are presumably accompanied by equally profound changes in the ecosystem. More important, however, the reduction of great stocks by fishing, particularly of an array of species, must have considerable effects of which we are unaware. Although fisheries biologists have had to restrict their attentions to the population dynamics of single stocks, they will have to look further afield if they are to understand the natural changes that may sometimes be the ultimate consequence of heavy fishing.

The planktonic changes in the North Sea

A planktonic change of a more restricted nature was noticed in the North Sea during the fifties (Burd & Cushing, 1962). Until 1950 small copepods like *Pseudocalanus* had predominated on the Flamborough Line and in that year the numbers of *Calanus* increased by a factor of three and the numbers of the smaller copepods subsequently declined considerably. The numbers of *Calanus* remained high until 1955 and subsequently they oscillated about a high level during the period up to 1961 when the line of sampling stations was abandoned. *Calanus* is the preferred food of the herring (in weight or in numbers, Hardy, 1924; Savage, 1937) and indeed the feeding efficiency, as wet weight per encounter, of the herring is greater when feeding on *Calanus* than when feeding on smaller or on larger animals (Cushing, 1964). Between 1930 and 1950, the herring of the southern North Sea had been growing slowly bigger. During this period, for example, a four-year old East Anglian herring increased from just below 24 cm to about 25cm but in 1951 these animals put on an additional centimetre in length. The increment of growth added in that year was equal to that in the previous twenty years. By 1964 a four-year-old herring was 4 cm longer than in 1930 (Burd, 1973). With some variation this increment was maintained in subsequent years. A consequence of this sharp increase in length was that the pattern of recruitment to the adult stocks of herring was changed. Before 1950–1 in the East Anglian herring fishery, a year class recruited partly at three years of age and partly at four, but after 1952 each year class recruited fully at three years of age (Cushing & Burd, 1957; Burd & Cushing, 1962) and so the stock became a little more vulnerable to fishing. During the early fifties the stock of herring increased in weight by 20–25% as the result of the growth change. In a preliminary study of the effect of fishing, the increased yield of weight masked some of the effects of fishing, especially that on recruitment during the late fifties and early sixties.

The information from the plankton recorder grid in the North Sea was used to extend the analysis. The temporal changes in *Calanus* in the southern North Sea were confirmed. The material was separated into two areas (west and east) and two periods in the year (April to June and July to September). The main nursery ground of the herring in the North Sea is on or near the Bløden ground, east of the Dogger Bank. The growth of the herring is conveniently estimated in length increments that are measured between the rings on the scale; with a scale length/fish length factor the scale measurements can be expressed in units of length, in cm. Such measurements are named l_1, l_2 and l_3 for the first three years of life, respectively. The increments, (l_2-l_1) and (l_3-l_2), which represent the growth in the second and third summers of life, were correlated positively with anomalies of *Calanus* density from a long-term mean in the period April–June, but not in July–September, in both regions, east and west. There was no correlation between the first year's growth of the herring with *Calanus* anomalies in any region or season. Iles (1968) demonstrated a negative correlation between the first year's growth and the number of larvae produced one year earlier, that is, the growth was more dependent upon density during the first year of life than upon available food. The later increments were correlated with food during the season of production,

April to June, but not with larval density. The juvenile fish grow in the eastern North Sea and migrate towards the north and west as they grow bigger (Burd & Cushing, 1962; Zijlstra, 1969) which accounts for the dependence on food in the second and third summers in both regions. Thus, the herring's growth is primarily density dependent in the first summer and primarily food dependent as they grow older; perhaps it would be truer to say that in the first year the effects of density masked the effects of food. The diminishing influence of density dependence with age is much as indicated in an earlier chapter. Hence, the increase in weight in the herring stocks during the fifties was probably due to the plankton change from small copepods to *Calanus*.

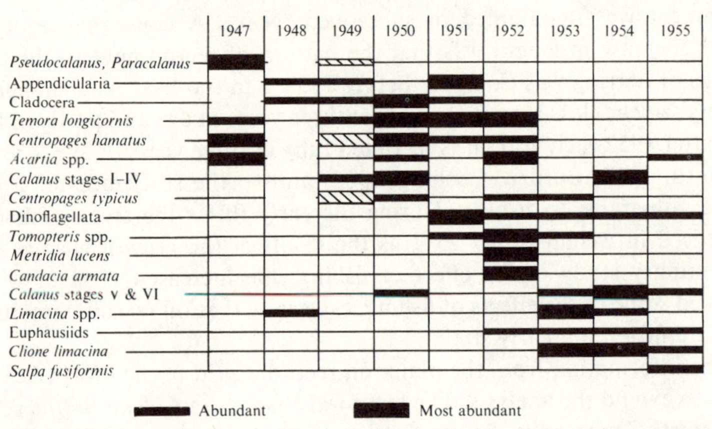

Figure 61. Changes in the planktonic regime in the North Sea (Glover, 1957).

The changes in the plankton in the southern North Sea were part of more general changes noticed in the material collected by the plankton recorder network. Glover (1957) showed that the North Sea plankton changed between 1947 and 1953 from a neritic type (exemplified by *Pseudocalanus*, *Paracalanus*, *Centropages hematus* (Lillj.) and *Acartia* spp.) to a more oceanic type (for example, *Clione limacina* Phipps, *Spiratella* spp., euphausids and *Centropages typicus* Kröyer) (Figure 61). Such changes might have followed from changes in the flow of Atlantic water into the North Sea; it is a restricted gulf in which the water is fully exchanged with the ocean every two or three years (Böhnecke, 1926) and so there is a continuous influx of oceanic animals most of which occurs in summer and autumn. Williamson (1961) analysed the changes for the decade 1949–59 by means of principal component analysis. In a matrix of rank correlation coefficients of the abundances of species on time, he distinguished four components of variance. More than half the total variance was correlated with a measure of vertical mixing, the difference between the temperature anomalies in the northern North Sea averaged for May, June and July and that for March; no such correlation was observed with indices of Atlantic inflow. The change in the planktonic regime from a neritic one to an oceanic one was probably associated with a change in the degree of vertical mixing. Such changes from year to year alter the timing of the production cycle and it is likely that the changeover from the neritic to the oceanic regime originated in the trend in differences of timing from year to year. Such a change is that which might be expected from a change in the wind strength and/or direction that might affect the production cycle for a period of time, as discussed earlier.

The increase of *Calanus* in the southern North Sea and the transformation of the plankton in the northern North Sea from a neritic type to an oceanic one were different facets of the same profound changes. A direct connection has been established with the changes in growth and recruitment to the North Sea herring. The changes in the western English Channel between 1924 and 1970 were also related to the numbers of herring and pilchard and they affected the ecosystem profoundly. It would not be surprising if temporal changes in fish populations were related to the changes described in the plankton. Analogous changes may have occurred off California, Japan and South Africa, if the dramatic variations in the abundances of pelagic species are reliable indicators. It follows that variation in growth and recruitment in pelagic species may originate in differences in the timing of the production cycle.

Changes in the fish populations

Records of catches of fishes are available for quite long time series. In many countries, records of total catch go back into the nineteenth century although records of all species caught by all countries are only available since 1946 and such statistics are incomplete (FAO Yearbooks). Records of the recruitment to fish stocks are available, but only for a small number of stocks.

Pelagic stocks

Four large stocks of pelagic fish have been heavily exploited and the fisheries that they supported have become extinct. Figure 62 shows dramatic changes in the catches of the Japanese sardine, the Californian sardine, the Hokkaido herring and the Norwegian herring (Figure 63). The catch of each reached about one million tons or so and then died away and the failure of each has been attributed to natural changes in the environment. Traditionally, fishermen expect stocks of herring-like fishes to appear and disappear (for example, those in Loch Fyne or off the Firth of Forth) and, as noted above, the Norwegian herring fishery fluctuated with a regular periodicity since the fourteenth century. The capacity for stabilization in numbers is low in the herring and so very large periodic fluctuations might be expected, particularly at the edge of the animal's range. In each fishery the effort exerted on the stock was high, particularly at the period of peak catches and in any case it must be high if very large catches are taken. There has often been a conflict in evidence between the possible effect of fishing and that of environmental factors, nowhere more noticeable than in the history of the pelagic fisheries.

The Hokkaido herring is a stock of Pacific herring (*Clupea pallasi* Cuvier et Valenciennes) and it spawned along the western coasts of Hokkaido (the northernmost of the Japanese islands) in spring. The eggs were laid on seaweeds and gravels very close inshore, just like the British Columbian herring or the White Sea herring, both of which belong to the same species. The fish were caught with set nets or traps (and some drift nets) and estimates of the numbers of nets are available from 1870 onwards (Motoda &

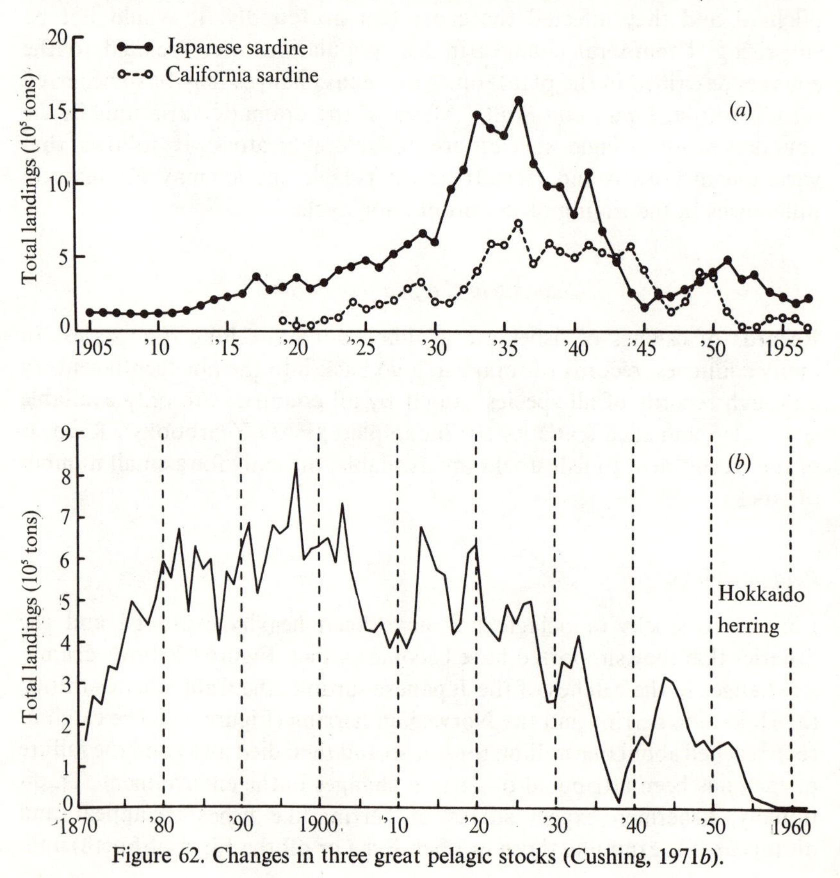

Figure 62. Changes in three great pelagic stocks (Cushing, 1971*b*).

Hirano, 1963). The peak catch was reached in 1897 (Figure 62) and it subsequently declined fairly steadily until it virtually disappeared in 1960. The stock probably extended into the Sea of Okhotsk and fish also spawned on the coasts of the island of Sakhalin (Ayushin, 1963). On the island of Hokkaido since 1900 (after the period of peak catch) the spawning runs have disappeared progressively from south to north. Motoda and Hirano showed that the greatest number of nets was reached in the period of 1899–1910, after the peak, and subsequently it declined. Examination of the catch and effort data showed that the 1939 year class provided a recovery with its successors between 1942 and 1952; subsequently catches again declined because effort increased between 1946 and 1956. Between 1910 and 1950 the annual survival rates were very low indeed. It is difficult to avoid the conclusion that in two periods, after 1897 and after 1952, fishing was heavy and that it reduced recruitment. In stocks of herring-like fishes, there can be no loss of catch due to growth overfishing and reduced catch must mean reduced recruitment due to either fishing or to natural causes. This happened when fishing was intense. Motoda and Hirano, however, suggested that the herring shifted its migration route in response to the warming of the sea in the thirties and forties.

The Japanese sardines (*Sardinops melanosticta* (Temminck and Schlegel)) were caught off Honshu and Hokkaido (Nakai, 1960) and in 1930 there were 700 purse seiners, 1000 in 1940 and the effort remained high in subsequent years (Kurita, 1960). Peak catches of two million tons were reached in 1936 (Figure 62), but after that date they fell dramatically with the failure of the year classes, 1938–41. During the period of prosperity the Japanese caught fish as two- and three-year olds, whereas the Koreans caught maturing fish. During the period of decline, after the collapse in the early forties, the Japanese concentrated on fish in their first summer of life and a further decline in catch followed a period of high and sustained effort. Yokoto (1951) suggested, probably correctly, that catches failed because of recruitment failure under the pressure of fishing. After thirty years the stock has not recovered and it has been replaced by anchovies. If such a dramatic collapse was due to fishing, it generated the competitive replacement of sardines by anchovies. Nakai (1960) associated this failure in recruitment with an anomaly in the flow of the Kuroshio current in the years 1938–45: cold water appeared off the coasts of Honshu and the axis of the current was diverted (Uda, 1952*b*) and it was suggested that the fish larvae might have starved. However, spawning occurred in water of normal temperature in 1941, south-east of the anomalous area (Nakai, 1960). The proposed environmental source of recruitment failure was disproved, but the fishing effort was high on pre-spawning fish.

The Californian sardine fishery flourished between 1920 and 1951. The peak catch was reached in 1936 and catches remained more or less steady until 1943, after which they declined. The year classes of 1949 and 1950 failed completely and with them the fishery became extinct. The fishing effort reached a plateau in 1936 and remained high until 1947 after the collapse had started. As stock declined growth in the first year increased and this density-dependent effect continued with the remaining year classes (Iles, 1973). Marr (1962) found a positive correlation between year-class strength and the cumulated temperatures off Scripps pier (in southern California); if low temperature delayed spawning, recruitment might have been adversely affected. However, changes in sea temperature in such time series can indicate climatic changes in a broad sense. But if expressed as deviations from a stock/recruitment curve, the thesis that

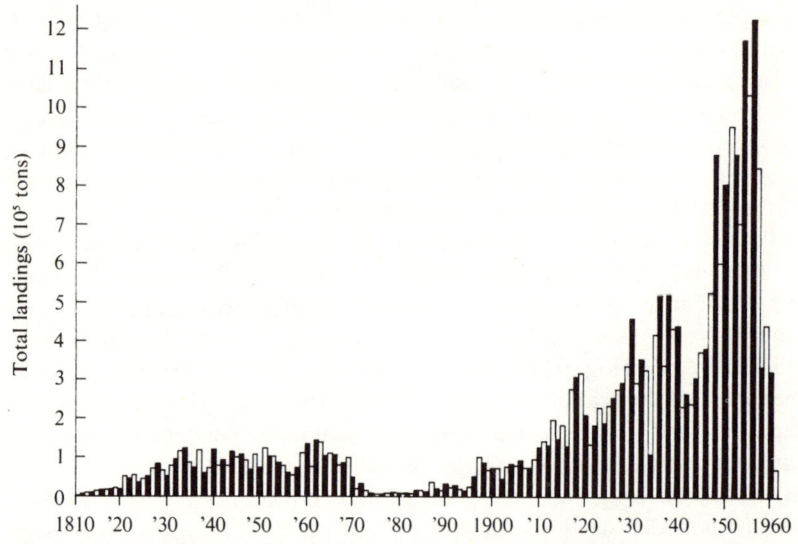

Figure 63. The catches of the Norwegian herring for a long period of time (Devold, 1963).

recruitment failure is due to fishing and that recruitment was correlated with climatic change, are not exclusive. Murphy (1966) has contrasted the the recruitment to the fished stock with that to the unexploited one and concluded that the reproductive capacity of the stock was reduced by fishing.

A time series of catches of the Norwegian herring throughout two of the famous periods is shown in Figure 63. The first extended from 1820 to 1870 and the second lasted from about 1910 to 1968. Between the two Norwegian periods, herring were caught on the Bohuslån coast of Sweden during the period of the Swedish herring. The fish were probably of North Sea stock and its fluctuations are perhaps governed by climatic changes out of phase with those favouring the Norwegian stock, as will be suggested in the next chapter. Towards the end of the first period, the spawning grounds appeared to shift north from Utsira to Stadt and from Stadt to the Lofoten Islands which may have been a progressive shift from stocklet to stocklet in three steps as described later by Runnstrom (1933). During the most recent herring period the peak catch of 1.5 million tons was recorded in 1956 with the 1950 year class. Before 1950 good year classes occurred quite frequently, but afterwards they were less frequent. The number of purse seiners increased from 273 in 1946 to 549 in 1955 and it did not decline until the early sixties when the catches had fallen by three times. Perhaps the high rate of exploitation at a low stock level has prevented the emergence of good year classes. Devold (1968) believes that recruitment failed naturally at the end of the last Norwegian period, but as will be pointed out in the next chapter, there is no evidence yet that the associated climatic change has taken place.

In the early stages of fisheries biology, 'man-made causes' and 'environmental causes' were opposed. So long as only changes in stock density were described, a decline could have been due to increased mortality or to decreased recruitment, because the two parameters were not distinguished. When decline in stock density was correlated with increasing effort, the cause could have been an increased mortality with constant recruitment, or a natural recruitment failure independent of the increased effort. The two possibilities could not be distinguished without a study of the age distributions and this was the crux of the argument between Thompson (1952) and Burkenroad (1948) on the interpretation of changes in the stock of Pacific halibut. In each of the papers that describe the failures of the four great pelagic stocks, the decline is attributed to environmental causes. The Hokkaido herring did not fail because the sea became warmer after its collapse. The Norwegian herring did not reach the end of its natural period in the sixties, because in the earlier Swedish periods European waters were very cold and the Baltic froze in winter; by 1971 this period had not yet been reached. The Japanese and Californian sardine stocks collapsed due to recruitment failure during periods of high fishing effort.

Cushing (1971*b*) and Cushing & Harris (1973) have suggested that the degree of density dependence was a function of fecundity. Consequently, the low fecundity pelagic fishes with poor capacity for stabilization are vulnerable to long-term climatic changes. Ricker (1958), in his analysis of stock and recruitment curves, showed that there is a limiting level of exploitation, which, if exceeded, reduces recruitment to zero and under such circumstances a stock can decline to such low levels that the fishery is extinguished. Cushing (1971) suggested that, for pelagic fishes the limiting level of exploitation (E_l) is comparatively low (0.35–0.50).

In the same publication the collapse of the Californian sardine fishery was examined in some detail. The rates of exploitation were estimated by two methods and the limiting level was exceeded on three occasions, in 1936 (when the subsequent recruitment did not fail), in 1942 and in 1946; in the latter years recruitment was reduced to very low levels indeed – with the effect that the fishery subsequently collapsed. It was also shown that the limiting level of exploitation was exceeded in the Norwegian herring stock after 1955, which explains why no large year class appeared after that of 1950. The limiting level was probably exceeded at the time of peak catch of the Japanese sardine and so the failure of the three subsequent year classes need not have been attributed to a shift in the axis of the Kuroshio current. The argument cannot be extended in detail to the decline of the Hokkaido herring, but recruitment must have declined during a period of intense fishing effort.

The effect of fishing upon the catches of the four great pelagic stocks is perhaps more dramatic than expected, perhaps because the very high yields tempted us to believe that the stocks were inexhaustible. If, however, the pelagic stocks are vulnerable to recruitment overfishing they need much more international care than has been lavished upon the more stable demersal stocks. The problem is really that of how recruit year classes are established, the degree of density dependence and the degree of environmental modulation. The Downs stock of herring was probably fished at a moderate rate for centuries and declined under heavy fishing between 1955 and 1965 (Cushing & Bridger, 1966; Cushing, 1968*b*). Because year classes were not very variable, the decline was slow. That of the Norwegian herring under the same conditions of fishing was more rapid, merely because the year classes were more variable. What is needed is a thoroughgoing investigation of the causes of variable recruitment or of how fish stocks regenerate themselves naturally.

Trends in other stocks

Changes in demersal stocks have also taken place. In the southern North Sea, haddock catches declined by two orders of magnitude between 1923 and 1960, whereas in the northern North Sea, during the same period, the catches of haddock doubled. A similar trend, but less well defined, can be observed from north to south in the densities of whiting. Sole catches increased by an order of magnitude in the German Bight, whereas those of cod and turbot remained more or less steady for the whole period of more than forty years (Beverton, 1962). Such trends in catches are really trends in recruitment. Between 1887 and 1953 the recruitment to the Karluk river salmon varied by a factor of about three, but it decreased over the whole period by very nearly an order of magnitude (Rounsefell, 1958). Recruitments to stocks of herring have been estimated. That of the Downs herring varied by a factor of three to five between 1924 and 1936, but between 1946 and 1961 it declined by a factor of three and in subsequent years by much more (Cushing & Bridger, 1966). The variability of the recruitment to the British Columbian Lower East stock ranged over an order of magnitude between 1915 and 1927, but between that date and 1952 the variation was reduced to a factor of three; within the period recruitment increased by a factor of about five (Taylor & Wickett, 1967). In the Norwegian herring, recruitment varied by a factor of about ten or twenty; between 1904 and 1950 there was a slow increase of about three times (Devold, 1963). In the sixties the fishery was extinguished as the stock was reduced to very low levels indeed. Garrod (1968) has published a series of data on recruitment to North Atlantic cod stocks. At West Greenland and at Iceland, between 1924 and 1959, recruitment varied by about an order of magnitude in the early part of the period, but the variation was somewhat less in the later part. Between 1931 and 1962 the year classes in the arctonorwegian cod stock varied by five to ten times, but in later years the recruitment was considerably reduced. The recruits to the stock of Georges Bank haddock between 1929 and 1963 were very variable, differing by a factor of twenty to forty before a dramatic collapse in the late sixties (Grosslein & Hennemuth, 1973). In the North Sea sole stock, recruitment varied by a factor of sixty between 1930 and 1966 (Anon., 1970).

In the examples quoted by Beverton the trends in catches were recorded for the most important stocks in the North Sea for nearly forty years. For a number of stocks that were examined for longish periods the variability and trends in recruitment were described. The variation ranged from a factor of two to three to one of sixty; the first was the environmentally protected stock of Karluk river salmon and the second was a sole stock at the northern end of its range. The year-to-year changes in year-class strength vary considerably, presumably in response to annual variations in the relevant climatic factors, wind strength or radiation. There are also long-term trends, slowly upward or downward over periods of twenty or thirty years in response to long-term trends. Collapses in recruitment also occur that are probably due to too much fishing; one hopes that as fisheries biologists think explicitly in terms of stock/recruitment problems, such collapses will disappear into history. For our present purpose the slow long-term trends in recruitment, upward or downward, are of greater interest. The examination of all the trends summarized in the previous paragraph shows that at least half of them change in time over a longish period. If such trends are modulated by climatic factors, as suggested, it becomes desirable to understand the nature of such changes. Weather can be forecast one, two or three days ahead, but fish stocks may rectify the longer term climatic variation, as will be indicated in the following chapter.

There is a paradox in the interpretation of the long-term events in the pelagic and demersal fisheries. The great fluctuations in the herring-like fisheries were held to be dominated by environmental causes, whereas in demersal fisheries changes are attributed to fishing. In this chapter environmental changes have been detected in the demersal stocks, whereas the great collapses of four pelagic stocks have been attributed to fishing. In the plaice stocks of the southern North Sea, the effect of fishing has been well established for a long time; when an increase in the German Bight was found it was detected by changes in the numbers of eggs, independently of the effect of fishing (Bückmann, 1961). In the pelagic stocks, the effect of fishing was appreciated, but its possible effect on recruitment was underestimated. The paradox is resolved in recalling that the vulnerability of herring-like fishes to fishing and to environmental change follows from the nature of their stock/recruitment curves.

Conclusion

The long-term changes in marine ecosystems that may result from climatic change can affect the recruitments to fish populations, not only as an upward or downward trend in a time series, but perhaps in detail from year to year. In addition, changes in fish populations can affect entire ecosystems. The events in the marine communities in the western English Channel during the twenties and thirties started when recruitment to the herring population there failed. Not only were the herring replaced by pilchards within about three generations, but consequential changes affected the whole ecosystem, the quantities of macroplankton, the numbers of planktonic predators, the elasmobranchs, the numbers of spring and summer spawning fishes, and also the maximum in winter phosphorus itself. This quantity represents the nutrient reservoir and its temporary reduction by about one-third for a period of forty years is an index of change that is considerable and to a considerable extent not understood. Equally far reaching in a different way were the changes in plankton and in herring growth noticed in the fifties in the North Sea. The increment in growth must have increased the stock throughout the North Sea by nearly one-fifth. Consequently, catches increased and the changes in numbers due to fishing were masked to some extent.

The recruitments to pelagic or demersal stock vary considerably between a factor of two or three to one of twenty to sixty. Differences in variability between stocks can be associated with differences in potential productivity. There are upward or downward trends in the recruitment to herring populations; those in the demersal stocks tend to be less noticeable or absent. This follows from the relation between density dependence and fecundity: a cod stock that is highly fecund is insulated from environmental changes in a way that a herring stock is not. Indeed such variability may be rectified by a herring stock (between abundance and extreme scarcity), but resisted by a cod stock.

The most important task for a fisheries biologist is to determine the death rate by fishing on any stock. If the fishing mortality is high enough to reduce recruitment during a period of declining year classes, the stock will collapse earlier than expected. Conversely, during a period of rising recruitment, collapse would be delayed. Whatever the trend is and whatever the results of its progress in terms of the stock, the biologist has failed if collapse is incorrectly predicted. Much of the discussion in the literature on the collapse of the four great pelagic stocks revolved around the question of whether the failure of recruitment was determined environmentally or by fishing. The argument was sterile because it could not proceed, but a development would be to ask whether fishing precipitated failure during a period of climatic deterioration, or conversely whether a stock was insufficiently exploited during a period of climatic amelioration. It might be suggested that it is too difficult to forecast a climatic change. However, as will be shown in the next chapter, long-term trends can be described and fish stocks appear to be peculiarly responsive to them.

11

Climatic changes and life in the sea

Introduction

In the last chapter the long-term changes in marine ecosystems and in recruiting year classes to fish stocks were described and a link between both and climatic changes was mentioned. There are two sorts of climatic variation that are relevant, the differences from year to year and the changes over long periods. The first have been correlated with a variety of climatic factors but the second has only been noticed in those time series that are long enough. In an earlier chapter a link between a fish stock and the local production cycle was proposed. The nature of the link was suggested in the match or mismatch of the production of fish larvae to that of their food; in temperate waters fish tend to spawn at fixed seasons in the same place from year to year, whereas the production of the larval food varies in timing. In the present chapter we investigate the nature of the correlation between year class strength and climatic factors.

The dependence of recruitment upon climatic factors

The dependence of recruitment in the Californian sardine stock upon the cumulated temperature in the year of hatching off Scripps pier in southern California (Marr, 1962) has already been referred to. There is no evidence that this correlation failed, but earlier Walford (1946) had established an analogous relationship based upon annual differences in salinity, that subsequently failed. Other early attempts to establish such relationships led to correlations that persisted for a long time and then failed. For a period of about a decade during the thirties, Carruthers (1938) correlated the year class strength of East Anglian herring with the wind strength in the area of spawning, that is, in the eastern English Channel and the southern North Sea. The year classes of the North Sea haddock were correlated with southerly winds (Carruthers, 1938), with westerly winds (Rae, 1957) and with the quantity of *Calanus finmarchicus* (Cushing, 1966*b*); Hermann & Hansen (1965) established a positive correlation between the recruitment to the West Greenland cod stock and temperature between 0 and 4° C; Martin & Kohler (1965) demonstrated a negative correlation between the annual cod catches off New Brunswick and temperature, between 5 and 9° C at St Andrews, New Brunswick. There is no evidence yet that such correlations with temperature in the north-west Atlantic have failed. But those in the North Sea were transient: that of the recruitment to the Downs herring did not survive the thirties and those of the North Sea haddock, save that based on the quantity of *Calanus*, failed during the fifties. The *Calanus* correlation was destroyed with the very strong year class of haddock in 1962, twenty-five times larger than any of its predecessors. If the magnitude of recruitment depended upon the match or mismatch of larval production to that of their food, it should depend on variation in radiation, wind strength or direction. As larval development is an inverse power function of temperature, cold water would delay larval production considerably as indicated in an earlier chapter; for the same reasons, however, the average differences in temperature might not be of very great importance. Differences in radiation from year to year may be of some importance, but differences in wind strength may well be predominant. The rationale of this explanation depends on the production ratio (D_c/D_m) referred to in an earlier chapter. The compensation depth, D_c, increases during the spring with the increasing altitude of the sun and upon the distribution of cloudiness and of these the first is more important than the second. The depth of mixing decreases during the spring as the wind strength slackens. Climate affects the rate of change of the production ratio during the spring and early summer. Thus there is good reason to correlate year class strength with variations in the strength of the wind. However, the depth of mixing depends also upon the wind's direction particularly if it shifts with respect to a coastline near which the fishes spawn. If the correlation were based upon wind strength only and took no account of a possible long-term shift in direction, it might be expected to fail. It is characteristic of correlations between recruitment and wind strength that they fail with the addition of information to quite a long series extant.

Evidence of quite a different character, on the relation between recruitment and climatic change, is found in the distribution of individual year classes. In the southern North Sea very strong year classes of plaice, cod and sole originated in the cold winter of 1962–3. Between 1962 and 1968 a number of very large year classes of cod, whiting and haddock appeared in the North Sea. Dickson *et al.* (1973) have correlated the recruitment of North Sea cod with temperature anomalies by months and by position on a line across the North Sea from the Humber to the Skaw; the highest correlation was obtained in April and May on those stations close to the English coast, i.e. the Flamborough Off Ground, where the cod are known to spawn. An extension of the argument suggests that the series of strong gadoid year classes during the sixties was associated with a decline in sea temperature, particularly in April and May, and it is possible that the match of larval production to that of their food was improved in those years.

Many years ago it was noticed (Tåning, 1953) that the outstanding year classes of cod were common throughout most of the North Atlantic. Templeman (1965) extended the study to cod, haddock and herring stocks in the Atlantic from 1902–62. Of ten stocks, nine were abundant in 1950; the 1904 year class was abundant in the arctic cod, Icelandic haddock and very abundant indeed in the atlantoscandian herring. In 1922 there were abundant year classes of cod at West Greenland and Norway and of haddock at Iceland and the same groups (with the addition of Icelandic cod) were abundant in 1934. All cod year classes and those of the Grand Bank haddock were abundant in 1942. In 1955 Icelandic cod and the Grand Bank haddock were abundant and in 1956 nine of the ten stocks listed were classed as abundant or very abundant. The classification as 'abundant', etc. is perhaps a little arbitrary, but more important is the fact that the phenomenon was so widespread. Climatic change is by its nature extensive; recent changes in the surface temperature have occurred across the whole Atlantic even if a rise in the east is often accompanied by a fall in the west. The most important point about Templeman's table is the implication that individual year classes are affected by climatic variability, within the limits imposed by the nature of the classification.

It is now known that in the years 1904–6 a number of biological events were noticed, the invasion of British waters by indicator species, the appearance of pilchard eggs in the western English Channel and, most important of all, it was a period of salinification, in Dickson's (1971) terms, which will be described below. The great year class of atlanto-scandian herring appeared in 1904 and the last notable one, that of 1950, appeared during a period of high salinity inflow. Figure 64 shows most of such periods during this century and it is obvious that both the changes described in the Channel and in the North Sea, in the previous Chapter, and the array of year classes of ten stocks during a period of sixty years are linked.

A third form of evidence is concerned with the effects of climatic change expressed in general terms. Ottestad (1942) correlated differences in total catches of the arctonorwegian cod stock for fifty-five years with periodicities in the widths of rings on pine trees in the area of the Vestfjord where the fish spawn. The time series was smoothed with a form of harmonic analysis in an attempt to detect cycles in climatic change. From the pine ring material four periodicities were calculated, each with a different phase in any given year. The average proportion of a year class in the age distribution is known in each year. Each fraction was raised by the sine of the phase value and then the four curves were added. The summed curve was fitted by least-squares to the catches ($r = 0.84$; $n = 55$). A clearer correla-

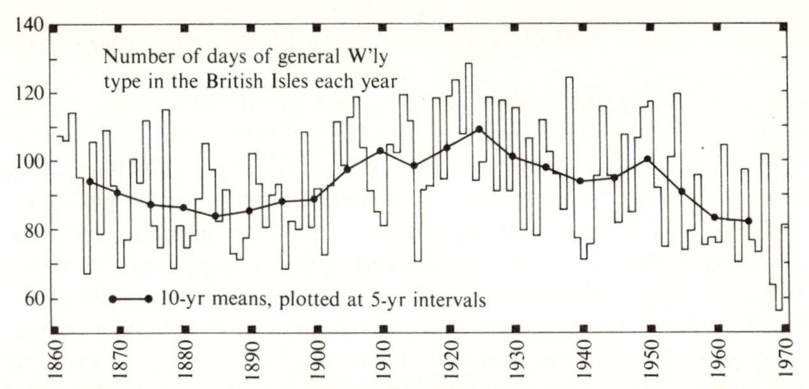

Figure 64. The number of days classified as 'westerly' (W) over the British Isles between 1863 and 1963 (Lamb, 1972).

tion might have been established had a series of year class strengths been available. The important point is that differences in catches are correlated with events that are themselves mediated by climatic changes. Because the effect on the growth of pine trees must be quite distinct from that on the cod year classes, so correlation is based on climatic changes in general terms. Such a general correlation may be contrasted with those on particular factors; for example, those on wind strength that failed perhaps when the direction changed. A much more recent form of general correlation was given by Dickson and Garrod (in Cushing, 1972*a*); they related true recruitments (as estimated by virtual population analysis) to the arctonorwegian cod stock to differences in surface salinity in the German Bight and to temperature differences on the Kola meridian. As will be shown below, both climatic indices are functions of the degree of southerliness of the windstream across Europe. The two correlations (that of Ottestad and that of Dickson and Garrod) cover distinct periods, which together extend for nearly eighty years. Hence, the year classes of the arctonorwegian cod stock are correlated with annual climatic factors in detail. Further, it is probable that a number of indices could be used to investigate the mechanisms by which recruitment is generated.

The evidence of a link between the magnitude of recruitment and climatic change is of three sorts: (1) the correlation with particular climatic indices, for example, wind strength from a given direction; (2) the appearance of year classes common to disparate stocks across the breadth of the Atlantic ocean, for example, the occurrence of nine abundant year classes in ten stocks both in 1904 and in 1950, both of which years were those of high salinity inflow; (3) the correlation between general indices of climatic change and the catches (and recruitments) of the arctonorwegian cod stock. The three forms of evidence are different and their disparity carries the conviction that the link between climate and recruitment really is rooted in the match or mismatch of larval production to that of their food.

One correlation between a fish stock and climatic change has already been noted, that between the alternation of Norwegian and Swedish herring periods and the degree of ice cover north of Iceland (Beverton & Lee, 1964). When the ice was close to Iceland, during the cooler period, the Swedish herring predominated whereas Norwegian herring were more abundant in the warmer one. Swedish herring were caught during the winter in pound nets on the rocky skerries that lie off the western (Bohuslån) coast of Sweden. According to Johansen (1924) the autumn spawning herring of the northern and central North Sea (in present-day terms, the Buchan and Dogger stocks; Anon., 1964) overwinter in the Skagerrak which is bordered by the Bohuslån coast. North Sea herring normally enter the Skagerrak in winter when the deep inflow in the Norwegian Rinne might be more intense (Dickson, 1971) and perhaps they reach the Swedish coast during the cooler periods. The fish exploited by the Swedish fishery were possibly North Sea spawners. Johansen examined the fish for ten years between 1909 and 1920 and concluded from the state of maturation and from meristic characters that they were North Sea autumn spawners; the Norwegian fish were spring spawners. More recently, Höglund (1972) has examined the skeletal remains of herring found in the middens of an oil reduction industry of an earlier Swedish herring period in the latter half of the eighteenth century. Rings on the cross sections of the vertebrae indicated age and the lengths of the fish of a given age corresponded closely to those expected today from the North Sea autumn spawners. Johansen listed the Swedish periods: 1307–62, 1419–74, 1556–87, 1660–89, 1748–1808, 1878–96 (or 1920).

Runnstrøm (1933) showed that the Norwegian herring spawned in three main regions off the coast of Norway, off Utsira near Haugesund, off Stadt (north of Bergen) and north of the Lofoten Islands and he may have distinguished three stocklets. Recently, Devold (1963) recounted the history of the fishery towards the end of the last Norwegian period in the decade 1860–70 as it was described at the time by Boeck (1871); the stocklets died out from south to north and so the fishery ended in the far north. If their fortunes depended upon the match of spawning time to that of the production cycle, then the decline of the stock may have taken place as the cycle became delayed in sequence in a northerly direction; if the wind had shifted from southerly to westerly on the Norwegian coast in spring such an effect would have been achieved. The shape of the coastline with respect to prevailing winds is important to such a concept. Such a shift in wind direction might account for the decline in the Norwegian herring and the rise of the Swedish stock at about the same time. Because the North Sea herring spawn close to the coasts of the British Isles near the Dogger and off the Scottish coasts, a westerly wind with a short fetch might generate a closer match of spawning to production than a southerly one with a longer fetch. In a purely speculative way, the Swedish fishery was rooted in an increase in the abundance of the Dogger and Buchan stocks or in a greater penetration of the fish of these stocks towards the Sound from the Skagerrak. The complex alternation between the two disparate fisheries is explained in a single cause, the shift of wind strength and direction. Devold (1963) used an equally simple hypothesis to account for the alternation between the spring-spawning Norwegian herring and the autumn-spawning Swedish herring by postulating a shift in the spawning ground by the Norwegian stock. However, more recent studies have suggested that such a change of ground with a gradual six months' shift in spawning time and a change of meristic characters is improbable (Cushing, 1966*b*, 1970; these papers describe the match of spawning time to time of production and the probably fixed time of spawning in stocks of fish in temperate waters). Devold's hypothesis, although simple, does not account for the observed changes. An alternative one is to suggest that the Norwegian stock and the Swedish stock are distinct and have remained distinct throughout the long history of the herring periods. Then the alternation of abundance in the two stocks is looked upon as the consequence of a long-term periodicity, as will be described in the next section.

Many of the correlations with environmental factors have been criticized because they failed or because the underlying theory was suspect. For example, Carruthers (1938) believed that the correlation of haddock year classes with westerly winds was generated when larval haddock were blown over deep water so that they could not settle to the bottom. Saville (1959), however, showed that the larval haddock never reached the deep water. The advantage of the thesis of match or mismatch of larval production to that of their food is that three forms of evidence can be all linked together. Further, the same hypothesis can explain the well-known periodicity of the Norwegian and Swedish stocks in a simple manner.

The nature of climatic change

After the last Ice Age in Europe, there was a post-glacial climatic optimum between 4000 and 2000 B.C., followed by a deterioration. Climatologists use the words 'amelioration' and 'deterioration' to describe changes in the weather; the first describes a period of pleasing weather and the second a nasty one. Between A.D. 400 and 1200, with a peak between A.D. 800 and 1000, there was a secondary optimum of dry, warm and storm-free weather. It was the period of the great Viking voyages and the settlements in Iceland and Greenland. Saxon civilization flourished in England and the saints travelled between Lindisfarne, Whitby and Iona. The high alpine passes were open and thirty-eight vineyards were recorded in the Domesday Book. Subsequently, there was a decline to variable weather of drought and storm, with floods in England and Holland. There was a partial recovery between 1400 and 1550 before a further decline into the Little Ice Age (1550–1850) when glaciers crept down the alpine valleys and there was considerable suffering in Scandinavia. In Britain, the traditional white Christmas was common and before 1831 the Thames was frozen every twelve or fifteen years, rather than once or twice a century which was the rule before the Little Ice Age. Mild summers appeared towards the end of the eighteenth century and they were followed by mild winters in the mid-nineteenth century. Towards the end of the nineteenth century in north-west Europe the climate deteriorated a little and then, except for the second decade, it slowly became very pleasant in the first half of the twentieth century, particularly during the thirties and forties. This account was taken from Lamb (1966) and Bray (1971).

The sun's energy is probably constant, or nearly so (indeed there may be slight trends in the solar constant), but its radiation at the earth's surface varies as its altitude, which itself varies daily and seasonally and with differences in dust, moisture in the air and snow on the ground. Hence, the input of heat varies seasonally and latitudinally and it is modified by mountain ranges and continents and markedly by snow. Thermal gradients in the atmosphere tend to increase in middle latitudes as an effect of the main icefields. The inequalities in heating generate density differences and hence winds. Warm air rises in low latitudes and cold air sinks nearer the poles and in middle latitudes the upper westerlies stream at the centre of the vertical circulation at a height of several kilometres. Variations in the trends of these westerly winds, as Rossby waves or in response to geographical features, generate the differences in the distributions of cyclone and anticyclones that cause the variations in wind strength and direction.

Much of the study of climatic change is concerned with long-term variation, such as might have started the Little Ice Age. However, the longest periodicity of interest to fisheries biologists so far is about fifty years or so, about that of a Norwegian or a Swedish herring period. Figure 64 shows the number of days classified as 'westerly' over the British Isles between 1873 and 1963. It was suggested in the last section that the alternation between Norwegian and Swedish periods was associated with a shift from westerly to southerly winds and back again over a long period. The last Swedish period occurred between 1870 and 1900 with catches at a lower level persisting until 1918. The last Norwegian period started just before 1920, and continued until the late sixties, when the fishery was extinguished. It is possible that the Swedish fishery occurred when the westerlies were increasing; the short fetch with westerly winds might have encouraged the stocks off the British coasts of the North Sea, whereas those that spawn on the Norwegian coasts might have declined in the face of the long fetch of the same winds across the Norwegian Sea. During the period of southerly winds the stocks spawning close to the British coasts were exposed to winds with a relatively long fetch, but the Norwegian fish were protected to some extent by the same wind because of the shape of the Norwegian coastline. If the shift from southerly to westerly wind and back again is associated with the herring periods, such a form of explanation can only make sense on the basis of the match of larval production to that of their food. The details of Figure 64 are equally interesting: the years of high southerliness relative to the smoothed curve tend to be some of those of high salinity inflow into the North Sea.

It is uncommon for events to be recorded for such long time periods. However, Dietrich (1954*b*) has published monthly anomalies of surface temperatures from 1849 to 1950 at Horn's Reef LV off Esbjerg in western Denmark. From these observations a period of cold winters and mild summers towards the end of the nineteenth century gave way to a period of mild winters and warm summers in the North Sea during the thirties and forties of the present century. Dietrich also charted wind strengths and directions from records at the light vessel; in the cooler periods, the spring winds have blown from the south and south-west rather than from the west. The summer winds blew from the west rather than from the west-north-west during the period of warming and such changes in wind strength and direction are perhaps those that mediate shifts in the timing of the production cycle.

Figure 65 shows the trends in surface temperature between 1880 and 1960 for sea areas off the British Isles (Smed, 1965). The chart shows the annual anomalies from a mean calculated for only part of the period, so the absolute values are biased but the temporal trend is not. In each area the sea was coolest during the first two decades of the twentieth century, but subsequently in the thirties, forties and fifties, the sea became warmer, by about a degree or so. Bjerknes (1963,1964) has explained the warming of the north-east Atlantic by showing that as westerly winds diminished in strength the quantity of heat remaining in the sea increased and the temperatures rose. The west wind slackened as the pressure difference between the Iceland low and the Azores high declined. In the mid-nineteenth century the Iceland low was weak but in the eighties it intensified giving low pressures, high pressure differences between it and the Azores high and hence strong westerly winds. Subsequently, in the thirties and forties of the present century, the pressure difference between the Iceland low and the Azores high decreased and the westerlies slackened again. Bjerknes demonstrated the effect in correlating pressure in the Iceland low with sea temperatures in the region 60°N.–50°N. for a long time period. Thus, the long-term variations in temperature indicate the trends in wind strength

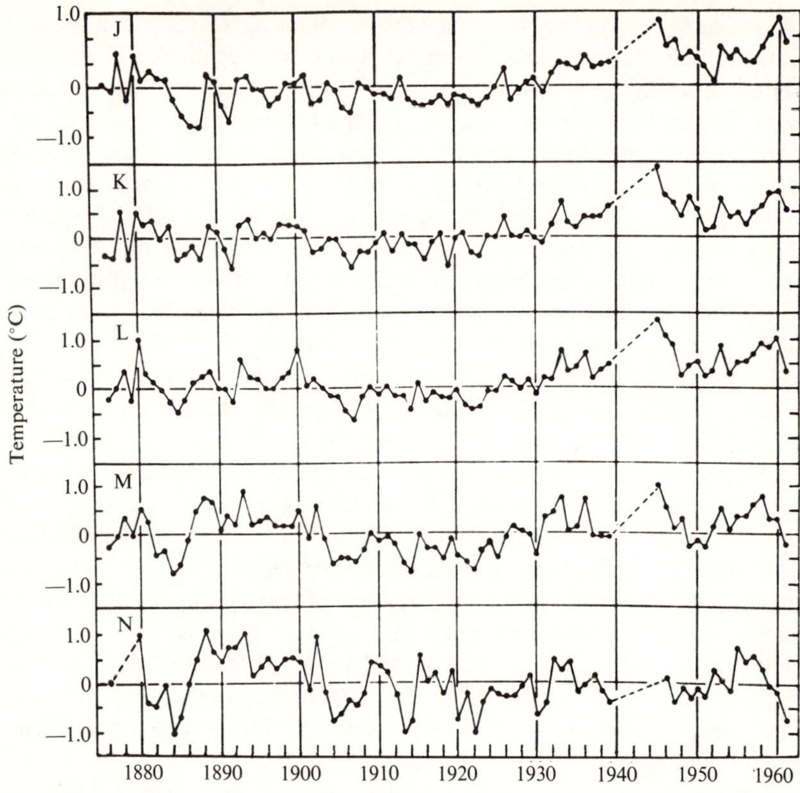

Figure 65. Trends in surface temperature between 1880 and 1960 for sea areas off the British Isles (Smed, 1965).

and direction that might be of considerable importance to the fortunes of the fish stocks.

Many papers have been published on the climatic changes in the ocean, both physical and biological. For fisheries biologists the most valuable is that of Dickson (1971) who has described the variations in the waters around the British Isles for a period of sixty-five years in a standard way. Further, he explained the variations in a particularly convincing manner. Anomalies in surface salinity from a grand monthly mean were calculated for each year by months in 293 standard fields. They were averaged by moving means and the fields were grouped, e.g. Western Approaches, German Bight, etc. Figure 66 shows a summary of the time series; 1905–20 in the English Channel, 1920–39 in the Central North Sea and German Bight, and 1945–70 in the German Bight. Although the same trends appear in all the areas examined, there are differences in degree between them and the reason for taking different areas is that there are gaps in the time series. There are peaks in high salinity anomaly in or near the following years:

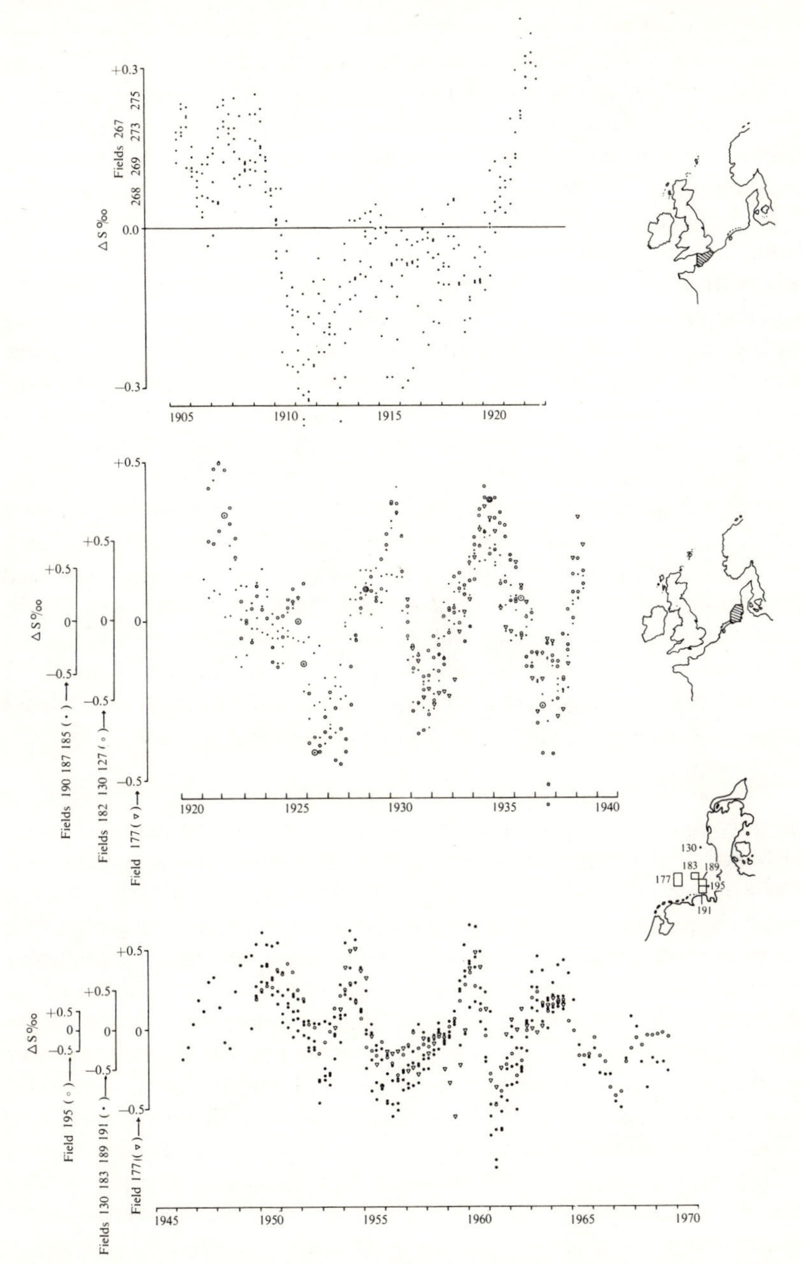

Figure 66. Salinity anomalies in the regions around the British Isles (Dickson, 1971): (*a*) 1905–20 in the western English Channel; (*b*) 1920–39 in the central North Sea and German Bight; (*c*) 1945–70 in the German Bight.

1904–6, 1920–1, 1925–6, 1930–1, 1934–5, 1939, 1950, 1954–5, 1959–60, 1964–5. These years have been referred to frequently in the last chapter and in an earlier section of this chapter. Records of the invasion of exotic or indicator species can be grouped in such periods, as might be expected.

A more important point is Dickson's form of explanation. He noted that there was a seasonal variation in the salinity anomaly and that long-term trends were of greater magnitude than the seasonal ones. Namias (1959, 1964) had examined the events in 1959, a long hot summer in the British Isles. He showed that the anomalous sea surface temperatures were related to anomalies in surface pressure, which is the same as Bjerknes' explanation. However, he also showed that the British waters were warmer in 1959 because of a protracted advection of warm and saline water throughout the eastern Atlantic from the south. At the same time there was an advection of cold water south-westerly along the coast of Labrador. Namias also pointed out that the augmented sea surface temperature, and other factors, will feed back into the meteorological system and there will be a tendency for these events to persist. Current flow increased and more saline water was drawn from the south and anti-cyclonic conditions over Europe decreased river flow and increased evaporation. By examining pressure distributions in the Atlantic for twenty-one years, Dickson was able to show that southerly winds occurred during years of high salinity inflow. Indices of meridionality (southerly winds) and cyclonicity were correlated with the salinity anomalies. Hence it was shown that the high salinity anomalies were explained by Namias' hypothesis.

In the waters off western Europe, the prevailing winds will shift between westerly and southerly and back during the long term and such shifts must be linked to the variability in the salinity anomalies shown in Figure 66. The wind system may contain within itself the form of variable system by which the production cycle might vary both in a short- and long-term time scale. As fish tend to spawn close to the coast (on an oceanic scale), differences in wind strength and/or direction can become amplified by the shape of the coastline, merely by modifying the degree of fetch. One of the most interesting points about Dickson's account is that the same pressure system generates warm water in the eastern Atlantic and cool water in the west. A particular year class of cod might be associated with an increment of temperature in the east and another one with a decrement in the west.

One of the consequences of Dickson's hypothesis is that events may be correlated at a considerable distance. For example, the year classes of the arctonorwegian cod stock might be quite properly correlated with events in the North Sea. In British waters the prevailing winds blow westerly or south-westerly between the Iceland low and the Azores high. In the Barents Sea the prevailing winds blow westerly and south-westerly between the

Iceland low and the Siberian high. The Iceland low is part of both systems and the same broad air stream that flows across north-west Europe affects both the Barents Sea and the waters around the British Isles. Differences in recruitments of stocks thousands of miles apart may originate in climatic events that bear upon them in a common fashion.

The winter of 1962–3 was not only very cold in Europe but also in Moscow, Japan and in the mid-west of the United States. The system of high and low pressure areas in the northern hemisphere moves in a more or less regular manner and there is no reason why Atlantic year classes should not be correlated with those in the Pacific. The herring periods in

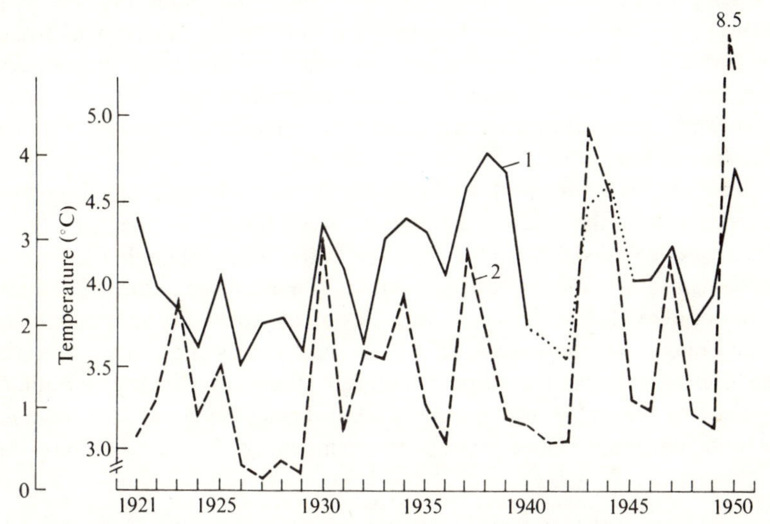

Figure 67. Correlation between year classes of the Norwegian herring and temperatures on the Kola meridian in the year of hatching (Ishevskii, 1964).

the Atlantic might well be associated with those of the tuna in the Pacific (Uda, 1960). The interest of such relationships would be high but they would be less useful than the expected closer correlation of recruitment on wind strength and direction in the region where the larvae drift.

Ishevskii (1964) suggested that the northern hemisphere could be divided into five sub-regions, Atlantic, Greenland and North America, Europe and Asia, Eastern Siberia and the Pacific; changes in any one are out of phase with those in an adjacent one. He believed that temporal differences in properties were generated by the effects of tidal periods, of up to eighteen or twenty years, upon the seasonal heat budgets. In the Atlantic, heat is transported north in the winter and the observations of temperature on the Kola meridian (a line of hydrographic observations in the Barents Sea) form a useful index of the heat budget. Catches of cod and herring were correlated with the temperatures on the meridian and the catches were set back in years by the modal age of the stock; in other words the crucial events were thought to have occurred during the first year of life. Figure 67 shows the relationship between catches of the Norwegian herring and temperature on the Kola meridian (observations in the Second World War were interpolated from a clear correlation between the temperature in the Barents Sea and the degree of Baltic ice cover). The inverse relation between Kola temperatures and Baltic ice cover is a periodic one, and probably the same as that reflected in the Norwegian–Swedish herring periods and that which Dickson might describe as southerly and westerly periods. Hill & Lee (1957) have correlated year classes of the arctonorwegian cod with southerly winds at Bear Island. Garrod (1967) has correlated the year classes of the same stock with temperature on the Kola meridian, as Ishevskii had done earlier. The changes in the environment must indicate the same broad processes. Perhaps the southerly wind off the Vestfjord, with a short fetch, generates an earlier production cycle than the westerly one, which has a long fetch.

There are two forms of variation in climatic change, the long-term periodicity and the annual differences. Recruitment to the fish stocks appears to be linked to both forms of variation, although the precise mechanisms have not yet been elucidated. The extensive nature of climatic events implies that variation in year classes of stocks far apart might be related and as we have seen, this is so right across the North Atlantic. The apparently detailed response of fish stocks to climatic variability implies that the major response must occur during the larval drift, because the food there is modifiable by the variation in timing of the production cycle.

The biological response to climatic variability

It was suggested earlier that recruitment to a fish stock was determined partly by density-dependent processes in juvenile life and partly by the match of larval production to that of their food. The natural mortality of the plaice was formulated as a partly density-dependent function of age, probably from the time of first feeding by the larvae. The grosser effects may take place during the larval drift, followed by lesser ones during life on the nursery grounds. Variation in the production cycle is effected through differences in the rate of development of the production ratio, governed by differences in radiation and in wind strength or direction. Our experience in north-west Europe is of fickle climate, but the important point is not so much the changeable character from day to day as the trends from week to week that are not quite as variable. In other parts of the world, where the sun stands in a brazen sky, the continuous production cycle probably generates food continuously and the variability of year classes in tropical seas may be less than that in high latitudes.

Temperature may have a direct effect upon the match or mismatch of larval production to that of their food. For example, during the cold winter of 1962–3 the rate of development of the plaice larvae was very much delayed, which is understandable because development is an inverse power function of temperature. Hence, cooler water will have a very much greater effect than would warmer. The 1963 year class of plaice was the largest recorded (Bannister, Harding & Lockwood, 1974). Not only was the production of larvae delayed but that of their food might have been advanced because the cold winter is associated with an anticyclone over the North Sea with relatively calm weather. Cold winters have been recorded in 1929, 1947 and 1963 during the history of the plaice study; the 1929 year class was also a large one, but that of 1947 was not outstandingly large. As noted above Dickson, Holden & Pope (1973) described a marked correlation between the year classes of cod on the Flamborough Off Ground and temperature in April and May. With the use of the observations of temperature from the 'bateaux routiers' a form of spatial correlation was shown on two lines across the North Sea. Because the spring temperature has dropped during the sixties as compared with the two or three decades before, it seems likely that the rise of gadoid stocks (haddock, cod and whiting) in a succession of year classes was associated with a climatic change.

The climate changes over long periods and within any one period a stock of fish might give the appearance of stability in numbers. A long-term trend in wind strength and direction at a particular position may alter the rate of change of the algal reproductive rate with time. If the production cycle tends to become abnormally late during a period of years, recruitment will decline on average. In Figure 68 is shown the variability of the year classes in a stock and recruitment curve of the Ricker type. In general recruitment will tend to return the stock to a replacement point or point of stabilization (whether the stock is exploited or not). If the recruitment falls persistently below the bisector when stock is low, the consequent recruitment must be low and so the stock will tend to be reduced further. If very low year classes appear in sequence, a new stock relationship might be established at a lower level; such events might have happened after the decline of the Californian sardine stock. At the other extreme, at high stock, if recruitment is persistently above the bisector, the condition for a considerable stock increase is fulfilled. When fish are transplanted from one environment to another, colonization may lead to high stock quite quickly as, for example, when the small mouthed bass were transplanted to San Francisco Bay.

The stock and recruitment curve used in Figure 68 implies that a stock stabilizes itself in a particular environment. We may recall the exploratory nature of a cohort and the conservative character of the stock in this

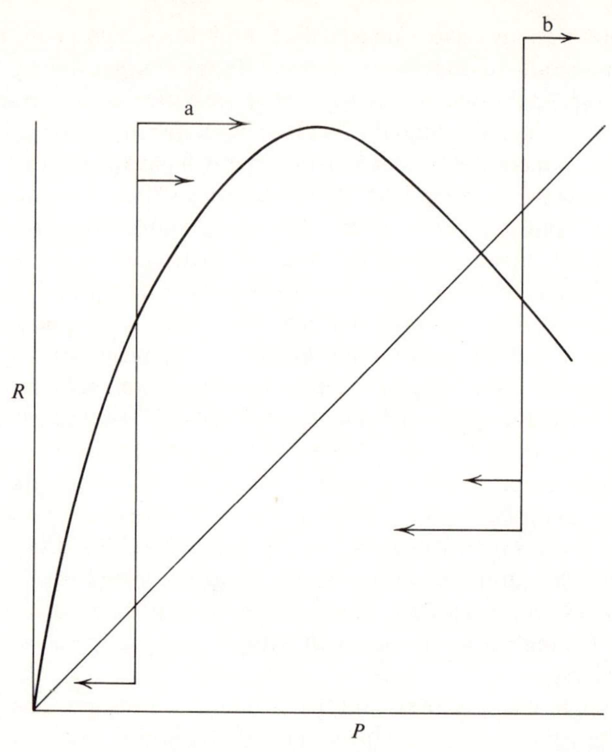

Figure 68. How the variability of recruitment can generate large decrements or increments in stock (Cushing, 1970); see text for explanations.

context. However, the variability of recruitment provides at one and the same time not only additional stabilization but also the capacity to adapt to climatic change by altering the stock level. The Norwegian herring live in an adverse environment in the deep water of the Norwegian Sea. The life span is long and the recruitment is very variable. The stock adapts to climatic change by maintaining high quantities during periods of amelioration and low levels during periods of deterioration; the changes in stock level between periods must be considerable, of many orders of magnitude. The ultimate cause of the changes in the Plymouth herring stock between the thirties and the seventies may have been a climatic change and the alternation between sardine and anchovy off the coast of California for about a thousand years may have the same basis. The mode of change from one stock level to another implies that it stabilizes within one range of variation and then switches sharply to another. The trends in climate, however, are not discontinuous: it is characteristic of some fish stocks to remain stable for one period and then for another at a different level when perhaps the climate itself is in a state of continuous change throughout both periods. Herring-like fishes may rectify such periodicities over a long period of time, whereas cod-like stocks may have sufficient capacity for stabilization to resist the variation in climatic change.

Upwelling areas may be sensitive to analogous changes in climate. For example, the present system off California is most intense south of Point Concepcion and that between Point Concepcion and Point Mendocino to the north is less important. However, the phosphatic deposits are greater in the latter area (Tooms, 1967) and considered over a very long period of time it may be the central zone of upwelling off California. The position of such coastal upwelling depends upon the angle which the wind makes to the coast and Yoshida (1967) has suggested that an angle of 21° to the coast is optimal. A small shift of wind direction may affect the position of upwelling quite considerably. On convex coasts like those of North America or North Africa, there may be shifts from region to region as the pressure systems move slightly. Variability in the positions of upwellings and in their intensities may be generated by the same factors that generate change in temperate seas.

In the deep tropical ocean, away from the upwelling areas, the situation may be quite different. In the North Pacific subtropical anticyclone tuna larvae are found in most areas at most seasons and food may be available everywhere most of the time. If food is available, but not necessarily abundant, the variability of year classes need not be high and the stocks of tuna-like fishes may be quite stable. The North Pacific anticyclone is very large and the oceanic regime associated with it may resist climatic change to some extent.

The simplest biological response by a fish stock is presumably that in the match or mismatch of larval production to that of their food. This response is modified by temperature particularly when it is cold enough to delay development considerably. In upwelling areas the spawning grounds are not fixed in position as in high latitudes and the time of spawning is highly variable. The areas may shift with climatic change, but from the dynamic structure of an upwelling area, the fish stocks are carried along in the shift. Any upwelling system generates a counter-current below 200 m and possibly one inshore at the surface. Together with the main eastern boundary current these currents form a structure within which a migratory circuit may become established. In the tropical ocean, however, it is possible that no structure of such a character is needed to control the variation of the stock. The stabilization mechanisms in higher latitudes ensure that the variations in the weather are either damped or rectified.

Conclusion

The long-term changes in climate of about fifty to seventy years in duration express themselves in periodic changes in wind strength and duration. The connection between such changes and production cycles and again with stock stability have been traced in a general way. There are two major applications of this work in fisheries biology: (1) in establishing the main trends in environmental conditions; and (2) in developing methods of predicting recruitment from the environmental conditions during the larval drift. Such methods may be of use in temperate seas and in upwelling areas but will probably be of little use in the deep tropical ocean where the tuna-like fishes live.

12

The regulation of fish populations by nature and by man

Introduction

The control of numbers in an animal population is one of the fundamental biological problems and it remains mysterious. Any description of the phenomenon reveals the stability and leaves the possible mechanisms indistinct. In his letter to the Linnean Society on 1 July 1858, after quoting Malthus, Darwin wrote: 'The amount of food for each species must, on an average, be constant, whereas the increase of all organisms tends to be geometrical, and in a vast majority of cases at an enormous ratio.' The absorption of the potential increase by innumerable checks (in the shape of mortality) is the struggle for existence. The survivors flourish by the deaths of their brothers and sisters and the stable numbers must be the result of a fine control of mortality, perhaps a density-dependent one.

Howard & Fiske (1911) suggested that the density-dependent control of numbers was mediated either through fecundity or mortality. Variations in the magnitudes of recruitment could originate in numbers of eggs with a fixed mortality or in a fixed number of eggs with variable mortality. In fishes, the required variation in fecundity is unlikely, because the growth of mature adults is not generally density-dependent. Perhaps the most important step was made by Lack (1954) who wrote that the reproductive rate was determined by natural selection and not by the magnitude of mortality. The development of the logistic curve in fisheries and in other fields implied that loss of stock with increased mortality was compensated by increased recruitment. In fishes recruitment increases relatively to stock with declining stock but absolutely only amongst the more fecund groups. The relative or absolute increment of recruitment probably arises from more food for the larvae as stock decreases. As stock is reduced further the increment of recruitment, relatively or absolutely, is also reduced. In the theory of balance it was considered unlikely that recruitment would be decreased either relatively or absolutely because stock would not be reduced so far. It is now well known that the reduction in stock can lead to both relative and absolute declines of recruitment.

If the reproductive rate is determined for a species, the density-dependent mortality is also determined, but within ecological constraints. If the numbers of an animal population remain stable for a very long time, the numbers of most of the components of the food web are similarly stable even if the numbers of the remainder change considerably. The natural mortality of fishes may be described as a density-dependent function of age (where the density-independent component is modulated in a density-dependent way) because the little animals grow in age through a series of predatory fields, each larger than its predecessor. The trend of natural mortality with age may be a function of the number of eggs laid and of predation in the early stages of the life cycle. In an evolutionary sense the number of eggs laid is linked to the mortality expected between egg and recruit. In this view the reproduction of the fish has become adapted to a permanent structure in the food webs. Within a cohort the variability of food may generate those differences in mortality that alter the year class strength and then the function of a multi-aged stock is to conserve the exploratory variability of recruitment. It is possible that such a mechanism is the long-term agent by which the reproductive capacity of the fish stock is linked to the permanent structure of the food web: fecundity varies as weight which is proportional to food taken. The same series of processes can accommodate changes in numbers anywhere in the system. A density-independent mortality may be modulated after the event in a density-dependent way and, if a predator enters or leaves the food web, the subsequent predatory fields take up the slack in the system.

The single process

The equation relating recruitment to parent stock used initially by Ricker and later by Beverton and Holt is a convenient summary of present opinion on the natural regulation of numbers in a fish population. Recruitment depends on stock modulated by density-independent and density-dependent mortality. It does not matter whether the density-independent mortality is subsequently modulated in a density-dependent manner or whether it is represented by the variance about a stock/recruitment curve or by the slope at zero stock. The equation is used empirically because neither coefficient can yet be developed independently.

There are three components of population control implicit in the equation. The most important is the magnitude of the stock that is the result of competition between species in evolution. From generation to generation the detailed control of numbers is obtained by stock-dependent mortality. The variation in recruitment is the result of factors both density-dependent and density-independent during the life history between hatching and recruitment. If the three components are effective at an early stage they might be the result of a single process, itself a consequence of the availability of food to the baby fish. Mortality may be density-dependent under such conditions and recruitment may depend on the match or mismatch of larval production to that of their food. The third component is the competitive one.

Of the three components the mode of competition in evolutionary history is inaccessible. The results, however, are manifest in the relative abundances of species in the same area. The differences are not fixed as if they were in fossil strata but have to be maintained from generation to generation. Competition in this sense may be most effective when food is shared between individual competitors and when the rates of growth and death are high. In an earlier chapter it was pointed out that adult fishes often lead isolated lives on foods that are specialized, exclusive and perhaps unlimited. Success in competition appears as an increase in numbers of one of the competing species and then greater chances of variation ensue. In such a way differences between species eventually become established although this principle does not bear upon the development of isolation between the incipient species of fish. It was suggested in an earlier chapter that competitive differences were generated or maintained in the ratio of growth to mortality at an early stage in the life history; how early a stage cannot be exactly determined, but the earlier the better because food is common and the rates are high. If numbers of a competitor are reduced, more food is available for the survivor, and its mortality slackens. With more food the ratio of specific growth rate to specific death rate increases and the effective competition is maintained.

In the Southern Bight of the North Sea, plaice larvae are hatched in the waters between the Thames and the Rhine and they drift towards Texel Island off the north coast of Holland. At the same time the larvae of cod, sandeels, dabs, whiting and herring are drifted on the same course. All the fish larvae feed on copepod nauplii, copepodites and on the appendicularian *Oikopleura*. The plaice and sandeels feed on *Oikopleura* and the other larvae feed on copepods. There is a simple food net in the plankton with five species competing for two or three forms of food. The opportunity for competition recurs each year in spring and the recent increase of cod in the southern North Sea may have originated in such a way.

The nature of competition between species of fish has not been described in any detail. However, the numbers of pelagic fish stocks can change by many orders of magnitude in a few years and a declining stock can be replaced by a rising one. In the thirties pilchards replaced herring in the western English Channel and in the seventies the change was reversed. Sardines off California gave way to anchovies as happened off Japan and off South Africa. The replacement of sardines by anchovies might have been started by heavy fishing but the events in the English Channel can only have been natural changes. An alternation between sardine and anchovy has been recorded for more than a thousand years from the anoxic sediments off California. The events in the western English Channel were associated with profound changes in the planktonic structure, which led to the supposition that the effects of competition become established during the period of larval drift.

The second component of population control is that of numbers from generation to generation. The most remarkable point about the time series of numbers of an animal population is not its variance or periodicity, but its constancy or stability. The stable numbers represent the degree to which the population is insulated from the variability of the environment. They are achieved by the density-dependent processes of growth and mortality predominantly during the period of larval drift and to some degree on the nursery ground. Both periods are early enough in the life history to be those of high density; indeed density on the nursery ground may be maintained in the face of mortality when the little fish settle progressively from the volume of the sea to the surface of the sea bed.

The stocks of herring in the north-east Atlantic segregate into three main spawning groups, autumn, winter and spring. The differences are much deeper than the mere separation of spawning time, important though that is. Between the groups the patterns of growth and reproduction are distinct. The differences in the asymptotic length, L_∞, can be as much as 5 cm in 35 cm and those in maximum age, T_{max}, are as much as ten years. Egg sizes and fecundity may differ between groups by a factor of two or three and the penultimate stage of maturation can range from periods of weeks in autumn spawners to many months in spring and winter spawners. Such differences in growth and fecundity are enough to imply that they may have a genetic origin. Differences in blood proteins between cod stocks in the North Atlantic have been established in three haemoglobins and seven transferrins (de Ligny, 1969); because the distribution of these proteins is almost certainly genetically determined, the chance of mixture between such stocks is very low indeed, $<1/10^4$, in most of the north-east Atlantic. The herring stocks resemble those of cod and perhaps they are also distinct to the same degree.

The most interesting point is that the profound differences between the three groups of herring (autumn, winter and spring spawners) are adapted to match three forms of production cycle, Central North Sea, Shelf and Oceanic. The studies of climatic change show that the range of variation is that generated between the trends from westerly to southerly air streams and back again. The associated trends in the timing of production cycles in the three groups are not yet known, but differences are unlikely to be much more than a month or so, which is a small part of the seasonal difference between groups. This adaptation to the timing of production cycles implies that selective pressures must occur during the planktonic phase of the life history. Selection must affect numbers and the control of numbers is achieved during the early life history. The regulation of numbers is thought to occur during the larval drift because the differences between the three spawning groups are adaptations to the match of spawning to production cycle, that is, the match of larval production to that of larval food.

The third component of population control is the generation of year classes. The recruitment of plaice, cod and herring in particular circumstances can be forecast from the numbers or densities during the first summers on the nursery grounds. By then, the density-dependent processes must be more or less complete, but could of course continue at a low level. Further, the metamorphosed plaice or cod feed on worms and bivalves, the numbers of which are only indirectly related to the nature and timing of the production cycle on the nursery ground, let alone that along the track of the larval drift. The production of larvae in temperate waters may be matched or mismatched to the production of their food. Larval production originates in spawning that is fixed in time and which is modified by extremes of temperature. The production of larval food is a function of the dynamics of the production cycle and is perhaps quite variable in timing. If the match is good, recruitment at a fixed level of stock may be high and if it is poor the year class may be low.

Thus, during this period of the planktonic life history, recruitment is perhaps in the main determined, numbers regulated and competition is established. Because it endures for a relatively short time, because the animals have little choice or capacity for movement and because their behaviour probably lacks the subtleties of that of higher animals, the three processes are probably mediated by the availability of food. It is not surprising for it is the basis of Lack's (1954) thesis on the control of numbers in animal populations. If natural mortality were a density-dependent function of age, in the particular sense described in an earlier chapter the trend of numbers in time could be modulated by the availability of food and by density-independent factors. Such a mechanism provides for the control of numbers and at the same time the generation of year classes (including the effects of density-independent factors); the same process might provide a possible mechanism for competition in the ratio of specific growth rate to specific death rate which is effected in numbers in the relative recruitments between species.

The single process, in which the three processes of population control may find their effects, is in the simplest terms, the process of feeding in the sea. The little fish that feeds well, grows well, swims quickly, evades predators and feeds better than ever. If the little animal fails to feed, the chance of death is greater and the initial failure to find food becomes greater. Such a mechanism may well form the basis of the density-dependent processes, which may determine the course of natural mortality with age. If the little fish grow through a succession of predatory fields, there are two constraints to the system, the fecundity of the species and the permanent structure of the food chain through which the fishes grow. Lack (1954) suggested that fecundity is determined by natural selection and the structure of the food chain must have a similar evolutionary history. It would then follow that perhaps the structure and function of food chains might be investigated in a study of fecundity throughout the system.

Complex processes are often analysed with model techniques and in fisheries biology they have been used for many years. The first was that of Baranov (1918) and it was followed by those of Russell (1931); Thompson & Bell (1934); Graham (1935); Ricker (1948,1954); Schaefer (1954,1957) and Beverton & Holt (1957). Models have a long scientific history which may extend as far back as the *Timaeus* of Plato and certainly as far as Copernicus, but biologists today are sometimes wary of them. There are three forms of test of a model. The first is to test the parameters; for example, fishing mortality might be estimated in more than one way and the estimates should not only correlate but an estimate of error should be available. The second test is to fit data to the model, as catches can be fitted to a Schaefer curve, with confidence limits. The third is the link between the model and the body of scientific knowledge; for example, the models of Beverton and Holt yielded greater understanding of the age structure of a fish population which represents an increment to knowledge quite independently of the use to which this model was put.

The place of fisheries research in marine ecology

The Victorian interest in marine biology arose in two ways. Holidays were spent at the seaside by the wealthier sections of the community after a network of railways had developed. A market developed in Britain during the second half of the century, for books describing the animals found in rock pools and on sandy beaches. The books of Philip Gosse are typical of the sea-shore manuals (*The Aquarium*, *The Canadian Naturalist*, *Tenby*, *Rambles on the Devonshire Coast*, etc.). They are well illustrated by engravings, with an occasional hand-painted plate. During the same period, expeditions put to sea to explore the depths of the ocean. Forbes (1843) had suggested that there were no animals in great depths, but work on the transatlantic cables belied this suggestion. The *Challenger* expedition was the first of many in the last three decades of the century. During the heyday of evolutionary and palaeontological studies, the diversity of life in the sea led to such beautiful works as Haeckel's Report on Radiolaria (1887) in the *Challenger Reports* and Savile Kent's exploration of the Great Barrier Reef.

The need to study fisheries did not arise until the last decades of the nineteenth century. During the century in Britain a number of Commissions had listened to evidence from fishermen on the failure of some fisheries and on the destructive effects of particular forms of gear. In the North Sea the decisive event was the introduction of the steam trawler in the eighties. Because a steam winch was used, and the ships were independent of wind and tide and because they used otter trawls rather than beam trawls, they were much more efficient than the sailing smacks. The initial catches were much greater, profits were higher and so more boats were built. During the nineties more than 600 steam trawlers were working from British ports alone and similar fleets were being built in Europe, particularly in Germany. Petersen (1894), Petersen *et al.* (1907) and Garstang (1900–3) showed that the stock-density of trawled fish in the North Sea was reduced by one-third during the last decade of the century. More important, the catches started to fail on the known grounds. Between 1870 and 1880 a purse seine fishery for mackerel and menhaden developed with steam boats off the eastern seaboard of the United States; in a fairly short period the first industrial fishery started, grew to its zenith and vanished. The catches were reduced to oil and soap to be used in the exploitation of the west.

In 1899 the Council for the Exploration of the Sea met in Christiania and recommended two lines of research, first on the problem of exploitation of the stocks and secondly on the effects of the environment on the variability of the stocks. As a consequence the International Council for the Exploration of the Sea was established in 1902. In subsequent years, right up to the present time, treaties of conservation have been signed between various groups of countries. In one ocean at least, the North Atlantic, there is a powerful structure of international action and co-operation including the power to stop fishing boats at sea, vested in the naval ships of those countries that contribute to it.

During the long history of about seventy years the science of fisheries biology developed slowly. There were two initial difficulties, first that the theory of the interaction of prey and predator had not been established at least in the form developed later by Ross (1910) and Volterra (1926) and secondly, the results of trawling trials in the first decade of the century revealed such high variance that the first practitioners of fisheries research saw little hope in studying fish stocks at sea. Although it was before the statistics of small samples had been fully developed, it was a fortunate accident because it led E. S. Russell to establish a method of sampling the catches on the quays and on the market which provided the most powerful system of estimating catch and stock-density of the exploited populations. Perhaps the most important advance in the early stage was the development of age determination in Scandinavia. Dahl (1907), working on herring, showed that the rings on the scales of the fish were correlated with the modes of the length distributions (Petersen's (1894) method) and showed that the rings were annual. A little later, Lea (1911) traced the history of the 1904 year class of the Norwegian herring as it passed through the fishery for many years; at the same time he deduced from the growth pattern of the scales that the fish left the inshore waters for the oceans at a fairly early stage in their life history. Reibisch (1899) and a few years later, Thompson (1915) showed that otoliths could be used for the same purpose. At a very early stage in the history of fisheries research the value of marking fish was established; Garstang (1908) and Petersen *et al.* (1909) showed that the fishing rate on plaice in a trawl fishery could be estimated. Further, Thompson & Herrington (1930) showed that the spread of fish in time after tagging could be estimated quite easily with good results from marked fish.

This short history of the early stages of fisheries biology before any theory was fully developed shows that the science was a little ingrown because it appeared to be concerned with techniques only. In a purely laboratory science the problems of technique and those of theory are usually linked in a dove-tailing development. The first theoretical advances occurred during the thirties, although it was discovered later that they were antedated by the work of Baranov (1918). Both the analytical and descriptive studies of fish populations were represented in the first steps of theoretical development. Russell's (1931) statement separated the vital parameters to be analysed in a fish population. Thompson & Bell (1934) reaffirmed the inverse relationship between stock density and fishing effort and developed a model from it. Graham (1935) generalized the description with the logistic curve to derive catch as a function of fishing rate and effectively it was the first formal statement that led to the solution of the problem of heavy fishing. One of the difficulties of such an approach was the estimation of fishing mortality, which led Graham (1938) to formalize the theory of a population of tagged fishes. With these two papers, Graham laid the foundations of modern formal fisheries dynamics, which requires a mathematical statement of principle and argument.

Ricker (1948) extended certain points of argument. For example, he established the nature of the catch equation and analysed more rigorously the errors that might arise in a tagging experiment. But his main achievement was the establishment of the dependence of recruitment on parent stock and raised the question of recruitment overfishing that remains to be solved for multi-aged stocks in rigorous terms. By historical accident the last piece of descriptive population dynamics was published at about the same time (Schaefer, 1954, 1957) as the rigorous statement of the solution of growth overfishing was published by Beverton & Holt (1957) in the yield/recruit formulation.

However, the work of Beverton and Holt was less the end of a historical process than the start of a new one. For example, they used the catch equation to prove that stock-density was a true index of stock. Their real achievement was to represent the population as the result of changes in growth and mortality with age and so the parameters could be estimated independently. They also formulated a self-regenerating stock/recruitment relationship (which, however, cannot represent the dome-shaped curve commonly found amongst the gadoid stocks).

These successes in fisheries biology were studies of single populations aimed at solving the problem of overfishing. During the fifties and sixties of the present century the much increased pressure of fishing has affected the recruitment to some fish stocks. The problem of the generation of recruitment, its magnitude or its failure, is linked to the availability of food for the larvae and so to the competition of other species. The study of a single population in isolation is no longer enough to solve the problems that face a fisheries biologist today. For example, the year classes of gadoids in the North Sea during the sixties increased considerably, as noticed earlier, but the reason for this upsurge is unknown. Solutions to this sort of problem will only appear from studies on a broader scale within the field of marine ecology. There are three groups of ecological work required. The first is a more rigorous formulation of the dependence of recruitment on parent stock, which should incidentally illuminate the study of the stability of food chains. A descriptive model like Ricker's will eventually be replaced by a more analytic one in which the processes involved are simulated. The second group of ecological problems is a fuller study of the density-dependent processes of growth and mortality at all ages. Enough has been done to show that both are of considerable magnitude in the early stages of the life cycle and that both decline with age. However, little is known of the mechanisms that generate such effects. The third group of problems involves the ecology of fish larvae, their growth and survival as they feed. Any model in this field must resemble those of the processes of production in the planktonic environment.

In a sense the set of three problems is really the same as those described in the last section on the single process. The recruitments to stocks of fish are not randomly distributed in time but follow trends that appear to be linked to climatic variation. Hence arises the hope that the trends can be predicted, at least coarsely. In the history of fisheries biology many scientists have expressed this hope even if it is only in the form of a failing correlation. The processes by which year classes are generated need to be understood and they are also food chain problems. Only when they are properly understood can the stock and recruitment relationship be fully formulated.

The purpose of fisheries biological studies within the field of marine ecology is to obtain safe and profitable yields from the stocks that support the fisheries. Some fisheries have tended to fail in the last decade or so because the problem of recruitment overfishing has been overlooked. At the present time the broader outlook of marine ecology is needed in fisheries science in order to solve the problems of the generation of recruitment and those of recruitment overfishing.

References

Adams, J. A. & Steele, J. H. (1966). Shipboard experiments on the feeding of *Calanus finmarchicus* (Gunnerus). In *Some Contemporary Studies in Marine Science*, ed. H. Barnes. pp. 19–35. Allen & Unwin, London.

Ahlstrom, E. H. (1966). Distribution and abundance of sardine and anchovy larvae in the California Current region off California and Baja California, 1951–64: a summary. *Spec. Sci. Rep. U.S. Fish and Wildlife Serv.* (*Fisheries*), pp. 534–71.

Allen, K. R. (1951). The Horokiwi Stream. A study of a trout population. *N.Z. Mar. Dept. Fish, Bull.* **10.**

Allen, W. E. (1941). Twenty years' statistical studies of marine plankton dinoflagellates off Southern California. *Amer. Mdl. Nat.* **26** (3), 603–35.

Alverson, D. L. (1967). Distribution and behaviour of Pacific hake as related to design of fishing strategy and harvest rationale. *FAO Fish. Rep.* **62** (2), 361–76.

Alverson, D. L. & Larkins, H. A. (1969). Status of knowledge of the Pacific hake resources. *Rep. Calif. coop. ocean. Fish. invest.* **13,** 24–31.

Alverson, D. L., Pruter, A. T. & Ronholt, L. L. (1964). *A study of demersal fishes and fisheries of the North eastern Pacific Ocean.* H. R. MacMillan Lectures in Fisheries, 190 pp. University of British Columbia, 1964.

Ancellin, J. & Nédelèc, C. (1959). Marquage de harengs en Mer du Nord et en Manche orientale (Campagne du 'Président Théodore Tissier', Novembre, 1957). *Rev. Trav. Inst. Pêches Marit.* **23,** 177–201.

Andrewartha, H. & Birch, L. (1954). *The Distribution and Abundance of Animals*, 782 pp. University of Chicago Press, Chicago.

Andrews, W. L., Cram, D. L. & Visser, G. A. (1970). An estimate of the potential production due to upwelling off the Cape Peninsula. *Sancor. Symp. Oceanogr. in South Africa*, paper B3, 14 pp.

Anon. (1953). *Rep. Calif. coop. ocean. Fish. invest.* 1952–53. 44 pp.

(1964). Fourteenth Report of the Commission. *Intern. Whaling Commn. London*, 122 pp.

(1965). The North sea herring. *Coop. Res. Rep. Intern. Council. Explor. Sea*, **4,** 57.

(1970). Report of the North Sea Flatfish working group. *Int. Council Explor. Sea*, CM1970, F14, 19 pp.

Antia, N. J., McAllister, C. D., Parsons, T. R., Stephens, R. & Strickland, J. D. H. (1963). Further measurements of primary production in coastal sea water using al arge volume plastic sphere. *Limnol. Oceanogr.* **6,** 237–58.

Apstein, C. (1896). *Das Süsswasserplankton*, Lipsius und Tischer Kiel and Leipzig. 196 pp.

Atkins, W. R. G. (1923). The phosphate content of fresh and salt water in its relation to the growth of algal plankton. Part I. *J. Mar. Biol. Ass. UK*, NS **13,** 119–50.

(1925*a*). Seasonal changes in the phosphate content of sea water in relation to the growth of the algal plankton. Part II. *J. Mar. Biol. Ass. UK*, NS **13,** 700–20.

(1925*b*). On the thermal stratification of sea water and its importance for the algal plankton. *J. Mar. Biol. Ass. UK*, NS **13** (3), 696–9.

Ayushin, B. N. (1963). Abundance dynamics of herring populations in the seas of the Far East and reasons for the introduction of fishery regulations. *Rapp. Procès-Verb. Cons. perm. int. Explor. Mer*, **154,** 262–70.

Bainbridge, R. (1960). Speed and stamina in three fish. *J. Exp. Biol.* **37,** 129–53.

Bainbridge, V. & McKay, B. J. (1968). The feeding of cod and redfish larvae. *Spec. Publ.* **7.** (1) *Int. Commn. North West Atl. Fish.* pp. 187–217.

Bannister, R. C. A., Harding, D. W. & Lockwood, S. J. (1974). Larval mortality and subsequent year class strength in plaice. *ICES, FAO, ICNAF Symp. on the early stages in the lives of fishes*, Springer-Verlag, Berlin.

Baranov, F. I. (1918). On the question of the biological basis of fisheries. *Nauchnyi issledovatelskii ikhtiologicheskii Institut, Izvestiia*, **1** (1), 81–128.

Barlow, J. P. & Bishop, J. W. (1965). Phosphate regeneration by zooplankton in Cayuga Lake. *Limnol. Oceanogr.* Suppl. X, R15–R24.

Batschelet, E. (1965). Statistical methods for the analysis of problems in animal orientation and certain biological rhythms. *Amer. Inst. Biol. Sci.* 57pp.

Baylor, E. R. & Sutcliffe, W. H. (1963). Dissolved organic matter in sea water as a source of particulate food. *Limnol. Oceanogr.* **8,** 369–71.

Becacos-Kontos, T. (1968). The annual cycle of primary production in the Saronicos Gulf (Aegean Sea) for the period November 1963–October 1964. *Limnol. Oceanogr.* **13** (3), 485–9.

Beklemishev, K. V. (1957). Superfluous feeding of marine herbivorous zooplankton. *Trud. Vsesoyuz. gidrobiol. Obshch.* **8,** 354–8.

Bernard, F. (1957). Research on phytoplankton and pelagic protozoa in the Mediterranean Sea from 1953–66. *Océanogr. Mar. Biol. Ann. Rev.* vol. 5, pp. 205–29.

Bernard, M. (1955). Étude préliminaire quantitative de la répartition saisonnière du zooplancton de la Baie d'Alger, 1950–1. *Bull. inst. Océanogr.* **52** (1065), 1–28.

Beverton, R. J. H. (1962). Long-term dynamics of certain North Sea fish populations. In *Exploitation of animal populations*, ed. E. D. Le Cren & M. W. Holdgate, pp. 242–64. Blackwell, Oxford.

(1963). Maturation, growth and mortality of clupeid and engraulid stocks in relation to fishing. *Rapp. Procès-Verb. Cons. int. Explor. Mer*, **154,** 44–67.

(1964). Differential catchability of male and female plaice in the North Sea and its effect on estimates of stock abundance. *Rapp. Procès-Verb. Cons. int. Explor. Mer*, **155,** 103–12.

Beverton, R. J. H. & Holt, S. J. (1957). On the dynamics of exploited fish populations. *Fish. invest. Lond. Ser.* 2, **19,** 533 pp.

(1959). A review of the life span and mortality rates of fish in nature and their relation to growth and other physiological characteristics. In *CIBA Foundation Colloquia on ageing*, ed. G. E. W. Wolstenholme & M. O'Connor, vol. 5, pp. 142–80. J. &. A. Churchill, London.

Beverton, R. J. H. & Lee, A. J. (1964). The influence of hydrographic and other factors on the distribution of cod on the Spitzbergen shelf. *Int. Commn. North West Atl. Fish. Environmental Symp.* Spec. Publ. **6,** 225–46.

Birkett, L. (1969). The nitrogen balance in plaice, soles and perch. *J. Exp. Biol.* **50** (2), 375–86.

(1970). Experimental determination of food conversion and its application to ecology. In *Marine Food Chains*, ed. J. H. Steele, pp. 261–4. Oliver & Boyd, Edinburgh.

Bishop, Y. M. M. (1959). Errors in estimates of mortality obtained from virtual populations. *J. Fish. Res. Bd Can.* **16** (1), 73–90.

Bjerknes, J. (1963). Climatic change as an ocean–atmosphere problem. In *Changes of climate*, Proc. Rome Symp. 1961, pp. 297–321. UNESCO/WMO.

(1964). Atlantic air sea interaction. *Adv. Geophys.* **10,** 1–82.

Blaxter, J. H. S. (1965). The feeding of herring larvae and their ecology in relation to feeding. *Rep. Calif. coop. ocean. Fish. invest.* **10,** 79–88.

Blaxter, J. H. S. & Hempel, G. (1963). The influence of egg size on herring larvae (*Clupea harengus* L.). *J. Cons. int. Explor. Mer*, **28,** 211–40.

Blegvad, H. (1916). Quantitative investigations of bottom invertebrates in the Limfjord with special reference for the plaice food. *Rep. Dan. Biol. Sta.* **34,** 33–52.

Boeck, A. (1871). *On silden og silde fiskerierne, naunlig om det norske varsildfisket.* Christiania.

Böhnecke, G. (1926). Der Jahrliche Gang des Salzgehaltes in der Nordsee. *Veröff. des Inst. f. Meereskunde*, NF Reihe, A. Heft, **17,** 5–22.

Bolster, G. & Bridger, J. P. (1957). Nature of the spawning area of herrings. *Nature (London)*, **179,** 638 pp.

Brandt, K. (1899). Über den Stoffwechsels im Meer. *Wiss. Meersuntersuch. Abt. Kiel*, NF **4,** 213–30.

Brandt, K. & Raben, E. (1919). Zur Kenntnis der chemischen Zusammensetzung des Planktons und einiger Bodenorganismen. *Wiss. Meeresuntersuch. Abt. Kiel*, NF **19,** 175–210.

Bray, J. R. (1971). Vegetational distribution, tree growth and crop success in relation to recent climatic change. *Adv. Ecol. Res.* **7,** 177–233.

Bridger, J. P. (1958). A modified high speed tow net. *J. Cons. int. Explor. Mer*, **23,** 357.

Brock, V. & Riffenburgh, R. (1960). Fish schooling: a possible factor in reducing predation. *J. Cons. int. Explor. Mer*, **25,** 307–17.

Brocksen, R. W., Davis, G. E. & Warren, C. E. (1970). Analysis of trophic processes on the basis of density-dependent functions. In *Marine Food Chains*, ed. J. H. Steele, pp. 468–99.

Buchanan-Wollaston. H. J. (1915). Report on the spawning grounds of the plaice in the North Sea. *Fish. Invest. Lond.* Ser. 2, **II** (4), 18 pp.

(1922). The spawning of plaice in the southern part of the North Sea in 1913–14. *Fish. Invest. Lond.* Ser. 2, **V** (2), 37 pp.

(1926). Plaice egg production in 1920–1, treated as a statistical problem, with comparison between the data from 1911, 1914 and 1921. *Fish. Invest. Lond.* Ser. 2, **9** (2), 36 pp.

Bückmann, A. (1932). Die Frage nach der Zweckmässigkeit des Schutzes untermassiger Fische und die Voraussetzungen für ihre Beantwortung. *Rapp. Procès-Verb. Cons. int. Explor. Mer*, **80** (VII), 1–16.

(1961). Über die Bedeutung des Schollenlaichens in der südöstlichen Nordsee. *Kurze. Mitt. Inst. Fisch. Univ. Hamburg*, **11,** 1–40.

Burd, A. C. (1973). The northeast Atlantic herring and the failure of an industry. *Sea Fisheries Research*, ed. F. R. Harden Jones, pp. 167–91.

Burd, A. C. & Cushing D. H. (1962). I. Growth and recruitment in the herring of the Southern North Sea. II. Recruitment to the North Sea herring stocks. *Fish. Invest. Lond.* Ser. 2, **23** (5), 71 pp.

Burkenroad, M. (1948). Fluctuations in abundance of Pacific halibut. *Bull. Bingh. Oceanogr. Coll. II*, **4,** 81 pp.

Burns, C. W. & Rigler, F. H. (1967). Comparison of filtering rates of *Daphnia rosea* in lake waters and in suspensions of yeast. *Limnol. Oceanogr.* **12** (3), 492–502.

Butler, E. I., Corner, E. D. S. & Marshall, S. M. (1969). On the nutrition and metabolism of zooplankton. VI. Feeding efficiency of *Calanus* in terms of nitrogen and phosphorus. *J. Mar. Biol. Ass. UK*, NS **49** (4), 977–1002.

Cahet, G., Fiala, M., Jacques, G. & Panouse, M. (1972). Production primaire au niveau de la thermocline en zone néritique de la Méditerranée Nord-Occidentale. *Mar. Biol*, **14,** 32–40.

Carruthers, J. N. (1938). Fluctuations in the herrings of the East Anglian Autumn fishery, the yield of the Ostend spent herring fishery and the haddock of the North Sea in the light of the relevant wind conditions. *Rapp. Procès-Verb. Cons. int. Explor. Mer*, **107,** 1–15.

Cattley, J. G. (1950). Zoo- and phytoplankton of the Flamborough Line, 1946–9. *Ann. Biol. Cons. perm. int. Explor. Mer*, **6,** 121–3.

(1954). Zoo- and phytoplankton of the Flamborough Line, 1950–3. *Ann. Biol. Cons. perm. int. Explor. Mer*, **10,** 101–3.

Clarke, G. L. & Denton, E. (1962). Light and Animal Life. In *The Sea*, ed. M. N. Hill, vol. 1, pp. 456–68. Wiley, New York.

Clayden, A. D. (1972). Simulation of the changes in abundance of the cod (*Gadus morhua* L.) and the distribution of fishing in the North Atlantic. *Fish. Invest. Lond.* Ser. 2, **27** (1), 58 pp.

Clemens, W. A., Foerster, R. E. & Pritchard, A. L. (1939). The migration of Pacific Salmon in British Columbian waters. *Publs Amer. Adv. Sci.* **8,** 51–9.

Clutter, R. I. & Whitesel, L. E. (1956). Collection and interpretation of sockeye salmon scales. *Bull. int. Pac. Salm. Fish. Commn.* **9,** 159 pp.

Colebrook, J. M. (1965). On the analysis of variation in the plankton, the environment and the fisheries. *Int. Commn North West Atl. Fish. Spec. Publ.* 6, 291–302.

Colebrook, J. M. & Robinson, G. A. (1961). The seasonal cycle of plankton in the North Sea and in the North Eastern Atlantic. *J. Cons. int. Explor. Mer*, **26,** 156–65.

(1965). Continuous plankton records; seasonal cycles of phytoplankton and copepods in the North East Atlantic and the North Sea. *Bull. Mar. Ecol.* vol. 6, pp. 123–9.

Conover, R. J. (1962). Metabolism and growth in *Calanus hyperboreus* in relation to its life cycle. *Rapp. Procès-Verb. Cons. int. Explor. Mer*, **153,** 190–7.

Conover, S. M. (1956). Oceanography of Long Island Sound. IV. Phytoplankton. *Bull. Bingh. Oceanogr. Coll.* **15,** 62–112.

Cooper, L. H. N. (1938). Phosphate in the English Channel 1933–8, with a comparison with earlier years, 1916 and 1923–32. *J. Mar. Biol. Ass. UK*, NS **23,** 181–95.

(1948). Phosphate and Fisheries. *J. Mar. Biol. Ass. UK*, NS **27,** 326–36.

Corner, E. D. S. & Davies, A. G. (1972). Plankton as a factor in the nitrogen and phosphorus cycles in the sea. *Adv. Mar. Biol.* **9,** 101–204.

Corner, E. D. S., Head, R. N. & Kilvington, C. C. (1972). On the nutrition and metabolism of zooplankton. VIII. The grazing of *Biddulphia* cells by *Calanus helgolandicus*. *J. Mar. Biol. Ass. UK*, NS **52,** 847–61.

Currie, R. I. (1962). Pigments in zooplankton faeces. *Nature* (*London*), **193** (4819), 956–7.

(1964). Environmental features in the ecology of Antarctic Seas. In *Biologie Antarctique*, pp. 89–94. Hermann, Paris.

Cushing, D. H. (1954*a*). Some problems in the production of oceanic plankton. Doc. 8. presented to Commonwealth Oceanographic Conference 1954.

(1954*b*). On the autumn spawned herring races in the North Sea. *J. Cons. int. Explor. Mer*, **21,** 44–60.

(1955). Production and a pelagic fishery. *Fish. Invest. Lond.* Ser. 2, **18** (7), 104 pp.

(1957). The number of pilchards in the Channel. *Fish. Invest. Lond.* Ser. 2, **21** (5), 27 pp.

(1959). On the nature of production in the sea. *Fish. Invest. Lond.* Ser. 2, **22** (6), 40 pp.

(1961). On the failure of the Plymouth herring fishery. *J. Mar. Biol. Ass. UK*, NS **41,** 799–816.

(1963). Studies on a *Calanus* patch. II. The estimation of algal reproductive rates. *J. Mar. Biol. Ass. UK*, NS, **43,** 339–47.

(1964). The work of grazing in the sea. In *Grazing in terrestrial and marine environments*, ed. D. J. Crisp, pp. 207–25, Blackwell, Oxford.

(1966*a*). *The Arctic cod.* 93 pp. Pergamon, Oxford.

(1966*b*). Biological and hydrographic changes in British Seas during the last thirty years. *Biol. Rev.* **41,** 221–58.

(1967*a*). The grouping of herring populations. *J. Mar. Biol. Ass. UK*, NS **47,** 193–208.

(1967*b*). The use of echo sounders and scanners in the study of fish behaviour. *FAO Conf. Fish Behaviour*, R.5, 14 pp.

(1968*a*). Grazing by herbivorous copepods in the sea. *J. Cons. int. Explor. Mer*, **32,** 70–82.

(1968*b*). The Downs stock of herring during the period 1955–66. *J. Cons. int. Explor. Mer*, **32,** 262–9.

(1968*c*). Direct estimation of a fish population acoustically. *J. Fish. Res. Bd Can.* **25** (11), 2349–64.

(1970). The regularity of the spawning season in some fishes. *J. Cons. int. Explor. Mer*, **33,** 81–97.

(1971*a*). Upwelling and the production of fish. *Adv. Mar. Biol.* **9,** 255–335.

(1971*b*). The dependence of recruitment on parent stock in different groups of fishes. *J. Cons. int. Explor. Mer*, **33,** 340–62.

(1972*a*). The production cycle and numbers of marine fish. *Symp. Zool. Soc. Lond.* (1972), **29,** 213–32.

(1972*b*). A comparison of production in temperate seas and in upwelling areas. *Trans. Roy. Soc. South Africa*, **40** (I), 17–33.

(1973*a*). The natural regulation of fish populations. *Sea Fish. Res.* ed. F. R. Harden Jones, pp. 399–411.

(1973*b*). Production in the Indian Ocean and the transfer from the primary to the secondary level. *Ecological studies Analysis and Synthesis*, **3,** 475–86.

Cushing, D. H. (1974). The possible density dependence of larval mortality and adult mortality in fishes. In *The early life history of fish*, ed. J. H. S. Baxter, pp. 21–38. Springer-Verlag, Berlin.

Cushing, D. H. & Bridger, J. P. (1966). The stock of herring in the North Sea and changes due to fishing. *Fish. Invest. Lond.* Ser. 2, **25** (1), 123 pp.

Cushing, D. H. & Burd, A. C. (1957). On the herring of the Southern North Sea. *Fish. Invest. Lond.* Ser. 2, **20** (11), 1–31.

Cushing, D. H. & Harris, J. G. K. (1973). Stock and recruitment and the problem of density dependence. *Rapp. Procès-Verb. Cons. int. Explor. Mer*, **164,** 142–55.

Cushing, D. H. & Nicholson, H. F. (1963). Studies on a *Calanus* patch. IV. Nutrient salts off the North East coast of England in the spring of 1954. *J. Mar. Biol. Ass. UK*, NS **43,** 373–86.

Cushing, D. H., Nicholson, H. F. & Fox, G. P. (1968). The use of the Coulter counter for the determination of marine productivity. *J. Cons. int. Explor. Mer*, **32,** 131–51.

Cushing, D. H. & Tungate, D. S. (1963). Studies on a *Calanus* patch. I. *The* identification of a *Calanus* patch. *J. Mar. Biol. Ass. UK*, NS **43,** 327–37.

Cushing, D. H. & Vućetić, T. (1963). Studies on a *Calanus* patch. III. The quantity of food eaten by *Calanus finmarchicus*. *J. Mar. Biol. Ass. UK*, NS **43,** 349–71.

Cushing, J. E. (1964). The blood groups of marine animals. *Adv. Mar. Biol.* **2,** 85–131.

Dahl, K. (1907). The scales of the herring as a means of determining age growth and migration. *Rep. Norw. Fish. Mar. invest.* **11** (6), 36 pp.

Dannevig, G. (1953). The feeding grounds of the Lofoten cod. *Rapp. Procès-Verb Cons. int. Explor. Mer*, **136,** 87–8.

Darbyshire, M. (1967). The surface waters off the coast of Kerala, south-west India. *Deep Sea Res.* **14** (3), 295–320.

Davidson, V. M. (1934). Fluctuations in the abundance of planktonic diatoms in the Passamaquoddy region, New Brunswick, from 1924 to 1931. *Contrib. Can. Biol. and Fish.* **8,** 357–407.

Deevey, E. S. (1947). Life tables for natural populations of animals. *Quart. Rev. Biol.* **22,** 283–314.

Deevey, G. B. (1952). A survey of the zooplankton of Block Island Sound, 1943–1946. *Bull. Bingh. Oceanogr. Coll.* **13,** 65–119.

(1956). Oceanography of Long Island Sound 1952–4. V. Zooplankton. *Bull. Bingh. Oceanogr. Coll.* **15,** 112–55.

Derzhavin, A. N. (1922). The stellate sturgeon (*Acipenser stellatus* Pallas), a biological sketch. *Bihll. Bakinskoi Ikhtiologicheskoi Stantsii*, **1,** 1–393.

Devold, F. (1963). The life history of the Atlanto-Scandian herring. *Rapp. Procès-Verb. Cons. int. Explor. Mer*, **154,** 98–108.

(1966). Vintersildinnsigene 1966. *Fiskets Gang.* **52** (16), 299–301.

(1968). The formation and disappearance of a stock unit of Norwegian herring. *Fisk. Dir. Skr. Ser. Havunders*, **15** (1), 3–15.

Dickie, L. M. (1971). Food chains and fish production. *Int. Commn North West Atl. Fish. Spec. Publ.* **8,** 201–21.

Dickson, R. R. (1971). A recurrent and persistent pressure anomaly pattern as the principal cause of intermediate scale hydrographic variation in the European shelf seas. *Dt. Hydrogr. Z.* **24** (3), 97–119.

Dickson, R. R. & Baxter, G. C. (1972). Monitoring deep water movements in the Norwegian Sea by satellite. *Int. Council Explor. Sea*, CM9, 9 pp. (mimeo).

Dickson, R. R., Holden, M. J. & Pope, J. G. (1973). Environmental influences on the survival of North Sea cod. In *The early life history of fish*, ed. J. H. S. Blaxter, pp. 69–80. Springer Verlag, Berlin.
Fish, Oban, May 1973.

Dietrich, G. (1950). Die anomale Jahresschwankung des Wärmeinhalts im Englischen Kanal, ihre Ursachen und Auswirkungen. *Dt. Hydrogr. Z.* **3,** 184–201.

(1954*a*). Verteilung, Ausbreitung und Vermischung der Wasserkörper in der südwestlichen Nordsee auf Grund der Ergebnisse der ' "Gauss"-Fahrt' im Februar–März 1952. *Ber. deutsch. wiss. Komm. Meeresf.* NF **B13** (2), 104–29.

(1954*b*). Ozeanographisch–meteorologische Einflusse auf Wasserstandsänderungen des Meeres am Beispiel der Pegelbeobachtungen von Esbjerg. *Küste*, **2** (2), 130–56.

Digby, P. S. B. (1953). Plankton production in Scoresby Sound, East Greenland. *J. Anim. Ecol.* **22,** 289–322.

Doty, M. & Oguri, M. (1957). Evidence for a photosynthetic daily periodicity. *Limnol. Oceanogr.* **2,** 37–40.

Droop, M. R. (1968). Vitamin B12 and marine ecology. IV. The kinetics of uptake, growth and inhibition in *Monochrysis lutheri. J. Mar. Biol. Ass. UK,* NS **48,** 689–733.

Dugdale, R. C. (1967). Nutrient limitation in the sea; dynamics, identification and significance. *Limnol. Oceanogr.* **12,** (4), 685–95.

Eggvin, J. (1964). Water movement in the central part of the Norwegian Sea based on recent material. *Int. Council Explor. Sea,* CM1964, **138,** 12 pp. (mimeo).

English, T. S. (1959). Primary production in the Central North Polar Sea Drifting station Alpha, 1957–8. *Preprints Intern. Oceanogr. Congress,* ed. Mary Sears, pp. 838–9.

(1964). Estimation of the abundance of a fish stock from egg and larval surveys. *Rapp. Procès-Verb. Cons. int. Explor. Mer,* **155,** 174–82.

Eppley, R. W. (1968). An incubation method for estimating the carbon content of phytoplankton in natural waters. *Limnol. Oceanogr.* **13** (4), 574–82.

Eppley, R. W., Holmes, R. W. & Paasche, E. (1967*a*). Periodicity in cell division and physiological behaviour of *Ditylum brightwellii,* a marine plankton diatom during growth on light and dark cycles. *Arch. Mikrobiol.* **56,** 305–23.

Eppley, R. W., Holmes, R. W. & Strickland, J. D. H. (1967*b*). Sinking rates of marine phytoplankton measured with a fluorometer. *J. Exp. mar. Biol. Ecol.* **1,** 191–208.

Eppley, R. W., Rogers, J. N. & McCarthy, J. J. (1969). Half saturation constants for uptake of nitrate and ammonium by marine phytoplankton. *Limnol. Oceanogr.* **14** (6), 912–20.

Eppley, R. W. & Sloan, P. R. (1966). Growth rates of marine phytoplankton; correlation with light absorption by cell chlorophyll *a. Physiol. Plant.* **19,** 47–59.

Eppley, R. W. & Thomas, W. H. (1969). Comparison of half saturation constants for growth and nitrate uptake of marine phytoplankton. *J. Phycol.* **5,** 375–9.

FAO *Yearbooks of Fishery Statistics, catches and landings.*

Fleming, R. H. (1939). The control of diatom populations by grazing. *J. Cons. int. Explor. Mer,* **14,** 210–27.

Fleminger, A. & Clutter, R. I. (1965). Avoidance of towed nets by zoo-plankton. *Limnol. Oceanogr.* **10** (1), 96–104.

Flores, L. A. (1968). Informe preliminar del crucero 6611 de la primavera de 1966 (Cabo Blanco-Punta Coles). *Informe Inst. Mar. Perú,* **14,** 16 pp.

Flores, L. A. & Elias, L. A. P. (1967). Informe preliminar del crucero 6608–9 de invierno 1966 (Mancora-Ilo). *Informe Inst. Mar. Perú,* **16,** 24 pp.

Foerster, R. E. (1934). An investigation of the life history and propagation of the sockeye salmon (*Oncorhynchus nerka*) at Cultus Lake in British Columbia. 4. The life history cycle of the 1925 year class with natural propagation. *Contrib. Can. Biol. Fish.* **8,** 349–55.

Foerster, R. E. (1936). An investigation of the life history and propagation of the sockeye salmon (*Oncorhynchus nerka*) at Cultus Lake, British Columbia. No. 5. The life history cycle of the 1926 year class with artificial propagation involving the liberation of free and swarming fry. *J. Biol. Bd Can.* **2,** 311–33.

(1937). The return from the sea of sockeye salmon (*Oncorhynchus nerka*) with special reference to percentage survival, sex, proportions and progress. *J. Biol. Bd Can.* **3,** 26–42.

Fogg, G. E. (1966). The extracellular products of algae. *Oceanogr. Mar. Biol. Rev.* **4,** 195–212.

Forbes, E. (1843). Report of the British Association for the Advancement of Science.

Ford, E. (1933). An account of the herring investigations conducted at Plymouth during the years from 1924 to 1933. *J. Mar. Biol. Ass. UK,* NS **19,** 305–84.

Friedrich, H. (1954). Die Planktonkunde des Meeres, ein Ergebnis integrierende Forschung. *Veröff. Inst. Meeresf. in Bremerhaven,* **3** (1), 1–8.

Frost, B. W. (1972). Effects of size and concentration of food particles on the feeding behaviour of the marine planktonic copepod *Calanus pacificus. Limnol. Oceanogr.* **17** (6), 805–15.

Fry, F. E. J. (1949). Statistics of a Lake Trout Fishery. *Biometrics,* **5** (1), 27–67.

(1957). Assessment of mortalities by the use of virtual populations. *ICNAF, ICES, FAO Spec. Sci. Meeting.* Lisbon 1957, paper P15 (mimeo).

Fukuda, Y. (1962). On the stocks of halibut and their fisheries in the north-east Pacific. *Intern. N. Pac. Fish. Commn Bull.* **7,** 39–50.

Fukuhara, F. M., Murai, S., Lalanne, J.-J., & Sribhibhadh, A. (1962). Continental origin of red salmon as determined from morphological characters. *Intern. N. Pac. Fish. Commn Bull.* **8,** 15–109.

Fujino, K. & Kang, T. (1968). Serum esterase groups of Pacific and Atlantic tunas. *Copeia,* **1,** 56–63.

Fulton, T. W. (1897). On the growth and maturation of the ovarian eggs of teleostean fishes. *16th Ann. Rep. Fish. Bd Scot.* **3,** 88–124.

Gardiner, A. C. (1939). Phosphate production by planktonic animals. *J. Cons. int. Explor. Mer,* **12,** 144–6.

Garrod, D. J. (1967). Population dynamics of the Arcto-norwegian cod. *J. Fish. Res. Bd Can.* **24** (1), 145–90.

(1968). Stock recruitment relationships in four North Atlantic cod stocks. *Int. Council Explor. Sea,* CM1968-F4, 6 pp.

Garstang, W. (1900–3). The impoverishment of the sea. *J. Mar. Biol. Ass. UK,* NS **6,** 1–70.

(1909). The distribution of the plaice in the North Sea, Skagerrak and Kattegat according to size age and frequency. *Rapp. Procès-Verb. Cons. int. Explor. Mer,* **11,** 65–134.

Gerking, S. (1959). Physiological changes accompanying ageing in fishes. *The Lifespan of Animals,* CIBA Colloquia on Ageing, vol. 5. Churchill, London.

Gilbert, C. H. (1916). Contributions to the life history of the sockeye salmon. 3. *Rep. Commn Fish. Brit. Columb.* 1915, 27–64.

Glover, R. S. (1957). An ecological survey of the drift net herring survey off the north-east coast of Scotland. II. The planktonic environment of the herring. *Bull. mar. Ecol.* **5,** 1–43.

Goldberg, E. D., Walker, T. J. & Whisenand, A. (1951). Phosphate utilization by diatoms. *Biol. Bull.* **101,** 274–84.

Goldman, C. R. (1968). The use of absolute activity for eliminating serious errors in the measurement of primary productivity with C^{14}. *J. Cons. int. Explor. Mer*, **32** (2), 172–79.

Gompertz, B. (1925). On the nature of the function expressive of the law of human mortality, and on a new mode of determining the value of life contingencies. *Phil. Trans. Roy. Soc. Lond.* **115** (1), 513–85.

Graham, G. M. (1934). The North Sea cod. *J. Cons. int. Explor. Mer*, **9** (2), 159–71.

(1935). Modern theory of exploiting a fishery and application to North Sea trawling. *J. Cons. int. Explor. Mer*, **10** (2), 264–74.

(1938). Rates of fishing and natural mortality from the data of marking experiments. *J. Cons. int. Explor. Mer*, **10,** 264–74.

(1958). Fish population assessment by inspection. In Some problems for biological fishery survey and techniques for their solution. *Int. Commn. North West. Atl. Fish.* Spec. Publ. **1,** 67–68.

Graham, J. J., Chenoweth, S. R. & Davis, C. W. (1972). Abundance, distribution, movements and lengths of larval herring along the western coast of the Gulf of Maine. *Fish. Bull. US Dept. Commerc.* **70** (2), 307–21.

Grainger, E. H. (1959). The annual oceanographic cycle at Igloolik in the Canadian Arctic. *J. Fish. Res. Bd Can.* **16,** 453–501.

Gran, H. H. & Braarud, T. (1935). A quantitative study of the phytoplankton in the Bay of Fundy and the Gulf of Maine (including observations on hydrography, chemistry and turbidity). *J. Biol. Bd Can.* **1,** 279–467.

Gray, J. (1926). The kinetics of growth. *J. Exp. Biol.* **6** (3), 248–74.

(1968). Animal Locomotion. 479 pp. Weidenfeld & Nicholson, London.

Greer-Walker, M. (1970). Growth and development of the skeletal muscle fibres of the cod (*Gadus morhua* L.). *J. Cons. int. Explor. Mer*, **33** (2), 228–44.

Grosslein, M. D. & Hennemuth, R. C. (1973). Spawning stock and other factors related to recruitment of Georges Bank haddock. *Rapp. Procès-Verb. Cons. int. perm. Explor. Mer*, **164,** 77–88.

Gulland, J. A. (1955). Estimation of growth and mortality in commercially exploited populations. *Fish. Invest. Lond. Ser.* 2, **18** (9), 46 pp.

(1961). Fishing and the stocks of fish at Iceland. *Fish. Invest.* Lond. Ser. 2, **23** (4), 52 pp.

(1967). Gulland's method of virtual population analysis. Appendix to Garrod (1967).

(1968*a*). The concept of the marginal yield from exploited fish stocks. *J. Cons. int. Explor. Mer*, **32** (2), 256–61. *In Marine Food Chains*, ed. J. H. Steele, pp. 296–318. Oliver & Boyd, Edinburgh.

(1968*b*). Recent changes in the North Sea plaice fishery. *J. Cons. int. Explor. Mer*, **31,** 305–22.

(1969). Manual of methods for fish stock assessment. I. Fish Population analysis. *FAO Manuals in Fisheries Science*, **4,** 154 pp.

(1970*a*). Food chain studies and some problems in world fisheries. In *Marine*

(1970*b*). The fish resources of the ocean. *FAO Fish. Tech. Rep.* **97,** 425 pp.

Gulland, J. A. & Williamson, G. R. (1967). Transatlantic journey of a tagged cod. *Nature* (*London*), **195**(4844), 921.

Hachey, H. B. (1936–7). Ekman's theory applied to water replacement on the Scotian Shelf. *Proc. Trans. Nova Scotian Inst. Sci.* **19** (3), 264–74.

Halldall, P. (1953). Phytoplankton investigations from Weather Ship M in the Norwegian Sea, 1948–9. *Hvalradets Skr.* **38**.

Haq, S. M. (1967). Nutritional physiology of *Metridia lucens* and *Metridia longa* from the Gulf of Maine. *Limnol. Oceanogr.* **12,** 40–51.

Harden-Jones, F. R. (1968). *Fish migration.* 325 pp. Edward Arnold, London.

Harding, D. W. & Talbot, J. W. (1973). Recent studies on the eggs and larvae of the plaice (*Pleuronectes platessa* L.) in the Southern Bight. *Rapp. Procès-Verb. Cons. int. Explor. Mer,* **164,** 261–9.

Hardy, A. C. (1924). The herring in relation to its animate environment. I. The food and feeding habits of the herring with special reference to the east coast of England. *Fish. Invest. Lond.* Ser. 2, **7** (3), 1–53.

Harris, E. (1959). The nitrogen cycle in Long Island Sound. *Bull. Bingh. Oceanogr. Coll.* **17,** 31–65.

Harris, J. G. K. (1968). A mathematical model describing the possible behaviour of a copepod feeding continuously in a relatively dense randomly distributed population of algal cells. *J. Cons. int. Explor. Mer,* **32** (1), 83–92.

(1975). The effect of density-dependent mortality on the shape of the stock and recruitment curve. *J. Cons. int. Explor. Mer* (in press).

Hart, T. J. (1934). On the phytoplankton of the Southwest Atlantic and the Bellingshausen Sea, 1929–1931. *Disc. Rep.* **8,** 1–268.

(1942). Phytoplankton periodicity in Antarctic surface waters. *Disc. Rep.* **21,** 261–356.

Hart, T. J. & Currie, R. I. (1960). The Benguela Current. *Disc. Rep.* **31,** 123–298.

Harvey, H. W. (1934). Annual variation of planktonic vegetation in 1933. *J. Mar. Biol. Ass. UK,* NS **19,** 775 pp.

Harvey, H. W., Cooper, L. H. N., Lebour, M. V. & Russell, F. S. (1935). The control of production. *J. Mar. Biol. Ass. UK,* NS **20,** 407–41.

Heincke, F. (1913). Untersuchungen über die Scholle, Generalbericht I. Schollenfischerei und Schonmassregeln. Vorläufige Kurze Übersicht über die wichtigsten Ergebnisse des Berichts. *Rapp. Procès-Verb. Cons. int. Explor. Mer,* **16,** 1–70.

Heinrich, A. K. (1961). Seasonal phenomena in plankton of the world ocean. I. Seasonal phenomena in the plankton of high and low temperature latitudes. *Trudy. Inst. Okeanol.* **15,** 57–81.

(1962). The life histories of plankton animals and seasonal cycles of plankton communities in the oceans. *J. Cons. int. Explor. Mer,* **27,** 15–24.

Hempel, G. (1955). Zur Beziehung zwischen Bestandschuchte und Wachstum in der Schollenbevölkerung der Deutschen Bucht. *Ber. dt. Wiss. Komm. Meeresforsch.* **15** (2), 132–44.

(1967). Egg weight in Atlantic herring (*Clupea harengus* L.). *J. Cons. int. Explor. Mer,* **31,** 170–95.

Hennemuth, R. C. (1961). Year class abundance, mortality and yield per recruit of yellowfin tuna in the eastern Pacific Ocean 1954–59. *Inter-Amer. Trop. Tuna Commn Bull.* **6** (1), 51 pp.

Henry, K. A. (1961). Racial identification of Fraser river sockeye salmon by means of scales and its applications to salmon management. *Bull. int. Pac. Salm. Fish. Commn,* **12,** 47 pp.

Hensen, V. (1887). Über die Bestimmung des Planktons oder das im Meere treibenden Materials an Pflanzen und Thieren. *Ber. Komm. Wiss. Unters. Meeres.* **5,** 1882–6. 13 pp.

Hentschel, E. & Wattenberg, H. (1931). Plankton und Phosphat in der Oberflächenschicht des Südatlantischen Ozeans. *Ann. Hydrogr. Berl.* **58,** 273–7.

Herdman, W. A. (1918). Spolia Runiana III. The distribution of certain diatoms and Copepoda throughout the year in the Irish Sea. *J. Linn. Soc.* **34,** 95–126.
(1921). Variation in successive vertical plankton hauls at Port Erin. *Trans. Liverpool Biol. Soc.* **35,** 161 pp.
Hermann, F. & Hansen, P. M. (1965) Possible influence of water temperature on the growth of the West Greenland cod. *Int. Commn North West Atl. Fish. Spec. Publ.* **6,** 557–663.
Herrington, W. C. (1948). Limiting factors for fish populations: Some theories and an example. *Bull. Bingh. Oceanogr. Coll.* **9** (4), 229–83.
Hidaka, K. & Ogawa, K. (1958). On the seasonal variations of surface divergence of the ocean currents in terms of wind stresses over the ocean currents in terms of wind stresses over the oceans. *Rec. oceanogr. Wks Japan.* **4** (2), 124–69.
Hill, H. W. & Lee, A. J. (1957). The effect of wind on water transport in the region of the Bear Island fishery. *Proc. Roy. Soc.* B. **148,** 104–16.
Hjort, J. (1914). Fluctuations in the great fisheries of Northern Europe. *Rapp. Procès-Verb. Cons. int. Explor. Mer*, **20,** 1–13.
Hjort, J., Jahn, G. & Ottestad, P. (1933). The optimum catch. Essays on Population. *Hvalr. Skr.* **7,** 92–127.
Höglund, H. (1955). Swedish herring tagging experiments, 1949–1953. *Rapp. Procès-Verb. Cons. int. Explor. Mer*, **140** (2), 19–29.
(1972). On the Bohuslån herring during the great herring fishery period in the eighteenth century. *Rep. Inst. mar. Res. Lysekil*, **20,** 1–86.
Holden, M. J. (1973). Are long-term sustainable fisheries for elasmobranchs possible? *Rapp. Procès-Verb. Cons. int. Explor. Mer*, **164,** 360–7.
Holling, C. S. (1959). The components of predation as revealed by a study of small mammal predation of the European sawfly. *Can. Ent.* **91,** 293–320.
(1965). The functional response of predators to prey density and its role in mimicry and population regulation. *Mem. Entom. Soc. Can.* **45,** 60 pp.
Holme, N. A. (1953). The biomass of the bottom fauna in the English Channel off Plymouth. *J. Mar. Biol. Assn. UK*, NS **32,** 1–49.
Holmes, R. W., Schaefer, M. B. & Shimada, B. M. (1957). Primary production, chlorophyll and zooplankton volumes in the tropical eastern Pacific Ocean. *Bull. Inter-Amer. Trop. Tuna Commn*, **2** (4), 129–69.
Howard, L. O. & Fiske, W. F. (1911). The importation into the United States of the parasites of the gipsy moth and the brown tail moth. *US Dept. Agr. Bur. Ent. Bull.* **91,** 1–312.
Hunter, J. G. (1959). Survival and production of pink and chum salmon in a coastal stream. *J. Fish. Res. Bd Can.* **16,** 835–86.
Hutchinson, G. E. (1950). Survey of existing knowledge of biogeochemistry. 3. The biogeochemistry of vertebrate excretion. *Bull. Amer. Mus. nat. Hist.* **96,** 554 pp.
(1967). A Treatise on Limnology. II. *Introduction to Lake Biology and the Limnoplankton.* 1115 pp. Wiley, London & New York.
Hylen, A., Middtun, L. & Saetersdal, G. (1961). Torskeundersokelsene i Lofoten og i Barents Havet 1960. *Fisken og Havet*, **2,** 1–14.
Idelson, M. (1931). Fish marking in the Barents Sea. *J. Cons. int. Explor. Mer*, **6,** 433 pp.
Iles, T. D. (1964). The duration of maturation stages in herring. *J. Cons. int. Explor. Mer*, **29,** 166–88.
(1968). Growth studies on North Sea herring. II. O-group growth of East Anglian herring. *J. Cons. int. Explor. Mer*, **32** (1), 98–116.

(1972). The interaction of environment and parent stock size in determining recruitment in the Pacific sardine as revealed by the analysis of density-dependent O group growth. *Rapp. Procès-Verb. Cons. int. Explor. Mer*, **164,** 228–40.

Ishevskii, G. K. (1964). The systematic basis of predicting oceanological conditions and the reproduction of the fisheries. Moscow, 165 pp.

Ivlev, V. S. (1939). Transformation of energy by aquatic animals. *Int. Rev. ges. Hydrobiol.* **38** (5/6), 449 pp.

(1945). The biological productivity of waters. *Usp. Sovrem. Biol.* **19,** 98–120.

(1961). Experimental ecology of the feeding of fishes, tr. D. Scott. 302 pp. Yale University Press, New Haven.

Jamieson, A. (1970). Cod transferrins and genetic isolates. *XII. Europ. Conf. Anim. Blood groups and Biochem. Aggreg.* Warsaw, pp. 533–8.

Jamieson, A. & Jones, B. W. (1967). Two races of cod at Faroe. *Heredity*, **22** (4), 610–12.

Jerlov, N. (1951). Optical studies of ocean waters. *Repts. Swed. Deep Sea Expdn*, **3** (1), 1–59.

Jitts, H. R., McAllister, C. D., Stephens, K. & Strickland, J. D. H. (1964). The cell division rates of some marine phytoplankters as a function of light and temperature. *J. Fish. Res. Bd Can.* **21,** 139–57.

Johannes, R. E. (1964). Phosphorus excretion and body size in marine animals; micro-zooplankton and nutrient regeneration. *Limnol. Oceanogr.* **9,** 235–42.

Johansen, A. C. (1924). On the summer and autumn spawning herring in the Southern North Sea. *Medd. Komm. Havunders. Fisk.* **7** (5), 119 pp.

Johnson, P. O. (1970). The Wash Sprat Fishery. *Fish. Invest. Lond.* Ser. 2, **26** (4), 75 pp.

Johnson, H. R., Backus, R. H., Hersey, J. B. & Owen, D. M. (1956). Suspended echosounder and camera studies of midwater sound scatterers. *Deep Sea Res.* **3,** 266–72.

Johnstone, J. (1908). *Conditions of life in the Sea.* 332 pp. Cambridge University Press, London.

Jones, P. G. W. & Folkard, A. R. (1968). Chemical oceanographic observations off the coast of North West Africa, with special reference to the process of upwelling. Paper 47, 13 pp. *ICES, FAO Symp. on Living Resources of the African continental shelf between the Straits of Gibraltar and Cap Verde.*

Jones, R. (1959). A method of analysis of some tagged haddock returns. *J. Cons. int. Explor. Mer*, **25,** 58–72.

(1973). The stock and recruitment relation as applied to the North Sea Haddock. *Rapp. Procès-Verb. Cons. int. Explor. Mer*, **164,** 156–73.

Joseph, J. J. (1957). Extinction measurements to indicate distribution and transport of water masses. *Proc. UNESCO Symp. Phys. Oceanogr. Tokyo* 1955, pp. 50–75.

Juday, C. (1940). The annual energy budget of an inland lake. *Ecology*, **21** (4), 438–50.

Kendall, M. G. & Stuart, A. (1966). *The Advanced Theory of Statistics*, vol. 3. *Design and analysis of time series.* 544 pp. Griffin, London.

Ketchum, B. H. (1939*a*). The development and restoration of deficiency in the phosphorus and nitrogen composition of unicellular plants. *J. Cell. Comp. Physiol.* **13,** 373–81.

(1939*b*). The absorption of phosphate and nitrate by illuminated cultures of *Nitzschia closterium. Amer. J. Bot.* **96,** 399–407.

Killick, S. R. (1955). The chronological order of Fraser River sockeye during migration, spawning and death. *Int. Pac. Salm. Commn*, **7**, 95 pp.

King, J. E. & Hida, T. S. (1957). Zooplankton abundance in the Central Pacific II. *US Fish and Wildlife Serv. Fish. Bull.* **57**, 365–95.

Kitou, M. (1958). Distribution of planktonic copepods at the ocean weather station X, May 1950 to April 1957. *Oceanogr. Mag.* **10** (2), 193–9.

Knox, C. & Brooks, R. E. (1969). Holographic motion picture microscopy. *Proc. Roy. Soc. Lond.* B. **174,** 115–21.

Koblentz-Mishke, O. I. (1965). The magnitude of the primary production of the Pacific Ocean. *Okeanologiya*, **5,** 325–37.

Koblentz-Mishke, O. I., Tsvetkova, A. M., Gromov, M. M. & Paramonova, L. I. (1973). Primary production and chlorophyll *a* in the West Pacific. In *Life activity of pelagic communities in the Ocean Tropics*, ed. M. E. Vinogradov, pp. 73–83. Navka, Moscow.

Koblentz-Mishke, O. J., Volkovinsky, V. V. & Kabanova, J. G. (1970). Plankton primary production of the world ocean. In *Scientific Exploration of the South Pacific*. pp. 183–93. National Academy of Sciences, Washington.

Krogh, A. (1931). Dissolved substances as food of aquatic organisms. *Biol. Rev.* **6,** 412–42.

Kurita, S. (1960). Causes of fluctuation of sardine population off Japan. *Proc. World. Sci. Meeting Biol. Sardines and related species*, **3,** 913–35.

Lack, D. (1954). *The natural regulation of animal numbers*. 343 pp. Oxford University Press, London.

Laevastu, T. (1960). Factors affecting the temperature of the surface layer of the sea. A study of the heat exchange between the sea and the atmosphere, the factors affecting temperature structure in the sea and its forecasting. *Commentat. physico-math.* **25** (1), 136 pp.

Lamb, H. H. (1972). *Climate, past, present and future*, vol. 1, 613 pp. Methuen, London.

Lanskaya, L. A. (1963). Fission rate of plankton algae of the Black Sea in cultures. *Symp. Mar. Microbiol. Springfield Ill.* pp. 127–32.

Lasker, R. (1970). Utilization of zooplankton energy by a Pacific sardine population in the Californian current. *Marine Food Chains*, ed. J. H. Steele, pp. 265–84. Oliver & Boyd, Edinburgh & London.

Lebour, M. V. (1918,1919). The food of post larval fish I–III. *J. Mar. Biol. Ass. UK*, NS **11,** 433–69; **12,** 22–47; **12,** 261–324.

(1921). The food of young clupeoids. *J. Mar. Biol. Ass. UK*, NS **12,** 458–67.

Le Cren, E. D. (1973). The population dynamics of young trout (*Salmo trutta*) in relation to density and territorial behaviour. *Rapp. Proces-Verb. Cons. int. perm. Explor. Mer*, **164,** 241–6.

Lee, A. J. (1952). The influence of hydrography on the Bear Island cod fishery. *Rapp. Procès-Verb. Cons. int. Explor. Mer*, **131,** 74–102.

Lee, A. J. & Folkard, A. (1969). Factors affecting turbidity in the Southern North Sea. *J. Cons. int. Explor. Mer*, **32,** 291–302.

Le Gall, J. (1935). Le hareng *Clupea harengus* Linné I Les populations de l'Atlantique Nord Est. *Ann. Inst. Ocean. Moncaco*, **15** (1), 215 pp.

Lea, E. (1911). A study of the growth of herrings. *Publ. Circ. Cons. int. Explor. Mer*, **61,** 35–57.

Letaconnoux, R. (1954). La morue de la Mer Celtique et de l'entrée de la Manche. *Rapp. Procès-Verb. Cons. int. Explor. Mer*, **136,** 30–2.

Li, C. C. (1955). *Population genetics*, 366 pp. Chicago.

de Ligny, W. (1969). Serological and biochemical studies on fish populations. *Oceanogr. Mar. Biol.* **7,** 411–513.

Lindeman, R. L. (1941). Food cycle dynamics in a senescent lake. *Amer. Midl. Nat.* pp. 636–73.

(1942). The trophic dynamic aspect of Ecology. *Ecology*, **23,** 399–418.

Lloyd, I. J. (1970). Primary production off the coast of Northwest Africa. *J. Cons. int. Explor. Mer*, **33,** 312–23.

Loftus, M. E. & Carpenter, J. H. (1971). A fluorometric method for determining chlorophylls *a*, *b*, *c*. *J. Mar. Res.* **29** (3), 319–38.

Lohmann, H. (1908). Untersuchungen zur Feststellung des Vollständigen Gehaltes des Meeres an Plankton. *Wiss. Meeresuntersuch. Abt. Kiel.* NF**10,** 129–370.

Longard, J. R. & Banks, R. E. (1952). Wind-induced vertical movement of the water on an open coast. *Trans. Amer. Geophys. Un.* **33** (3), 377–80.

Lorenzen, C. J. (1966). A method for the continuous measurement of *in vivo* chlorophyll concentration. *Deep Sea Res.* **13** (2), 223–7.

(1967). Vertical distribution of chlorophyll and phaeopigments: Baja California. *Deep Sea Res.* **14** (6), 735–46.

Lotka, A. J. (1925). *Elements of Physical Biology*. Williams & Wilkins, Baltimore, Ltd.

Lumb, F. E. (1964). The influence of cloud on hourly amounts of total solar radiation at the sea surface. *Quart. J. Meteorol. Soc.* **90** (386), 493–5.

Lund, J. W. G. (1949). Studies on *Asterionella formosa* Hass. I. The origin and nature of the cells producing seasonal maxima. *J. Ecol.* **37** (2), 389–419.

(1950). Studies on *Asterionella formosa* Hass. II. Nutrient depletion and the spring maximum. Part 1. Observations on Windermere, Esthwaite Water and Blelham Tarn. Part 2. Discussion. *J. Ecol.* **38,** 1–35.

(1964). Primary production and periodicity of phytoplankton. *Verh. Intern. Verein. Limnol.* **15,** 37–56.

Lund, J. W. G., Mackereth, F. H. J. & Mortimer, C. H. (1963). Changes in depth and time of certain chemical and physical conditions and of the standing crop of *Asterionella formosa* Hass in the North Basin of Windermere in 1947. *Phil. Trans. Roy. Soc.* B. **731** (246), 255–90.

McAllister, C. D. (1961). Zooplankton studies at Ocean Weather Station 'P' in the North east Pacific Ocean. *J. Fish. Res. Bd Can.* **18** (1), 1–29.

(1970). Zooplankton rations, phytoplankton mortality and the estimation of marine production. In *Marine Food Chains*, ed. J. H. Steele, pp. 419–451. Oliver & Boyd, Edinburgh & London.

McAllister, C. D., Parsons, T. R., Stephens, K. & Strickland, J. D. H. (1961). Measurements of primary production in coastal sea water using a large-volume plastic sphere. *Limnol. Oceanogr.* **6,** 237–58.

MacArthur, R. (1960). On the relative abundance of species. *Amer. Nat.* **94,** 25–36.

McNeil, W. J. (1963). Redd superimposition and egg capacity of pink salmon spawning beds. *J. Fish. Res. Bd Can.* **21,** 1385–96.

Mackereth, F. J. (1953). Phosphorus utilization by *Asterionella formosa* Hass. *J. exp. Bot.* **4,** 296–313.

Mare, M. (1942). A study of a marine benthic community with special reference to the microorganisms. *J. Mar. Biol. Ass. UK*, NS **25** (3), 517–54.

Margolis, L., Cleaver, F. C., Fukuda, Y. & Godfrey, H. (1966). Salmon of the North Pacific. VI. Sockeye salmon in offshore waters. *Int. N. Pac. Fish. Commn.* **20,** 1–68.

Marr, J. C. (1951). On the use of the terms *abundance*, *availability* and *apparent abundance* in fishery biology. *Copeia*, **2,** 163–9.

(1962). The causes of major variations in the catch of the Pacific Sardine, *Sardinops caerulea* (Girard). *FAO Proc. World. Sci. Meeting Biol. Sardines and related species*, **3,** 667–791.

Marshall, P. T. (1958). Primary production in the Arctic. *J. Cons. int. Explor. Mer*, **23,** 173–7.

Marshall, S. M. & Orr, A. P. (1927). The relation of the plankton to some chemical and physical factors in the Clyde Sea area. *J. Mar. Biol. Ass. UK*, NS **14** (4), 837–68.

(1930). A study of the spring diatom increase in Loch Striven. *J. Mar. Biol. Ass. UK*, NS **16,** 853–73.

(1952). On the biology of *Calanus finmarchicus*. VII. Factors affecting egg production. *J. Mar. Biol. Ass UK*, NS **30,** 527–48.

(1958). On the biology of *Calanus finmarchicus*. X. Seasonal changes in oxygen consumption. *J. Mar. Biol. Ass. UK*, NS **37,** 459–72.

Marshall, S. M., Nicholls, A. G. & Orr, A. P. (1935). On the biology of *Calanus finmarchicus*. VI. Oxygen consumption in relation to environmental conditions. *J. Mar. Biol. Ass. UK*, NS **20,** 1–28.

Martin, J. H. (1965). Phytoplankton-zooplankton relationships in Narrangansett Bay. *Limnol. Oceanogr.* **10,** 185–91.

Martin, W. R. & Kohler, A. C. (1965). Variation in recruitment of cod (*Gadus morhua* L) in southern ICNAF waters, as related to environmental change. *Spec. Publ. Int. Commn North West Atl. Fish.* **6,** 833–46.

Marty, J. J. (1959). The fundamental stages of the life cycle of Atlanto-Scandinavian herring. *U.S. Fish and Wildlife Serv. Fisheries* (Translation), **327,** 5–68.

Maslov, N. A. (1944). Bottom fishes of the Barents Sea. V. Migrations of the cod. *Trans. Knip. Inst. Murmansk.* **8,** 3–186.

Mather, F. J., III. (1962). Transatlantic migration of two large bluefin tuna. *J. Cons. int. Explor. Mer*, **27,** 325–7.

Matsumoto, W. (1966). Distribution and abundance of tuna larvae in the Pacific Ocean. *Proc. Governor's Conf. Centr. Pac. Res. Honolulu 1966* ed. T. A. Manier, pp. 221–30.

Medawar, P. B. (1945). Size, shape and age. In *Essays on Growth and Form presented to D'Arcy Wentworth Thompson*, 157–87, Clarendon Press, Oxford.

de Mendiola B. R. (1971). Some observations on the feeding of the Peruvian Anchoveta *Engraulis ringens* J. in two regions of the Peruvian coast. *Fertility of the Sea*, ed. Costlow, **2,** 417–40.

Menzel, D. W. & Ryther, J. H. (1960). The annual cycle of primary production in the Sargasso Sea off Bermuda. *Deep Sea Res.* **6,** 351–7.

(1961*a*). Annual variations in primary production of the Sargasso Sea off Bermuda. *Deep Sea Res.* **7,** 282–8.

(1961*b*). Zooplankton in the Sargasso Sea off Bermuda and its relation to organic production. *J. Cons. int. Mer*, **26,** 250–8.

Merriman, D. (1941). Studies on the striped bass (*Roccus saxatilis*) of the Atlantic coast. *Fish. Bull. US. Fish and Wildlife Serv.* **50,** 77 pp.

Milne, D. J. (1957). Recent British Columbia spring and coho salmon tagging experiments and a comparison with those conducted from 1925 to 1930, *Bull. Fish. Res. Bd. Can.* **113,** 56 pp.

Møller, D. (1968). Genetic diversity in spawning cod along the Norwegian coast. *Hereditas*, **60,** 1–32.

Moore, H. B. (1949). The zooplankton of the upper waters of the Bermuda area of the North Atlantic. *Bull. Bingh. oceanogr. Coll.* **12,** 1–97.

Motoda, S. & Hirano, Y. (1963). Review of the Japanese herring investigations. *Rapp. Procès-Verb. Cons. int. Explor. Mer*, **154,** 249–61.

Mullin, M. M. (1963). Some factors affecting the feeding of marine copepods of the genus *Calanus*. *Limnol. Oceanogr.* **8,** 239–50.

Mullin, M. M., Sloan, P. R. & Eppley, R. W. (1966). Relationship between carbon content cell volume and area in phytoplankton. *Limnol. Oceanogr.* **11** (2), 307–11.

Munk, W. H. & Riley, G. A. (1952). Absorption of nutrients by aquatic plants. *J. Mar. Res.* **11** (2), 215–40.

Murphy, G. I. (1965). A solution of the catch equation. *J. Fish. Res. Bd Can.* **22** (1), 191–202.

(1966). Population biology of the Pacific sardine (*Sardinops caerulea*). *Proc. Calif. Acad. Sci.* **34,** 1–84.

Nakai, Z. (1960). Changes in the population and catch of the Far East sardine area. *Proc. World Sci. Meet. Sardines and related species*, **3,** 807–53.

Nakamura, H. (1969). *Tuna distribution and migration.* 76 pp. Fishing News (Books) Ltd.

Namias, J. (1959). Recent seasonal interactions between North Pacific waters and the overlying atmospheric circulation. *J. geophys. Res.* **64,** 631–46.

(1964). Seasonal persistence and recurrence of European blocking during 1958–1960. *Tellus*, **16** (3), 394–407.

Neess, R. & Dugdale, R. C. (1959). Computation of production for populations of aquatic midge larvae. *Ecology*, **40,** 425–30.

Nielsen, E. S. (1952). The use of radioactive carbon for measuring organic production in the sea. *J. Cons. int. Explor. Mer*, **18,** 117–40.

(1958). A survey of recent Danish measurements of the organic productivity in the sea. *Rapp. Procès-Verb. Cons. int. Explor. Mer*, **144,** 92–5.

(1965). Investigations of the rate of primary production at two Danish light ships in the transition area between the North Sea and the Baltic. *Medd. Danm. Fisk. Hav.* **4** (3), 31–77.

Nielsen, E. S. & Hansen, K. V. (1959). Measurements with the carbon-14 technique of the respiration rates in natural populations of phytoplankton. *Deep Sea Res.* **5,** 222–33.

Odum, H. T. (1956). Primary production inflowing waters. *Limnol. Oceanogr.* **1,** 102–17.

Olson, F. C. W. (1964). The survival value of fish schooling. *J. Cons. int. Explor. Mer*, **29,** 115–16.

Ostvedt, O. J. (1955). Zooplankton investigations from Weather Ship M in the Norwegian Sea, 1948–9. *Hval. Skr.* **40,** 93 pp.

Ostwald, W. (1903). Zur Theorie der Schwebevorgange sowie der spezifischen Gewichts bestimmungen schwebender Organismen. *Arch. ges. Physiol.* 94 pp.

Otsu, T. (1960). Albacore migration and growth in the north Pacific Ocean as estimated from tag recoveries. *Pacific Sci.* **14** (3), 257–60.

Ottestad, P. (1942). On periodical variations in the yield of the great sea fisheries and the possibility of establishing yield prognoses. *Fisk. Dir. Skr. Ser. Havundersøk*, **7** (5), 1–11.

Paloheimo, Y. & Dickie, L. M. (1965). Food and growth of fishes. 1. A growth curve derived from experimental data. *J. Fish. Res. Bd Can.* **22,** 521–54.

(1966*a*) Food and growth of fishes. 2. Effects of food and temperature on the relation between metabolism and growth. *J. Fish. Res. Bd Can.* **23,** 869–908.

(1966*b*). Food and growth of fishes. 3. Relations among food, body size and growth efficiency. *J. Fish. Res. Bd Can.* **23,** 1209–48.

Parker, R. R. & Larkin, P. A. (1959). A concept of growth in fishes. *J. Fish. Res. Bd Can.* **16** (5), 721–45.

Parrish, B. B. (1957). The cod, haddock and hake. In *Sea Fisheries*, ed. G. M. Graham, 251–331. Edward Arnold, Edinburgh.

Parrish, B. B. & Saville, A. (1965). The biology of the north east Atlantic herring population. *Oceanogr. Mar. Biol. Ann. Rev.* **3,** 323–73.

Parrish, B. B., Saville, A., Craig, R. E., Baxter, I. G. & Priestley, R. (1959). Observations on herring spawning and larval distributions in the Firth of Clyde in 1958. *J. Mar. Biol. Ass. UK*, NS **38,** 445–53.

Parsons, T. R., Le Brasseur, R. J. & Fulton, J. D. (1967). Some observations on the dependence of zooplankton grazing on cell size and concentration of phytoplankton blooms. *J. oceanogr. Soc. Japan.* **23** (1), 10–17.

Pearcy, W. G. (1962). Ecology of young winter flounder in an estuary. *Bull. Bingh. Oceanogr. Coll.* **18,** 1–78.

Pearl, R. (1930). *The Biology of Population growth.* 330 pp. Knopf.

Pella, J. J. & Tomlinson, P. K. (1969). A generalized stock production model. *Bull. Inter-Amer. Trop. Tuna Commn*, **13** (3), 421–58.

Petersen, C. J. G. (1894). On the biology of our flatfishes and on the decrease of our flatfisheries. *Rep. Dan. Biol. Sta.* **4,** 146 pp.

(1918). The sea bottom and its production of fish food. *Rep. Dan. Biol. Sta.* **25,** 1–62.

Petersen, C. J. G., Garstang, W. & Kyle, H. M. (1907). Summary Report on the present state of our knowledge with regard to the plaice and plaice fishery. A. of the Kattegat; B. of the North Sea. *Rapp. Procès-Verb. Cons. int. Explor. Mer*, **7** (C), 54–150.

Petipa, T. S. (1964). Diurnal rhythm of the consumption and accumulation of fat in *Calanus helgolandicus* (Claus) in the Black Sea, *Dokl. Akad. Nauk. SSSR*, **156** (6), 1440–3.

(1965). The food selectivity of *Calanus helgolandicus* (Claus). In *Investigations of the Plankton of the Black Sea and Sea of Azov. Acad. Sci. Ukrain. SSR*, 102–10.

Petipa, T. S. & Makarova, N. P. (1969). Dependence of phytoplankton production on rhythm and rate of elimination. *Mar. Biol.* **3** (3), 191–5.

Petipa, T. S., Pavlova, E. V. & Mironov, G. N. (1970). The food web structure, utilization and transport of energy by trophic levels in the planktonic communities. In *Marine Food Chains*, ed. J. H. Steele, pp. 143–67. Oliver & Boyd, Edinburgh & London.

Platt, T. (1970). Spatial heterogeneity of phytoplankton in a near shore environment. *J. Fish. Res. Bd Can.* **27** (8), 1453–73.

Pomeroy, L. R., Mathews, H. M. & Min, H. B. (1963). Excretion of phosphate and soluble organic phosphorus compounds in zooplankton. *Limnol. Oceanogr.* **8** (1), 50–3.

Pope, J. G. (1972). An investigation of the accuracy of virtual population analysis using cohort analysis. *Int. Commn North West Atl. Fish. Res. Bull.* **9,** 65–75.

Poulsen, E. M. (1931). Biological investigations upon the cod in Danish waters. *Medd. Komm. Danmarks. Fisk. og Havundersok Fiskeri*, **9** (1), 149 pp.

Pritchard, A. L. (1939). Homing tendency and age at maturity of pink salmon (*Oncorhynchus gorbuscha*) in British Columbia. *J. Fish. Res. Bd Can.* **4,** 233–51.

Purdom, C. E. & Wyatt (1969). Racial differences in Irish and North Sea plaice (*Pleuronectes platessa*). *Nature* (*London*), **222,** 780–8.

Raben, E. (1905). Über quantitative Bestimmung von Stickstoffverbindungen im Meerwasser, nebst einem Anhang über die quantitative Bestimmung der im Meerwasser gelösten Kieselsäure. *Wiss. Komm. Kiel*, **8,** 81–101.

Rae, K. S. M. (1957). A relationship between wind, plankton distribution and haddock brood strength. *Bull. Mar. Ecol.* **4,** 247–69.

Raitt, D. F. S. (1968). The population dynamics of the Norway pout in the North Sea. *Dept. Agric. Fish. Scotland. Mar. Res.* **5,** 24 pp.

Raitt, D. S. (1939). The rate of mortality of the haddock of the North Sea stock, 1919–1938. *Rapp. Procès-Verb. Cons. int. Explor. Mer*, **110,** 65.

Ramster, J. W. (1973). The residual circulation of the Northern Irish Sea with particular reference to Liverpool Bay. *Fish. Lab. Lowestoft Tech. Rep.* **5,** 21 pp.

Rashevsky, N. (1959). Some remarks on the mathematical theory of nutrition in fishes. *Bull. Math. Biophys.* **21,** 161–83.

Raymont, J. E. G. (1963). *Plankton and productivity in the Oceans*, 660 pp.

Reeve, M. R. (1970). The biology of Chaetognatha. I. Quantitative aspects of growth and reproduction in *Sagitta hispida.* In *Marine Food Chains*, ed. J. H. Steele, pp. 168–92. Oliver & Boyd, Edinburgh & London.

Reibisch, J. (1899). Über die Eizahl bei *Pleuronectes platessa* und die Alterbestimmung aus den Otolithen. *Wiss. Meeresunters Kiel*, IV.

Reid, J. L. (1962). On circulation, phosphate phosphorus content, and zooplankton volumes in the upper part of the Pacific Ocean. *Limnol. Oceanogr.* **7** (3), 287–306.

Rich, W. H. (1937). Homing of Pacific Salmon. *Science*, **85,** 477–8.

Richardson, I. D., Cushing, D. H., Harden Jones, F. R., Beverton, R. J. H. & Blacker, R. W. (1959). Echo sounding experiments in the Barents Sea. *Fish. Invest. Lond.* **2,** 22 (9), 55 pp.

Richman, S. (1958). The transformation of energy by *Daphnia pulex. Ecol. Monogr.* **28,** 273–91.

Ricker, W. E. (1945). A method of estimating minimum size limits for obtaining maximum yield. *Copeia*, **2,** 84–94.

(1948). Method of estimating vital statistics of fish populations. *Indiana univ. Publ. Sci. Serv.* **15,** 101 pp.

(1954). Stock and recruitment. *J. Fish. Res. Bd Can.* **11** (5), 559–623.

(1958). Handbook of computations for biological statistics of fish populations. *Fish. Res. Bd Can. Bull.* **119,** 300 pp.

(1969). Food from the Sea. In *Resources and Man*, ed. Cloud. Freeman & Co., Chicago.

(1973). Critical statistics from two reproduction curves. *Rapp. Procès-Verb. Cons. int. Explor. Mer*, **164,** 333–40.

Ricker, W. E. & Foerster, R. E. (1948). Computation of fish populations. In A symposium of fish populations. *Bull. Bingh. Oceanogr. Coll.* **11** (4), 173–211.

Rigler, F. H. (1961). The uptake and release of inorganic phosphorus by *Daphnia magna* Straus. *Limnol. Oceanogr.* **6** (2), 65–174.

Riley, G. A. (1946). Factors controlling phytoplankton populations on George's Bank. *J. Mar. Res.* **6,** 54–73.

(1949). Phytoplankton of Block Is. Sound. *Bull. Bingh. Oceanogr. Coll.* **13** (3), 40–64.

(1957). Phytoplankton of the North Central Sargasso Sea 1950–2. *Limnol. Oceanogr.* **2** (3), 252–70.

(1963). Organic aggregates in sea water and the dynamics of their formation and utilization. *Limnol. Oceanogr.* **8,** 372–81.

(1965). Theory of food chain relations in the ocean. *The Sea*, ed. M. N. Hill, vol. 2, pp. 438–63.

Riley, G. A. & Conover, S. M. (1956). Oceanography of Long Island Sound. III. Chemical Oceanography. *Bull. Bingh. Oceanogr. Coll.* **12,** 1–169.

Riley, G. A. & Schurr, H. M. (1959). Oceanography of Long Island Sound. III. Transparency of Long Island Sound waters. *Bull. Bingh. Oceanogr. Coll.* **17** (1), 66–82.

Riley, G. A., Stommel, H. & Bumpus, D. F. (1949). Quantitative ecology of the plankton of the western North Atlantic. *Bull. Bingh. Oceanogr. Coll.* **12,** (3), 1–169.

Riley, G. A., van Hemert, D. & Wangersky, P. J. (1965). Organic aggregates and aggregates in surface and deep waters of the Sargasso Sea. *Limnol. Oceanogr.* **10** (3), 354–63.

Riley, J. D. (1966). Marine fish culture. VII. Plaice (*Pleuronectes platessa* L) post larval feeding on *Artemia salina L* nauplii and effects of varying feeding levels. *J. Cons. int. Explor. Mer.* **30,** 204–21.

Robinson, G. A. (1970). Continuous plankton records; variation in the seasonal cycle of phytoplankton in the North Atlantic. *Bull. Mar. Ecol.* **6** (9), 333–45.

Rodhe, W. (1963). Report of discussion. In *Marine Biology*, vol. 1. ed. G. A. Riley, *Proc. 1st Intern. Interdiscipl. Conf., Wash. D.C., Amer. Inst. Biol. Sci.* 286 pp.

Rollefsen, G. (1954). Observations on the cod and cod fisheries of Lofoten. *Rapp. Procès-Verb. Cons. int. Explor. Mer*, **136,** 40–7.

(1955). The arctic cod. *Proc. UN Sci. Conf. Conserv. Utiliz. Resources, Rome*, pp. 115–17.

Ross, R. (1910). *The Prevention of malaria.* London.

Rounsfell, G. N. (1958). Factors causing decline of sockeye salmon of Kalka river, Alaska. *U.S. Fish and Wildlife Serv. Fish. Bull.* **58** (130), 83–169.

Royce, W. (1964). A morphometric study of yellowfin tuna, *Thunnus albacares* Bonnaterre. *US Fish and Wildlife Serv. Fish. Bull.* **63** (2), 395–444.

Runnstrøm, S. (1933). Sildeundersøkelser 1930–31. *Arsberet vedkomm. Norges Fisk* 1931, **1,** 65–77.

(1934). The pelagic distribution of the herring larvae in the Norwegian waters. *Rapp. Procès-Verb. Cons. int. Explor. Mer*, **88** (5), 6 pp.

(1936). A study of the life history and migrations of the Norwegian spring herring based on the analysis of the winter rings and summer zones of the scale. *Fisk. Dir. Skr. Hav.* **5** (2), 103 pp.

Russell, E. S. (1931). Some theoretical considerations on the 'overfishing' problem. *J. Cons. int. Explor. Mer*, **6** (1), 3–20.

(1932). Is the destruction of undersized fish prejudicial to the stock? *Rapp. Procès-Verb. Cons. int. Explor. Mer*, **80** (8), 13 pp.

(1937). Fish migrations. *Biol. Rev.* **12,** 320–37.

Russell, F. S. (1930). The seasonal abundance and distribution of the pelagic young of teleostean fishes caught in the ring trawl in offshore waters in the Plymouth area. *J. Mar. Biol. Ass. UK*, NS **16,** 707–22.

(1969). On the seasonal abundance of young fish. XI. The year 1966. *J. Mar. Biol. Ass. UK*, NS **49,** 305–10.

Russell, F. S. & Colman, J. S. (1934). The zooplankton. II. *The Great Barrier Reef Expdn Sci. Rep. BM* (*NH*), **2,** 159–76.

Russell, F. S. & Demir, N. (1971). On the seasonal abundance of young fish.

XII. The years 1967, 1968, 1969 and 1970. *J. Mar. Biol. Ass. UK*, NS **51,** 127–30.

Ryland, J. S. (1963). The swimming speeds of plaice larvae. *J. Exp. Biol.* **40,** 285–99.

Ryther, J. H. (1954). Inhibitory effects of phytoplankton upon the feeding of *Daphnia magna* with reference to growth, reproduction and survival. *Ecology*, **35** (4), 522–33.

Ryther, J. H., Hall, J. R., Pease, A. K. & Jones, M. M. (1966). Primary organic production in relation to the chemistry and hydrography of the western Indian Ocean. *Limnol. Oceanogr.* **11,** 371–80.

Ryther, J. H. & Menzel, D. W. (1960). The seasonal and geographical range of primary production in the western Sargasso Sea. *Deep Sea Res.* **6,** 235–8.

Ryther, J. H., Menzel, D. W. & Vaccaro, R. F. (1961). Diurnal variations in some chemical and biological properties of the Sargasso Sea. *Limnol. Oceanogr.* **6,** 149–53.

Saila, S. B. (1961). A study of winter flounder movements. *Limnol. Oceanogr.* **6,** 292–8.

Saila, S. B. & Flowers, J. M. (1969). Elementary applications of search theory to fishing tactics as related to some aspects of fish behaviour. *FAO Fish. Rep.* **62** (2), 343–56.

Saila, S. B. & Shappy, R. A. (1963). Random movements and orientation in salmon migration. *J. Cons. int. Explor. Mer*, **28,** 153–66.

Saunders, R. L., Kerswill, C. J. & Elsdon, P. F. (1965). Canadian Atlantic salmon recaptured near Greenland. *J. Fish. Res. Bd Can.* **22,** 625–9.

Savage, R. E. (1937). The food of North Sea herring 1930–1934. *Fish. Invest. London*, **2,** 15 (5).

Saville, A. (1959). The planktonic stages of the haddock in Scottish waters. *Mar. Res. Scot. 1959* (3), 23 pp.

(1964). Estimation of the abundance of a fish stock from egg and larval surveys. *Rapp. Procès-Verb Cons. int. Explor. Mer*, **155,** 164–70.

Schaefer, M. B. (1954). Some aspects of the dynamics of populations important to the management of the commercial fish populations. *Inter. Amer. Trop. Tuna. Commn Bull.* **1** (2), 27–56.

(1957). A study of the dynamics of the fishery for yellowfin tuna in the eastern tropical Pacific Ocean. *Bull. Inter. Amer. Trop. Tuna Commn*, **2,** 245–85.

Schmidt, J. (1909). The distribution of the pelagic fry and the spawning regions of the gadoids in the North Atlantic from Iceland to Spain. *Rapp. Procès-Verb. Cons. int. Explor. Mer*, **10** (4), 158 pp.

(1922). The breeding places of the eel. *Phil. Trans. Roy. Soc.* B, **211,** 179–208.

Sette, O. E. (1943). Biology of the Atlantic Mackerel (*Scomber scombrus*) of North America. I. Early life history including the growth, drift and mortality of egg and larval populations. *US Fish and Wildlife Serv. Fish. Bull.* **50** (38), 149–234.

Sewell, R. B. S. (1925). Geographic and Oceanographic research in Indian waters. II. A study of the nature of the sea bed and of the Deep Sea deposits of the Andaman Sea and the Bay of Bengal. *Mem. Asiatic. Soc. Bengal* IX (2), 27–50.

(1935). Geographic and oceanographic research in Indian waters. VII. The topography and bottom deposits of the Laccadive Sea. *Mem. Asiatic Soc. Bengal* XI (7), 425–60.

Shapovalov, L. (1937). Trout and salmon marking in California. *Calif. Fish Game*, **23,** 205–7.

Shelbourne, J. E. (1957). The feeding and condition of plaice larvae in good and bad plankton patches. *J. Mar. Biol. Ass. UK*, NS **36,** 539–52.

Sheldon, R. W. & Parsons, T. R. (1967). *A practical manual on the use of the Coulter Counter in marine science.* 66 pp. Coulter Electronics Sales Co., Canada, Toronto.

Shushkina, E. A. (1968). Calculation of copepod production based on metabolic features and the coefficient of the utilization of assimilated food for growth. *Okeanologiya*, **8** (1), 126–38.

Sick, K. (1961). Haemoglobin polymorphism in fishes. *Nature* (*London*), **192,** 894–6.

Simpson, A. C. (1949*a*). Notes on the occurrence of fish eggs and larvae in the Southern Bight of the North Sea during the winter 1946–7. *Ann. Biol. Cons. int. Explor. Mer*, **4,** 90–5.

(1949*b*). Notes on the occurrence of fish eggs and larvae in the Southern Bight of the North Sea during the winter of 1947–8. *Ann. Biol. Cons. int. Explor. Mer*, **5,** 90–7.

(1959). The spawning of the plaice in the North Sea. *Fish. Invest. Lond.* Ser. 2, **22** (7), 111 pp.

Sindermann, C. J. & Mairs, D. F. (1959). The C blood group system of Atlantic sea herring. *Anat. Rec.* **134,** 640 pp.

(1961). A blood group system for spiny dogfish, *Squalus acanthias. Biol. Bull.* **120** (3), 401–10.

Slobodkin, L. B. (1960). Ecological energy relationships at the population level. *Amer. Nat.* **94,** 213–36.

(1966). *Growth and regulation of animal populations.* 184 pp. Holt, Rinehart & Winston, New York,

Smayda, T. H. (1970). The suspension and sinking of phytoplankton in the sea. *Oceanogr. Mar. Biol. Ann. Rev.* 1970, **8,** 353–414.

Smayda, T. H. & Boleyn, B. J. (1965). Experimental observations on the flotation of marine diatoms. I. *Thalassiosira nana*, *Thalassiosira rotula* and *Nitszchia seriata. Limnol. Oceanogr.* **10,** 449–509.

Smayda, T. H. & Boleyn, B. J. (1966). Experimental observations on the flotation of marine diatoms. III. *Skeletonema costatum* and *Rhizosolenia setigera. Limnol. Oceanogr.* **11,** 18–34.

Smed, J. (1965). Variation of the temperature of the surface water areas of the northern North Atlantic, 1876–1961. *Spec. Publ. Int. Commn North West Atl. Fish.* **6,** 821–6.

Smith, E. L. (1936). Photosynthesis in relation to light and carbon dioxide. *Proc. Nat. Acad. Sci. Wash.* **22,** 504 pp.

Solomon, M. E. (1949). The natural control of animal populations. *J. Anim. Ecol.* **18,** 1–35.

Southward, A. J. (1963). The distribution of some plankton animals in the English Channel and approaches. III. Theories about long term biological changes including fish. *J. Mar. Biol. Ass. UK*, NS **42,** 431–29.

Southward, G. M. (1967). Growth of Pacific halibut. *Rep. Int. Pac. Halibut Commn*, **43,** 40 pp.

Sprague, L. M. & Vrooman, A. M. (1962). A racial analysis of the Pacific sardine (*Sardinops caerulea*) based on studies of erythrocyte antigens. *Ann. NY Acad. Sci.* **97,** 131–8.

Sproston, N. G. (1947). *Icthyosporidium hoferi* (Plehn and Mulsow, 1911) and internal fungoid parasite of the mackerel. *J. Mar. Biol. Ass. UK*, NS **26,** 72–98.

Soutar, A. & Isaacs, J. D. (1969). History of fish populations inferred from fish scales in anaerobic sediments off California. *Ref. Calif. coop. oceanic Fish. Invest.* **13,** 63–70.

Steele, J. H. (1958). Plant production in the northern North Sea. *Mar. Res. Scot.* 1958, **7,** 36 pp.

(1961*a*). The environment of a herring fishery. *Mar. Res. Scot.* **6,** 19 pp.

(1961*b*). Primary production. *Oceanography Amer. Adv. Sci.* **67,** 519–38.

Steele, J. H. (1962). Environmental control of photosynthesis in the Sea. *Limnol. Oceanogr.* **7** (2), 137–50.

(1965). Notes on some theoretical problems in production ecology. *Mem. Ist Ital. Idrobiol. Dott. Marco de Marchi 18 Suppl.* pp. 383–98. Edinburgh & London.

(ed) (1970). *Marine Food Chains.* 552 pp. Oliver & Boyd, Edinburgh & London.

(1974). *The structure of marine ecosystems.* Yale University Press, Newhaven.

Steele, J. H. & Edwards, R. R. C. (1969). The ecology of 0-group plaice and common dabs in Loch Ewe. IV. Dynamics of the plaice and dab populations. *J. Exp. Mar. Biol. Ecol.* **4,** 174–88.

Steele, J. H. & Menzel, D. W. (1962). Conditions for maximum primary production in the mixed layer. *Deep Sea Res.* **9,** 39–49.

Steuer, A. (1910). *Plankton Kunde.* 723 pp. Leipzig & Berlin,

Strathmann, R. R. (1967). Estimating the organic carbon content of phytoplankton from cell volume or plasma volume. *Limnol. Oceanogr.* **12,** 411–18.

Strickland, J. D. H. (1960). Measuring the production of marine phytoplankton. *Fish. Res. Bd Can. Bull.* **122,** 172 pp.

Strickland, J. D. H., Eppley, R. W. & Rojas, B de M. (1969). Phytoplankton populations, nutrients and photosynthesis in Peruvian coastal waters. *Bol. Inst. Mar. Peru,* **2** (1), 45 pp.

Strickland, J. D. H. & Parsons, T. R. (1968). Practical handbook of sea water analysis. *Fish. Res. Bd Can. Bull.* **167,** 311 pp.

Strickland, J. D. H. & Terhune, L. D. B. (1961). The study of *in situ* marine photosynthesis using a large plastic bag. *Limnol. Oceanogr.* **6** (1), 93–5.

Stubbs, A. R. & Lawrie, R. G. G. (1962). Asdic as an aid to spawning ground investigations. *J. Cons. int. Explor. Mer.* **27** (3), 248–60.

Subramanyan, R. (1959). Studies on the plankton of the east coast of India. I. Quantitative and qualitative fluctuations of total phytoplankton crop, the zooplankton crop and their interrelationships with remarks on the magnitude of the standing crop and production of matter and their relationship to fish landings. *Proc. Ind. Acad. Sci.* **50,** 113–87.

Suda, A. (1963). Structure of the albacore stock and fluctuation in the catch in the north Pacific area. *FAO Fish. Rep.* **6** (3), 1237–77.

(1971). Tuna fisheries and their resources in the IPFC area. *Far Seas Fish. Res. Lab.* 5, Ser. 5, 58 pp.

Sverdrup, H. U. (1938). On the process of upwelling. *J. Mar. Res.* **1,** 155–64.

(1953). On conditions for vernal blooming of phytoplankton. *J. Cons. int. Explor. Mer,* **18,** 287–95.

Taft, A. C. & Shapovalov, L. (1938). Homing instinct and straying amongst steel head trout (*Salmo gairdnerii*) and silver salmon (*Oncorhynchus kisutch*). *Calif. Fish and Game,* **24,** 118–25.

Tåning, A. V. (1937). Some features in the migration of cod. *J. Cons. int. Explor. Mer,* **12,** 1–35.

(1953). Long term changes in hydrography and fluctuation in fish stocks. *Int. Comm. North West Atl. Fish. Commn Ann. Proc.* **3,** 69–77.

Taylor, P. H. C. & Wickett, W. P. (1967). Recent changes in abundance of British Columbia herring and future prospects. *Circ. Fish. Res. Bd Can. (Biol. Sta. Nanaimo)*, **80,** 17 pp.

Teal, J. M. (1957). Community metabolism in a temperate cold spring. *Ecol. Monogr.* **27,** 283–302.

Templeman, W. (1965). Relation of periods of successful year classes of haddock on the Grand Banks to periods of success of year classes of cod, haddock and herring to areas to the North and East. *Spec. Publ. Int. Commn North West Atl. Fish.* **6,** 523–34.

Thomas, W. H. (1966). Effects of temperature and illuminance on cell division rates of three species of tropical oceanic phytoplankton. *J. Phycol.* **2** (2), 17–22.

Thompson, W. F. (1915). A preliminary report on the life history of the halibut. *Rep. Commn Fish Brit. Columbia*, 76–99.

(1952). Condition of stocks of halibut in the Pacific. *J. Cons. int. Explor. Mer*, **18** (2), 141–6.

Thompson, W. F. & Bell, F. H. (1934). Biological statistics of the Pacific Halibut fishery. 2. Effect of changes in intensity upon total yield and yield per unit of gear. *Rep. Int. Fish. Commn* no. 8, 49 pp.

Thompson, W. F. & van Cleve, R. (1936). Life history of the Pacific halibut. 2. Distribution and early life history. *Rep. Int. Fish. Commn* No. 9, 184 pp.

Thompson, W. F. & Herrington, W. C. (1930). Life history of the Pacific halibut. *Rep. Int. Fish. Commn* **2,** 137 pp.

Thrailkill, J. R. (1959). Zooplankton volumes off the Pacific coast, 1957. *Spec. Sci. Rep. US Fish and Wildlife Serv.* **326,** 57 pp.

(1961). Zooplankton volumes off the Pacific coast, 1958. *Spec. Sci. Rep. US Fish and Wildlife Serv.* **374,** 70 pp.

(1963). Zooplankton volumes off the Pacific coast, 1959. *Spec. Sci. Rep. US Fish and Wildlife Serv.* **414,** 77 pp.

Tooms, J. (1967). Marine minerals in perspective. *Hydrospace*, **1** (1), 40–6.

Townsend, C. H. (1935). The distribution of certain whales as shown by the log book records of American whale ships. *Zoologica, NY*, **19,** 1–50.

Trevallion, A., Edwards, R. R. & Steele, J. H. (1970). Dynamics of a benthic bivalve. In *Marine Food Chains*, ed. J. H. Steele, pp. 285–95. Oliver & Boyd. Edinburgh & London.

Trout, G. C. (1957). The Bear Island cod; migrations and movements. *Fish. Invest. Lond.* Ser. 2, **21** (6), 51 pp.

Tutin, W. (1955). Preliminary observations on a year's cycle of sedimentation in Windermere. *Mem. Ist Ital. Idrobiol. Dott. Marco de'Marchii, Suppl. 8*, 467–84.

Uda, M. (1952*a*). On the relation between the variation of the important fisheries conditions and oceanographical conditions in the adjacent waters of Japan. *J. Tokyo Univ. Fish.* **38** (3), 363–89.

(1952*b*). On the fluctuation of the main stream axis and its boundary line of Kuroshio. *Bull. Tokai. Reg. Fish. Res. Lab.* **3,** 44 pp.

(1960). The fluctuation of the sardine fishery in oriental waters. *Proc. World Sci. Meet. Biol. Sardines and related species*, III. 937–47.

Utermohl, H. (1933). Neue Wege in der quantitativen Erfassung des Planktons. *Verh. d. intern. Vereinig. Limnol.* **5,** 567–95.

(1958). Zur Vervollkommnung der quantitativen Phytoplankton methodik. *Intern. Verein. Theor. v. angew. Mitt. int. Ver. Limnol.* no. 9, 38 pp.

van Campen, W. C. (1960). Translation. Japanese summer fishery for albacore (*Germo alalunga*). *US Fish and Wildlife Ser. Res. Rep.* **52,** 29 pp.

van Oorschot, J. L. P. (1953). Conversion of light energy in algal culture. *Medd. v. Landbouwhoge school te Wageningen Nederland*, **55** (5), 225–76.

de Veen, J. F. (1961). The 1960 tagging experiments on mature plaice in different spawning areas in the southern North Sea. *Int. Council Explor. Sea*, CM-1961 (44), 7 pp.

(1962). On the subpopulations of plaice in the southern North Sea. *Int. Council. Explor. Sea*, CM1962 (94), 6 pp.

(1970). On the orientation of the plaice (*Pleuronectes platessa* L). 1. Evidence for orientating factors derived from the ICES transplantation experiments in the years 1904–08. *J. Cons. int. Explor. Mer*, **33** (2), 192–227.

de Veen, J. F. & Boerema, L. K. (1959). Distinguishing southern North Sea populations of plaice by means of otolith characteristics. *Int. Council Explor. Sea*, CM1959 (191), 5 pp.

Verhulst, P. F. (1838). Notice sur la loi que la population suit dans son accroissement. *Corv. Math et Phys.* **10,** 113 pp.

Vinogradov, M. E. (1970). *Vertical distribution of the oceanic zooplankton.* Jerusalem 1 PST 1970, 339 pp.

Volterra, V. (1926). Variazioni e fluttuazione del numero d'individui in specie animali conviventi. *Mem. Accad. Naz. Linza*, **6** (2), 31–113.

Walford, L. A. (1946). Correlation between fluctuations in abundance of the Pacific sardine (*Sardinops caerulea*) and salinity of sea waters. *J. Mar. Res.* **6** (1), 48–53.

Warren, C. E. & Davis, G. E. (1967). Laboratory studies on the feeding, bioenergetics and growth of fish. In *The Biological Basis of Freshwater Fish Production*, ed. S. Gerking, 175 pp.

Watt, K. E. F. (1959). A mathematical model for the effect of densities of attacked and attacking species on the number attacked. *Can. Ent.* **91** (12), 29–44.

Webb, P. W. (1971). The swimming energetics of trout. I. Thrust and power output at cruising speeds. *J. Exp. Biol.* **55** (2), 489–520.

Wiborg, K. (1954). Investigations on zooplankton in coastal and offshore waters of western and north western Norway. *Fisk. Dir. Skr.* XI, 1.

Widrig, T. M. (1954). Method of estimating fish populations with application to the Pacific sardine. *US Dept. Int. Fish. Bull.* **94** (56), 139–66.

Williamson, M. H. (1961). An ecological survey of a Scottish herring fishery. IV. Changes in the plankton during the period 1949–1959. *Bull. Mar. Ecol.* **5,** 207–29.

Wimpenny, R. S. (1944). Plankton production between the Yorkshire coast and the Dogger Bank. *J. Mar. Biol. Ass. UK*, NS **26** (1), 1–6.

(1953). *The plaice, being the Buckland lectures for 1949.* 144 pp, Edward Arnold, London.

(1966). *The plankton of the Sea.* 426 pp. Faber, London.

Winberg, G. C. (1956). Intensivnost obmena i pisheheuye potrebnosti ryb. *Nauchnye Trudy Belorusskovo Gosuderstvennovo Universitata Imeni VI Lenina Miusk*, 235 pp.

Wolfe, D. A. & Schelske, C. L. (1967). Liquid scintillation and geiger counting efficiencies for carbon-14 incorporated by marine phytoplankton in productivity measurements. *J. Cons. int. Explor. Mer.* **31,** 31–7.

Wood, H. (1937). Movements of herring in the northern North Sea. *Fish. Scotland Sci. Invest.* 1937, **3,** 49 pp.

Woodhead, A. D. & Woodhead, P. M. J. (1959). The effects of low temperature on the physiology and distribution of the cod *Gadus morhua* L. in the Barents Sea. *Proc. Zool. Soc.* **133** (2), 181–99.

Wooster, W. S. & Reid, J. L. (1962). Eastern boundary currents. In *The Sea*, ed. M. N. Hill, pp. 253–80. Wiley, London & New York.

Wyatt, T. (1971). Production dynamics of *Oikopleura dioica* in the Southern North Sea and the role of fish larvae which prey on them. *Thalassa Yugoslavica*, **7** (1), 435–44.

Wyrtki, K. (1964). Upwelling in the Costa Rica Dome. *Fish. Bull. US Fish and Wildlife Serv.* **63** (2), 355–72.

Yarrell, W. (1836). *A history of British fishes.* van Voorst, London, 472 pp.

Yentsch, C. S. (1965). Distribution of chlorophyll and phaeophytin in the open ocean. *Deep Sea Res.* **12,** 653–66.

Yokoto, T. (1951). Studies on the sardine stock in the Hykganad. II. Estimation of the amount of sardine stock (from 1935 to 1941). *Bull. Jap. Soc. Sci. Fish.* **17** (1), 5–8.

Yoshida, H. O. & Otsu, T. (1963). Synopsis of biological data on albacore *Thunnus germo* La cepede 1800 (Pacific and Indian Oceans). *FAO Fish. Rep.* **6** (2), 274–318.

Yoshida, K. (1967). Circulation in the eastern tropical oceans with special reference to upwelling and undercurrents. *Jap. J. Geophysics*, **4** (2), 1–75.

Zeuthen, E. (1947). Body size and metabolic rate in the animal kingdom with special regard to the marine microfauna. *Comptes. Rendues. Lab. Carlsberg*, **26** (3), 161 pp.

Zijlstra, J. J. (1969). On the recruitment of the Downs herring. *Rapp. Procès-Verb. Cons. int. Explor. Mer*, **160,** 158–60.

(1972). On the importance of the Wadden Sea as a nursery area in relation to the conservation of the Southern North Sea fishery resources. *Symp. Zool. Soc. Lond.* (1972) no. 29, 233–58.

Index